# Farm Animal Welfare Law

This book introduces the various aspects of international farm animal protection and wildlife conservation through the lenses of food safety and environmental protection law. Bite-sized chapters focus on a wide range of topics from agrobiodiversity, fishing, and aquaculture to pollinators and pesticides, soil management, industrial animal production, and transportation, as well as international food trade.

Animal welfare and biodiversity conservation sit at the core of the selected chapters, each one providing real-world examples to make the complex field easy to understand. Current developments including food safety modernization, blockchain, and COVID-19 considerations are addressed head-on.

*Farm Animal Welfare Law* provides a primer for law school courses and masters' programs, for practitioners, advocates, and animal enthusiasts alike. Through its emphasis on sustainable food production, this book offers a cutting-edge selection of evolving topics at the heart of the pertinent discourse.

# Essentials in Animal Law and Welfare

*Series Editor*
Gabriela Steier

**Farm Animal Welfare Law:** International Perspectives on Sustainable Agriculture and Wildlife Regulation, *edited by Gabriela Steier*

For more information about this series, please visit: https://www.routledge.com/Essentials-in-Animal-Law-and-Welfare/book-series/CRCANIMLAW

# Farm Animal Welfare Law
## International Perspectives on Sustainable Agriculture and Wildlife Regulation

Edited by
Gabriela Steier

Junior Editors
Sahana Ramdas
Christian Mulgrew de Laire
Audrey Amescua
Amanda G. Verkest
Morgan Boutilier

CRC Press
Taylor & Francis Group
Boca Raton London New York

CRC Press is an imprint of the
Taylor & Francis Group, an **informa** business

First edition published 2024
by CRC Press
6000 Broken Sound Parkway NW, Suite 300, Boca Raton, FL 33487-2742

and by CRC Press
4 Park Square, Milton Park, Abingdon, Oxon, OX14 4RN

*CRC Press is an imprint of Taylor & Francis Group, LLC*



ISBN: 978-0-367-68477-8 (hbk)
ISBN: 978-0-367-68476-1 (pbk)
ISBN: 978-1-003-13773-3 (ebk)

DOI: 10.1201/9781003137733

Typeset in Times
by codeMantra

## EDITORS' DISCLAIMER

The editors make no representations or warranties whatsoever as to the accuracy, completeness, or suitability for any purpose of the content, citations, or attributions of this publication or any publications cited therein. Any opinions and views expressed in this publication are the opinions and views of the authors and are not necessarily the views of or endorsed by the editors. All contributing authors agreed to the edits of the final version of this book and released the editors from any responsibility or liability directly or indirectly connected to this work, the supplements, and works cited, quoted, or otherwise used therein. Much of the work in this book is based on a previously published edited volume, https://link.springer.com/book/10.1007/978-3-319-18002-1, which is duly cited herein. This current publication is intended for educational purposes only, and any materials cited and statements made are not to be construed as or understood to be legal advice.

*For my Sons*

*May you always cherish, respect, and marvel*

*in the enchanting creatures of our world.*

*May your inner compass steer you toward*

*Conservation, protection, and compassion.*

# Contents

# Preface

My sweet little baby boy is sleeping next to me as I am typing these words. Dear Readers, this is the second time I have the pleasure and privilege to put such an extraordinary volume together. Last time, my older son, who is now six years old, was the baby sleeping next to me. Last time, dear Readers, the book took on a life of its own with over 800 pages and over two dozen contributors. We had to use dictionary paper to manage the contents in one volume binding. This time, I had a team of junior editors who helped me to adapt, update, and condense the material for our day and age.

Much has happened in the world of farm animal welfare law – and in some ways nothing has happened. The ongoing Covid-19 pandemic has taught people little even though it has offered manifold lessons, studies, and data to learn from. From zoonoses to health complications from diets rich in animal protein, the human race continues to exploit, enjoy, and embrace the precious creatures whom we share the planet with for delectable, complicated, and convenient foods. This relationship of eating our fellow Earth dwellers causes indescribable pain, environmental burdens, and culinary delight. Some developments, such as rainforest stewardship and ag-gag strike-downs, have brought animal law to the foreground.

We hope to make some of the advances, the shortcomings, and the gaps in regulation a bit more palatable in this work. With the textual explanations and summaries of previous work, as well as the appendix of global legislation in place, we are giving you, dear Readers, the Tools for Change.

We stand on the shoulders of giants and all those who helped me to forge the field of farm animal law in *International Farm Animal, Wildlife, and Food Safety Law* (Springer 2007). The contributors to my previous book on farm animal welfare represented every continent and wrote about every taxonomic group of creatures used in food and agriculture. Even though the book stands on its own, I am proud to expand upon the field with this volume.

As the first book of the *Essential in Animal Welfare Series*, edited by my team for Taylor and Francis, many topics summarize grand subjects of our field and provide merely overviews. The series is expanding quickly and shall benefit many readers around the globe. Hopefully, when my sons grow up, they will have conscientious and animal-friendly diets at their fingertips, surrounded by lush agroecology for a resilient and sustainable future food system, guided by integrity, prosperity, and kindness to all beings on our beautiful planet.

This volume provides a summary, an issue-spotting exercise ripe for law school courses and masters' programs, for practitioners, advocates, and animal enthusiasts alike. It is not as comprehensive as the subject, but it is the primer for the series.

Every word is written for you, dear Readers. On behalf of my junior editors and myself, it is our hope that you will find this book eye-opening. May the conservation and protection of animals rise to the challenges of consolidating appetites for meat, dairy, and eggs and profit with the soul-searching of protecting animal welfare, habitats, and ensuring sustainable futures.

**Gabriela Steier**
*Boston, MA*

# Acknowledgments

The editor thanks the junior editors, Audrey, Christian, Sahana, Amanda, and Morgan, for their diligent work and sleepless nights to ensure the success of this book. Thanks to their energetic and youthful enthusiasm, this book is brimming with hope for a better future.

Special thanks go to the contributing authors, including the junior editors, and Autumn, Gerardo, and Paulo. Thank you for sharing your expertise with the world via this book.

For their continued encouragement, support, and pride in my work, I thank my parents, sons, and husband.

Thanks on behalf of all of us go to our publishing and editing team, Alice Oven and Shikha Garg, and everyone who helped us to put this book together. Thank you for believing in us!

This book has been such a pleasure to work on with all parties involved that seeing its completion is bitter-sweet. It is wonderful to see it completed, and I will hold the memories of this amazing collaboration in my heart for years to come.

# Editor

**Dr. iur. Gabriela Steier, LL.M., Esq.** is a lawyer, educator, speaker, and scholar focusing on food law and regulation. Her areas of focus are food system regulations in the EU and US, and environmental and climate change laws. She is a part-time lecturer at Northeastern University, where she teaches several courses on food regulation, food safety modernization, and international food trade. She also teaches as an adjunct professor at the Duquesne University School of Law in Pittsburgh, and as a visiting professor at the University of Perugia, Italy. As founder of FoodLawInternational.com, she manages a vast network, hosts lecture series, and promotes academic scholarship in her field. Her recent online course on Udemy and her *International Perspectives* Column in *Quality Assurance Magazine* give her far-reaching traction. She has published widely in both textbooks and peer-reviewed journals. Dr. Steier has a BA from Tufts University, a JD from the Duquesne University School of Law, an LLM from the Vermont Law School, and a Doctorate in Comparative Law from the University of Cologne, Germany. She is based in Boston, MA.

# Junior Editors

**Sahana Ramdas** is an animal and environmental advocate. She has specialized in three disciples, with an LLM in Animal Law, LLM in Global Environmental law, and an LLM in General Law. She is a practicing attorney in the State of New York with over four years of legal experience in the field of Animal Law and Environmental Law. Sahana is a published author in the two public interest fields and has also collaborated with the Food Law International on scholarly work that highlights the intersectionality of animal law and environmental law.

**Christian Mulgrew de Laire** is a graduate student at Northeastern University School of Law working to earn a dual JD/MBA. Before Northeastern University School of Law, he graduated from The College of the Holy Cross where he studied Economics and French. His concentration consisted of anti-trust, sports law, and case studies. Christian is fluent in English and French. As Junior editor of this volume, he has edited several chapters and created figures.

**Audrey Amescua** completed a Master of Science in Regulatory Affairs of Food from Northeastern University and is poised to become a future leader in the food industry. Through her work experience, Audrey has gained substantial insight into international food affairs and is well-versed in the areas of food safety and food safety regulation. Audrey is passionate about stronger, smarter regulatory policies and hopes that through enhanced cooperation between global stakeholders and by embracing technological innovation, we can develop a safer, more sustainable, and equitable global food supply. Audrey intends to continue her flourishing career in global food policy, focusing on human and animal safety and sustainability. She holds a Bachelor of Science in Food Science and Technology from Texas A&M University.

**Amanda G. Verkest** is an experienced food scientist working as an international consultant in food regulation and sustainability and an operations manager at a medicinal herb farm and retreat center. Amanda is also involved in philanthropic work regarding sustainable food and agriculture in southeast Tennessee. Amanda holds a Bachelor of Science in Nutrition and Food Science with a minor in Chemistry from Middle Tennessee State University and a Master of Science in Global Studies and International Relations from Northeastern University. Ultimately, Amanda is passionate about global sustainable agriculture, plant medicine, and education and intends to continue her international consulting work while studying herbal medicine.

**Morgan Boutilier** holds a Bachelor of Arts in Sociology from UCLA, where she is pursuing her law degree. In addition to her education, she draws upon experience gained from working with a range of nonprofit animal welfare organizations such as Animal Equality, Mercy for Animals, and Animal Place. Morgan is passionate about Farm Animal Welfare Law and plans to dedicate her career to creating a more compassionate and equitable food system. She can be reached at morganmboutilier@gmail.com.

# Contributing Authors

**Autumn T. Johnson** is the CEO of Tierra Strategy, a firm focused on public interest policy advocacy and sustainability consulting. She is also the founder of Autumn Acres, a micro sanctuary for farm and domestic animals. She holds a BA from the University of Arizona, a JD from the University of Oregon, an MBA from Seattle University, and is a PhD Candidate at Boise State University in the Public Policy and Administration program.

**Gerardo Figueiredo Junior** is a Brazilian lawyer, graduated from Faculdade Milton Campos, postgraduate in Business Law from Universidade Gama Filho and specialist in Food Law from Michigan State University. He chaired the Regulated Sectors Study Commission of the Brazilian Bar Association in São Paulo and is the author of chapters published in specialized works and magazines in several countries, in addition to participating in courses and lectures. He is currently a founding partner of Brazilian Law Firm Zeigler Advogados, heading the Food Law and Regulatory Law team. Gerardo can be contacted at gerardo@zeigler.com.br.

**Paulo Bellonia** is System Analist graduated from FUMEC University, entrepreneur and founder of SaveCerrado, a non-governmental organization dedicated to the preservation and sustainable development of the Cerrado, one of the Brazilian biomes that brings together the largest variety of animal and plant species, as well as one of the largest water reserves from the country. Paulo can be reached by email paulo@savecerrado.org.

# *Part 1*

## *Marine Animals*

# Section 1

# Editor Notes – Aquaculture Section

The world is 75 percent ocean. Cultures have developed around sea travel, exploration, and economy. Many coastal cultures have incorporated the oceans into their culture, and for good reason. The oceans provide resources, passage, and food. Fishing is a staple that provides nourishment and supports the local economy. In France, for example, many coastal cities focus on fishing and opportunities present from being next to the ocean. In the Pacific, the many Polynesian tribes created trade networks, trained master navigators, and incorporated the oceans into their lives. And the foremost example dates back to the Vikings who would sail to discover new lands and search for resources. There are many more examples ranging from the ancient Phoenicians and Greeks to the colonial and modern British and Americans.

## SYMBOLISM AND CUSTOM ANCHOR FISHERIES AND ECONOMICS

There is custom and symbolism in the connection between fisheries and the economy. For instance, a trade network known as the "kula ring"[1] was formed when men from one island would go to another island to barter attempting to obtain unique white or red shells. These shells were considered signs of importance and that a man

[1] https://www.encyclopedia.com/social-sciences/dictionaries-thesauruses-pictures-and-press-releases/kula-ring.

DOI: 10.1201/9781003137733-2

is a strong suitor for women. The routes were predetermined, creating routes that resembled a ring or cycle. The only way to obtain the shells was to sail to the partner island and barter with men who had shells. Then, in turn, men from another island would come to give them the gifts in exchange for shells, so that the cycle could continue. Conversely, the "kula ring" was formed because of the access to shells, the oceans, and a need for trade. Without the ocean or shells, the culture surrounding the barter system for shells would not have developed the same. In fact, the culture relies on the ocean and its ecosystem. Major changes to either would have an impact on the culture of the islands involved in the ring.

Further east, the Maori, Samoans, and other groups were creating trade routes and exploring islands using master navigators. Given the lack of resources present, the cultures became reliant on other means to survive.[2] As such, they traded, and part of the culture revolved around the ocean and seafaring. The tribes in Polynesia are all slightly different but resemble each other. Through exploration they were able to settle new islands and create homes.

Further examples for the connection between fisheries and culture come from the people of Polynesia's "haka." While each country has its own unique version, the chants resemble each other and are used for the same purpose. Furthermore, these cultures created trade networks that extended miles to obtain any missing resources. The culture's identities are based on seafaring and the opportunities from the ocean. In both cases, the ocean was a part of the culture that allowed for society and trade to grow. Without the ocean, these cultures would not be the same. Beyond culture, there are other ways that the oceans have become integral to coastal societies.

## FOOD SOURCING AND SUSTAINABILITY OF FISHERIES

Another major use of the ocean is for food procurement. Fish are a major source of food for coastal states. With their nutritional value and presence when other sources of meat were not available, it is not surprising that fish became an essential part of certain diets. Coastal villages could easily send out boats to catch fish in the morning to sell. Furthermore, given the large variety of fish and nutrients that are found in the fish, it is notable that fishing has become a critical part of meals. As a result, when the human population grew, so did the number of fisheries. The resulting increased pressure from fishing on the ocean biosphere meant that the sustainability of fishing came into question. Without any regulation of these "common goods," fish could go extinct which would significantly impact cultures, economies, and health (Figure 1).

The international community has created regulation that allows for sustainable fishing and protection of the environment. These rules are made considering longevity and leaving something for the next generation. The focus of the regulation is ending negative and environmentally harmful practices, while still allowing people to get what they need and use the resources available. Creating regulation is a veritable balancing act that is not perfect but must consider many different aspects to be successful. This section looks at the ocean and fish analytically, analyzing how the current situation arose and what is being done to fix or prevent further harm.

[2] https://www.britannica.com/place/Polynesia.

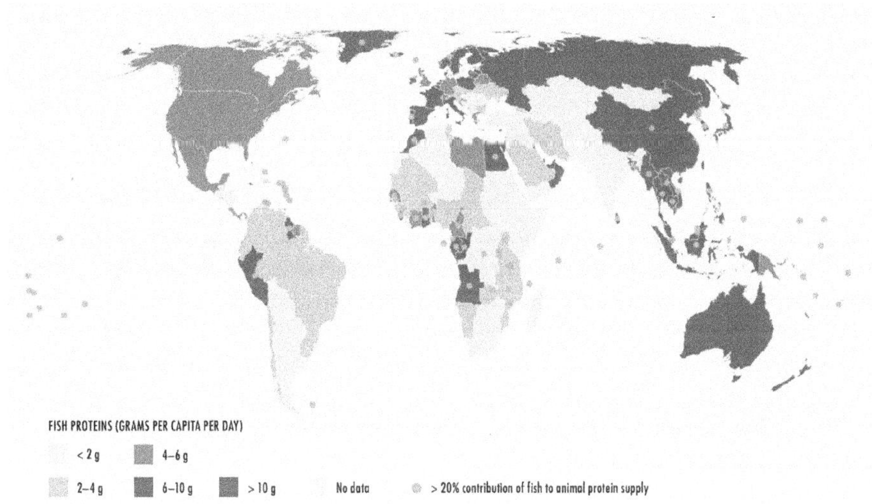


**FIGURE 1** Fish contributions to animal protein intake (averages 2015–2017). (FAO. 2020. *The State of World Fisheries and Aquaculture 2020.* Sustainability in action. Rome. https://doi.org/10.4060/ca9229en.)

The principal notion to remember is that fish are critical to the ocean ecosystem and human health.

> For the last decade or two, the Food and Agriculture Organization of the United Nations (FAO) and other reporting bodies in Europe, North America, and other nations have stated that the world's fish stocks are being overfished.[3]

Overfishing is an issue that must be resolved. The effects could be catastrophic on the environment and ocean ecosystem. Without fish, snorkeling trips, nature vacations, and the ocean would all change. The unfortunate fact is that there is overfishing and ending it will take large amounts of effort and time. As is the case in other industries, when there is money to be made, someone will find a way to make the money. This mantra holds for multiple types of fishery activities, but results in negative environmental effects and more money for the fisheries that are successful. Obviously, there is money to be gained by selling fish, even if the long-term effects are grim at best. Unfortunately, ending overfishing poses a challenge since it went unregulated for so long.

An additional negative practice for some fisheries is Illegal, Unreported, and Unregulated (IUU) fishing. The practice, which is clearly legally questionable, creates numerous difficulties when trying to rejuvenate species of fish. Most countries strive to combat this exploitation. Historically, there are examples that demonstrate the life cycle of the current situation. A prime example of a possibly successful rejuvenation is whales. Whales were subject to heavy hunting for their fat. Since whaling came to prominence and was banned before the turn of the 21st century,[4] we have the

[3] Introduction Fishing and Aquaculture Chapter.

[4] https://www.pbs.org/wgbh/americanexperience/features/whaling-history-whaling-america/.

luxury of hindsight when analyzing the phenomenon. During its peak years, whaling proved to be an important economic catalyst. The U.S. first treaty with Japan centered around the idea of helping stranded sea men, re-fueling American ships, and opening Japanese ports to American ships.[5] At that point in history, the American whaling industry had expanded to the Northern Pacific and would benefit from friendly ports. While one of the multiple reasons to create a relationship with the Japanese, the U.S. did hope to aid American whalers. This interaction shows the influence that whaling had at that time. Consequently, whales were hunted to near extinction and currently many are on the endangered species list.[6] Whaling had a significant impact on the whale population, and in turn the aquatic ecosystem. Luckily, countries ratified a treaty that banned whaling and has resulted in a steady growth in the whale population.[7] Something worth noting is that Japan recently withdrew from the international whaling ban and has legalized whaling in its territorial waters. As a result, the recent progress has been slowed.[8] Despite the efforts of many, the actions of a few still carry significant influence on the larger scale.

Overall, the whaling example shows how human actions can be economically beneficial, yet detrimental to a species. The choices that governments, companies, and individuals make can have a large impact on the world. Fortunately, the whaling example also shows that negative trends can be reversed and hopefully undone with time. Whaling also shows the importance of international support and cooperation in reversing bad practices. The actions of a few countries are hindering the beneficial actions of many. Currently, fisheries are facing a similar dilemma. Many countries updated their laws to create sustainable fisheries; however, the actions of a few through IUU undermine the positive impact of the many. Currently, an issue with fishing is bycatch, trawling, and long-line. These methods can destroy ecosystems and harm species of fish that are not the primary targets. While the effects are unintentional, they must still be addressed.

Given the issues facing aquaculture and environment, many countries and international organizations have taken strides to reverse the worrisome trends. The U.N. has ratified treaties and passed charters allowing countries to enforce national law in the ocean and setting standards for fisheries. The Food and Agriculture Organization of the United Nations (FAO) created binding agreements for sustainable and environmentally safe fishing and collecting aquatic resources. Since fish are not the only resource in the ocean, it is important to protect all parts of the ocean. When an oil rig leaks, or a boat tears up the ocean floor, the consequences extend beyond fish. Human activities affect the environment, as well as the fish populations. When an oil rig leaks, fish, birds, and terrestrial life is impacted. The oil kills the fish, makes it hard for birds to fly and find food, and land-based species lose part of their habitats to the oil. One accident harms many parts of the environment. These are the incidents that while are regrettable and unintended, still occur and need to be curbed. One of the following chapters analyzes the current situation and possibilities going

[5] https://history.state.gov/milestones/1830-1860/opening-to-japan.
[6] https://www.worldwildlife.org/species/whale.
[7] https://iwc.int/estimate.
[8] https://www.worldwildlife.org/species/whale.

forward. There have been actions taken by the U.N., FAO, and groups of countries to combat problems facing the ocean. The actions taken are recent and not all have been enforced for enough time to judge their impacts. Nevertheless, action is important and will eventually be measurable. At that time, the cultures, economies, and animals that rely on the ocean will be able to thrive and people will have what they need.

As is evident by the growing information and statistics, the ocean needs protecting. There have been actions taken to preserve the ocean and all that rely on it. Protecting the ocean and fish is a noble and critical goal, human needs are still important. A balancing act comes from protecting the ocean and supplying people with what they demand.

Obviously, demand is in a way a source of the situation. Furthermore, as the global population grows, the quantity demanded of fish will not decrease. The cultures, people, and economies that relied on fish and that are being preserved through the international regulation cannot abandon their habits. The question of solutions and how to achieve a sustainable solution becomes relevant. Evidently, creating regulation for sustainable fishing is important, but does not necessarily resolve the increasing quantity demanded. In our modern world, there are solutions, some of which revolve around new technologies.

Technology is a blessing that has created opportunities and allowed for advancements once believed to be impossible. Technology has shaped the late 20th and early 21st centuries. Since the 1980s the upgrades and mass production of computers, phones, and transportation have changed the daily lives of many people. Technology has also changed medicine, research, and food. Recently, Genetically Engineered (GE) animals have been introduced as possible food options. The new development raises many questions, possibilities, and risks (Figure 2).

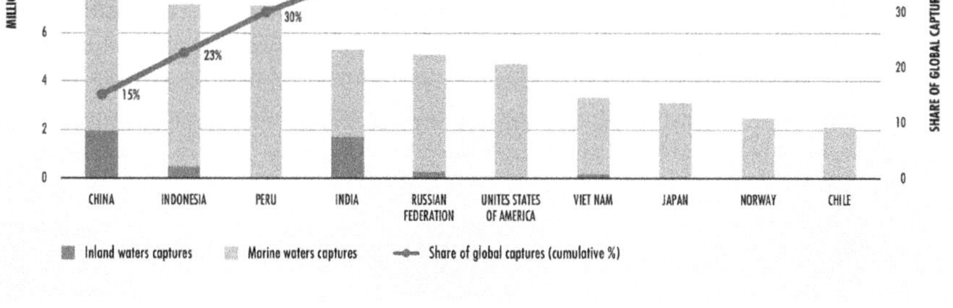


**FIGURE 2** FAO data on global fish capture. (FAO. 2020. *The State of World Fisheries and Aquaculture 2020.* Sustainability in action. Rome. https://doi.org/10.4060/ca9229en.)

## GE FISH ARE ANIMAL DRUGS, NOT FOOD

"Notably, the biotechnology industry plays a vital role in the U.S economy, making the development of GE products relatively favorably regarded in the U.S from both an economic and political perspective."[9] What a grave and short-sighted mistake! The economic perspective in allowing GE fish to be marketed is far outweighed by the risks to the sustainability of food systems around the world.

A new market offers economic growth, new source of tax revenue, new job opportunities, and in this case a consumer good. But the political benefits and pitfalls to GE salmon are more challenging to measure.

Politically, the salmon can create a tool for trade and a way to promote environmental change. Countries trade with each other to boost their economies and gain access to goods previously inaccessible. Trade has also been used as a negotiating tool to sway (or bully) a country. An embargo is an example of how trade can be used politically. WTO disputes are rich in examples of how GE foods have been forced onto countries for economic and political reasons, setting a dangerous precedent for GE fish overrunning those countries with a conscience against these Frankenfish.

GE animals are new and pose risks. There are the environmental, health, and regulatory risks that must be addressed before GE animals can be a complete positive – if ever. The risks involved could also result in GE animals causing more harm than benefits, and this harm can cross species and geographic borders. Notably, the environmental and regulatory risks involve the safety of the environment. There are many unanswered questions about what could happen if GE animals escape from captivity. Assuming they escape, they would become invasive species since many have been modified and are different from wild animals. Such an incident would be detrimental to the overall goal of environmental and economic boosts.

GE fish, the first accepted by the FDA, is not deemed a food at all. Seeing that GE animals are considered food, they are certified by the FDA before being distributed. However, the FDA does not certify the animals as food, but rather an animal drug. The process does still account for safety but is different from food certification. As a result, GE animals do not need to meet the same standards as other food products. Fortunately, the producers will respect and strive to make a product that is healthy for humans. The producer can only create revenue and stay in business if the GE salmon is bought. While the FDA may not be regulating the salmon, the free markets will impact the product. The combination of certification and market power provides a level of safety and confidence in a product. However, confidence is not equivalent to knowing what is in your food. The best way to reassure customers is transparency.

In reality, there are opportunities for GE animals to be served without the consumer being aware. Some of the situations do not need transparency from a legal perspective but certainly from an ethical one. The potential lack of transparency is worrisome as many effects of GE animals remain unknown, but the science suggests that GE animals are more trouble than they are worth. Furthermore, having knowledge of what is in food is paramount for people with allergies. There are concerns

---

[9] GE Salmon Chapter 1.2.1 in *International Farm Animal, Wildlife, and Food Safety Law* (Steier & Patel, Eds.) (Springer 2017).

with the modern fishing industry and GE animals. Both must be resolved to maximize any long-term rejuvenation of aquaculture and human health. GE animals can be viewed as a solution to the questions concerning fishing, but with the unanswered questions, the new technology could also create further issues.

To safely combat both issues, governments have introduced regulation to control fisheries, create sustainable fishery practices, and protect the environment and humans from potential GE animal issues. In the following section, the current aquaculture situation is analyzed in depth. Specific observations are made regarding international fishing regulation and sustainable fishing. There is room for improving the sustainability of global fishing efforts, their legality, as well as the animal welfare concerns across the oceans and rivers.

# 1 Marine Fishing and Aquaculture

## *A Global Perspective*

*Audrey Amescua*
Northeastern University

**CONTENTS**



DOI: 10.1201/9781003137733-3

## 1.1 INTRODUCTION

For the last decade or two, the Food and Agriculture Organization of the United Nations (FAO) and other reporting bodies in Europe, North America, and other nations have stated that the world's fish stocks are being overfished. The alarming statistic, reported by many media outlets, regulatory bodies, and academics, has left some to argue that the state of the marine fisheries is being exaggerated and not encompassing the many national and international efforts in the past decade to improve the world's fish stocks. Despite these recent improvements, the fact is that global fish production reached 179 million tons in 2018, and global capture fisheries production reached the highest level ever recorded (96.4 million tons).[1] Demand for fish has grown with a 122 percent rise in total fish consumption between 1990 and 2018, and the world's estimated population growth will continue to put pressure on the world's fisheries.[2] Though the state of the world's fisheries is improving, there is still a dire need for intervention and improved regulations and policies for marine wildlife and resource management. A prime example of poor management is the shrimp industry, where the high demand is expected to increase. Shrimp fisheries rely heavily on the use of bottom trawlers, which leads to high amounts of bycatch (catch of non-target species) and waste.[3] The top fish-producing countries include regions in South Asia,[4] where there are also high instances of illegal, unreported, and unregulated (IUU) fishing activities.[5] The primary issues that need to be addressed through the proper management and regulatory oversight activities are overfishing, IUU fishing, bycatch, and overcapacity of fishing vessels.

### 1.1.1 OVERFISHING

The many economic incentives in fisheries have historically led to a race to exploit the oceans as quickly as possible. The results were overfishing and a reduction in fish stock capacity. The FAO's 2020 report on the state of world fisheries and aquaculture put the estimate of overfished stocks at more than one-third,[6] specifying that 34.2 percent of marine fish stocks are overfished (biologically unsustainable) and 59.6 percent are maximally sustainably fished (biologically sustainable). Only 6.2 percent of marine fish stocks are underfished.[7] This means that only 65.8 percent of fish stocks are at biologically sustainable levels.[8] Given these facts, the prospects of marine fisheries are worrying. This problem is felt particularly hard in developing countries where the primary source of protein is fish. Many also rely on the fishing industry as a source of livelihood, stemming from the strong global demand for high-value fisheries exports like shrimp. For example, in Bangladesh, Cambodian

[1] Food and Agricultural Organization of the United Nations (2020a), p. 2.
[2] Food and Agricultural Organization of the United Nations (2020b).
[3] Gillett (2008), p. 47.
[4] Food and Agricultural Organization of the United Nations (2020a), p. 80.
[5] Food and Agricultural Organization of the United Nations Regional Office for Asia and the Pacific (2007).
[6] Food and Agricultural Organization of the United Nations (2020a), p. 2.
[7] Food and Agricultural Organization of the United Nations (2020a), p. 47.
[8] Ibid, p. 7 and p. 47.

Gambia, Ghana, Indonesia, Sierra Leone, and Sri Lanka, fish accounts for half of the population's intake of animal protein.[9] Poor fishery management, short-sighted practices, overfishing, and unregulated fishing have been the primary drivers in the decrease in catches and rising costs of operations.[10]

Between 1990 and 2018, there has been a 14 percent growth in global capture fisheries production, a 527 percent increase in global aquaculture production, and a 122 percent rise in total food fish consumption.[11] Total global captures fisheries production reached 96.4 million tons, the highest level ever reached; aquaculture also reached an all-time high of 114.5 million tons.[12] The primary species that drove global capture were anchoveta, Alaska pollock, and skipjack tuna.[13] Aquaculture had the highest rise in aquatic animals, aquatic algae, and ornamental seashells and pearls.[14]

Technological advancements in marine fishing equipment have allowed for increased catch efficiency. These advancements, while economically beneficial, have taken a toll on fishery and coastal stocks. As humans became more efficient at predatory fishing practices, fisheries became increasingly exploited. Between 1961 and 2016, global fish consumption outpaced population growth twofold, with fish increasing 3.2 percent, while population growth increased by only 1.6 percent.[15] This increased consumption has been driven partially by increased production, particularly in aquaculture and thanks to technology.

If managed correctly, fish stocks can be consistently above target levels or rebuilt. The key to sustainability in aquaculture and fisheries is strong management. The key has been proven in recent initiatives from the Global Environmental Facility (GEF) and the FAO to improve the state of the world's tuna stocks. Thanks to the joint GEF-FAO Program, 8 of 13 major commercial tuna stocks are no longer considered overfished; before this program, 13 were experiencing overfishing; the number is now at 5.[16] Programs and joint efforts such as the joint GET_FAO program are ways in which fisheries' management is improving. In areas where there is poor management or total lack of management, fish stocks are depleting quickly. These areas tend to be in developing nations where hunger, poverty, and conflict exist.[17]

### 1.1.2 IUU Fishing

IUU fishing is a term that attempts to capture many independent yet related factors and is one of the greatest threats to the sustainability of the world's fisheries. Craftily adopted by the international community, the term IUU fishing is an attempt to clearly differentiate sustainable fishing from unsustainable fishing. "Illegal" refers to fishing

---

9 Hunt (2020).
10 Garcia and Rosenberg (2010).
11 Food and Agricultural Organization of the United Nations (2020b).
12 Food and Agricultural Organization of the United Nations (2020a), p. 6.
13 Food and Agricultural Organization of the United Nations (2020b).
14 Ibid.
15 Food and Agricultural Organization of the United Nations (2018a).
16 Food and Agricultural Organization of the United Nations (2020c).
17 Food and Agricultural Organization of the United Nations (2020a), p. 115.

activities that violate national or international laws or measures. "Unreported" refers to vessels that have not reported or have misreported catches to the relevant national authority. "Unregulated" means that the vessel is operating in a manner that is inconsistent or contradicting the conservation and management measures put in place by regional fisheries management organization (RFMO) when the flag state of a given vessel is not a member of the RFMO when fishing activities occur in areas where there is no applicable conservation or management measures, or where fishing activities are conducted inconsistently with conservation of living marine resources under international law.[18]

The resulting consequences from IUU fishing are crucial to identify provided their level of potential and actual ecologic, economic, and social harm. The collapse of the Bluefin tuna stocks in the 1990s and early 2000s has been one of the worst ecological consequences of illegal fishing activities. In combination with overfishing, illegal fishing caused Bluefin tuna in the Pacific, Atlantic, and Indian Oceans to decline dramatically. The Eastern Atlantic Bluefin has significantly recovered over the last decade thanks to better management, low conservative quotas, and the crackdown on IUU fishing, but there is still much work that needs to be done to ensure the sustainability of the species.[19] In 2018, Spanish authorities discovered an extensive operation of illegal Bluefin tuna trade. The illegal catch, originating in Italy and Malta, entered the EU market through Malta, where illegal tuna trade exceeded legal trade twofold, resulting in an annual profit of 12.5 million euros.[20] Economically, IUU fishing undervalues global fishery resources and harms legitimate business. Each year, 7.7 to 14 million metric tons of unreported fish catches might be linked to illicit trade.[21] These fish taken out of the legitimate, legal food system pose a threat to food security and the livelihood of many, negatively impacting the annual income of seafood workers by an estimated $6.8 to 13.3 billion.[22] The exact economic loss due to IUU fishing is difficult to value due to the illicit nature; however, it is agreed that the number is in the billions[23]; one study estimates the loss to be anywhere from $25.5 billion to as high as 49.5 billion USD annually. Socially, the illegal and illicit nature of IUU fishing lures in other organized crimes. According to one report, IUU fishing was found to be the sixth most valuable black-market industry in 2017.[24] Studies show that fishing vessels that are involved in IUU fishing have also been linked to forced labor, labor abuses, modern slavery, and human trafficking.[25] In addition, tuna and swordfish are highly profitable industries and ones in which IUU fishing is challenging to detect due to the commingling of legal and illegal groups. These industries could provide another profit-generating strategy for organized crime syndicates that would be involved in the trafficking of drugs, arms, or people.[26]

18 Food and Agricultural Organization of the United Nations (n.d.).
19 Cheney (2020).
20 WWF (2018).
21 Sumaila (2020), p. 2.
22 Ibid.
23 NOAA (n.d.a).
24 Mavrellis (2017).
25 International Labor Organization (2013).
26 UN Office of Drugs and Crime (2011).

Primary channels for illicit trade include using trans-shipments, exporting bulk in large, refrigerated containers, and transporting illegally caught seafood into local and regional markets. Trans-shipments involve offloading catches from multiple vessels into large freezers and processing ships at sea, making traceability of the catch nearly impossible.[27] Many, however, say that trans-shipment activity is essential to operations, largely due to economic factors.[28] Exports in large containers are generally subjected to less stringent reporting and inspection requirements, making it an excellent way to ship the illegally caught seafood. Furthermore, transporting illegally caught fish into local and regional markets where demand is strong and exports are high also creates a lack of transparency because the fish move quickly through the market.[29]

### 1.1.3 Bycatch

Bycatch refers to marine discards of non-target fish (both commercial and non-commercial) and may include sea birds, mammals, and turtles accidentally trapped in nets. Animals caught as bycatch are generally undesired species or unintended species that fishermen dispose of at sea either because they do not want to be accused of damaging already endangered species (regulatory discards) or because keeping the discards would impact economic bottom lines by occupying needed cargo space (economic discards). Most unintentionally caught animals suffer injuries or die.[30] The FAO's 2020 annual report found that between 2010 and 2014, annual discards were estimated at 9.1 million tons (10.8 percent of global catch), of which 4.2 million were from bottom trawls, 1.0 million tons from purse seines, 0.9 million tons from midwater trawls, and 0.8 million tons from gillnet fisheries.[31]

The last numbers show that some fisheries are more prone to bycatch than other fisheries and partly dependent upon the type of fishing gear, some of which are particularly problematic. For example, the shrimp fisheries have notoriously high levels of bycatch, with trawlers catching 3–15 times more bycatch than actual shrimp. The bycatch results are due to the use of bottom trawling (dragging a net through the bottom of the seafloor to catch fish and other aquatic species for tropical shrimp in particular). Trawling not only harms marine life but also damages the environment and ecosystem.[32] Another example of problematic equipment includes the use of gillnets, which are often associated with the unwanted catch of large marine species such as sharks, dolphins, whales, and other protected marine species.[33] The FAO's 2020 report revealed that the number of interactions between fisheries and endangered, threatened, and protected species included at least 20 million individual animals.[34] On the whole, the extent of bycatch is not fully understood since many states do

---

27 Sumaila (2020).
28 Food and Agricultural Organization of the United Nations (2020d).
29 Sumaila (2020).
30 NOAA (n.d.b).
31 Food and Agricultural Organization of the United Nations (2020a).
32 Food and Agricultural Organization of the United Nations (n.d.).
33 Reeves, et al. (2013).
34 Food and Agricultural Organization of the United Nations (2020a).

not have comprehensive statistics during coherent periods of discard rates, nor does this exist at the global level. One study suggested that fishery-specific estimates of discard rates are not available for over two-thirds of global fisheries.[35] It is clear that bycatch contributes to overfishing, slowing efforts to replenish fish stocks, and harming marine wildlife and the environment, ultimately causing a major threat to sustainability efforts and the biodiversity of the world's oceans.

### 1.1.4 Overcapacity of Vessels in Industrial Fleets

The use of the world's ocean resources exemplifies the economic concept of the Tragedy of the Commons. Common pool resources are often characterized by excess capacity and economic inefficiency, a fundamental problem in global fisheries. Fishing capacity is the number of fish that can be caught within a given period by a vessel or fleet if that vessel or fleet is fully utilized. When the maximum harvest that a fleet could catch given set levels of inputs (e.g., fuel, gear, ice, bait, engine power, vessel size) exceed the desired harvest level, the vessel or fleet has reached overcapacity. Vessels or fleets operating at overcapacity catch more fish than there are fish available.[36] Overcapacity of fishing fleets is largely perceived as a major impediment to achieving economically productive fisheries. Nevertheless, overcapacity is driven by numerous factors, including heavy subsidization and high global demand. The subsidization of the fleets personifies as mismanagement of the fisheries industry from a political standpoint. Programs such as the Capital Construction Fund and the Fisheries Obligation Guarantee Program lead to more vessels being produced than can possibly fish in a given fishery.

Globally, the number of fishing vessels was at 4.56 million in 2018, a reduction of 2.8 percent compared to 2016.[37] Asian countries have the largest fleet with 3.1 million vessels (68 percent of the global total). Next to Asia, Africa holds the second largest number of fishing vessels (20 percent of the global total), followed by the Americas (10 percent), Europe (2 percent), and Oceania (less than 1 percent). China's capacity reached peak numbers in 2013 but has declined between 2016 and 2018, from 1,071,000 vessels to 864,000 vessels (20 percent).[38] Overcapacity is about the number of vessels in the ocean and the size of those vessels. Small vessels dominate the overall number of boats in the ocean; large vessels make up a much smaller percentage of the global fleet, with the highest proportions of these boats in Oceania, Europe, and North America. Small vessels tend to lack as much scrutiny and reporting as large vessels.[39] Large fleets tend to be more technologically advanced and efficient, resulting in an increased catch. Too many large vessels deployed leads to overall less fish to catch and fewer profits for all stakeholders. Buybacks and allocations of individual fishing quotas are often used as a way to manage overcapacity. Ultimately, overcapacity leads to lower economic performance and fuels IUU fishing.

[35] Gilman, et al. (2020).
[36] Pew Environment Group (2009).
[37] Food and Agricultural Organization of the United Nations (2020a).
[38] Food and Agricultural Organization of the United Nations (2020a), pp. 41–42.
[39] Ibid, pp. 44–46.

## 1.2 WATER AND MARINE ANIMAL LAW

### 1.2.1 International Efforts

Even more so than land-based animals, fish move freely through the world's oceans and do not swim strictly in specific international geographic and political borders or boundaries. Therefore, it is increasingly necessary that countries work together to promote proper management practices in international fisheries.

#### 1.2.1.1 United Nation's Law of the Sea

The oceans have historically been subject to the Freedom-of-the-Seas Doctrine, a principle emerging in the 17th century that essentially limited all national and jurisdictional rights over the ocean to merely the sea surrounding a nation's coastline. The rest of the ocean's waters were to be "free to all and belong to none" – a heavily romanticized notion captured in movies, television, and books. Over time, as nations built ever-expanding fleets and began exploiting the vast resources offered by the open ocean, concern began to grow over coastal fish stocks and the pollution caused by transport ships and oil tankers. As pollution grew and tension over rights to the world's resources grew, so did the threat of conflict and instability. In 1945, amid pressures from domestic oil interests, U.S. President Harry Truman declared jurisdiction over all natural resources on the nation's continental shelf, the first direct contradiction to the freedom-of-the-seas doctrine. Other nations quickly began making similar claims as many promptly realized the vast resources, such as oil, tin, diamonds, metals, and fish, that could be exploited thanks to new and emerging technologies.[40]

Due to these growing concerns and the clear need for international regulation and intervention, in 1957, United Nations (UN) Member States held the United Nations Convention on the Law of the Sea (UNCLOS) where they negotiated "a legal order for the seas and oceans which will facilitate international communication, [to] promote the peaceful uses of the seas and oceans, the equitable and efficient utilization of their resources, the conservation of the living resources, and the study, protection and preservation of the marine environment."[41] The Convention met three times, and in 1982, formally adopted the United Nations Law of the Sea (LOS) and created the first international framework for regulating the Earth's water resources. This body of customs, treaties, and international agreements was ground-breaking for international law and policy.[42]

The LOS covers several primary issues, including limits to states' maritime zones (establishing territorial seas), navigation rights, continental shelf rights, deep-sea mining, protection of the marine environment, and mechanisms for conflict resolution. These issues established several key regulatory and oversight committees, including the International Tribunal for the LOS, International Seabed Authority, and the Commission of the Limits on the LOS.

---

40 United Nations (n.d.).

41 Statute of the International Tribunal for the Law of the Sea, Dec. 10, 1982

42 United Nations (n.d.).

**Limits to States Maritime Zones (Territorial Seas)**: The first task of the UNCLOS was to set territorial limits (who gets what and who owns what). Without this, no other topic could be negotiated. A territorial sea is a coastal state's right to completely control a certain amount of distance off its shoreline. This right had been recognized by international law, but the extent of how this territorial sea should extend and how wide it should be was not clear. At the start of the convention, the amount of territorial sea claimed by states varied – 25 states maintained a 3-mile territorial sea, 66 had extended their territorial seas to a 12-mile limit, 15 others claimed limits between 4 and 10 miles, and 8 states claimed 200 nautical miles of territorial sea. Probably not surprisingly, smaller coastal states wanted a vast territorial sea to protect themselves better. In contrast, larger nations with formidable naval and marine power wanted a much more limited territorial sea to allow for more freedom of movement and easier navigation of the ocean's waters. Ultimately, the move toward a 12-mile territorial sea gained acceptance. Within this range, the controlling nation would have the rights to enforce their laws and regulate the use and exploitation of marine resources. The right of innocent passage was permitted to naval and merchant ships, meaning states would not have to detour 3,000 miles to avoid territorial seas of coastal states. To address the concerns of coastal states of protecting themselves, preventing certain violations, and enforcing police powers, they were granted a "contiguous zone" which extended certain rights 24 nautical miles beyond their shore. This zone may be used by a coast guard or other naval agencies to protect against smugglers, illegal immigrants, or evaders of the law. Another special right was given to archipelagic states (states made up of a dense cluster of islands such as Indonesia and the Philippines). These states have a territorial sea of 12 miles extending from a line drawn connecting the outermost points of the perimeter islands of the group that are near each other. The waters between the islands are archipelagic waters, where ships of all states have the right of innocent passage and to established sea lanes and air routes to ensure speedy and unobstructed passage.[43]

**Exclusive Economic Zone (EEZ)**: EEZs recognize the right for coastal states to have jurisdiction over the waters extending 200 miles from their shores, giving each state sovereignty to exploit, develop, manage, and conserve all resources within their zone. The majority of offshore resources, including the most lucrative fish stocks, are located within EEZs, making them some of the most economically beneficial areas of the oceans. It is also one of the most critical areas for fishing and conservation of the marine environment and was recognized in the 1958 Convention on Fishing and Conservation of the Living Resources of the High Seas. Today, 86 coastal states have economic jurisdiction of up to 200 miles limit, causing 99 percent of the world's fisheries to fall under the purview of some nation's jurisdiction.[44]

**Navigation Rights**: One of the significant points of contention during the UNCLOS was in the differing interests of the coastal states and the major naval powers regarding navigational rights. Traditionally, most countries claimed some part of the sea just beyond their shores as part of their territory to protect against smugglers, warships, and other intruders. The rights of this territory were primarily

[43] UNCLOS; United Nations (1998).
[44] United Nations (1998).

for protection, though other interests such as the use of resources did begin to emerge. By the late 1960s, many countries began to declare a territorial sea stretching 12 miles from their coasts. These declarations were a significant issue for major naval powers, especially given the tensions between the United States and the Soviet Union. Having a 12 mile territorial sea essentially closed off access to straights used for international navigation, including the straights of Gibraltar (8 miles wide and only open access to the Mediterranean), Malacca (20 miles wide, main route between Pacific and Indian Oceans), Hormuz (21 miles wide, only passage to oil-producing areas of the Gulf States), and the Bab el Maneb (14 miles wide, connects the Indian Ocean and the Red Sea). The United States and the Soviet Union wanted free passage through the straights, effectively giving these waters the same rights as international waters of the high seas, while the coastal states, not wanting to be involved in the affairs and conflicts of the outside Powers, felt this was a significant threat to national security. The coastal states were willing to grant the right of "innocent passage," meaning naval powers controlling submarines would have to rise to show their flags, a critical concern for these powers during the Cold War. Also not covered in the right of innocent passage was the right of passage over the straights for airships.[45]

This issue of straights, overflight, and width of the territorial sea was one of the last major negotiations needed and was the focus of the third UNCLOS conference. A compromise was reached when the parties settled on a new concept of transit passage, which legally accepted the provisions of innocent passage and freedom of navigation on the high seas. Essentially, it retains the international status of the straights and gives naval powers the right of unimpeded navigation and overflight. However, in all other matters, the sovereignty of the straights belonged to the coastal states, and they could maintain the territorial sea.[46]

**Continental Shelf Rights**: Increased exploration and new technologies have allowed for the seafloor to be explored, exposing a vast number of natural resources ranging from natural gas to minerals, from sand and gravel to oil, diamonds, and gold. The use of the ocean to exploit resources offered by the continental shelf is a significant departure from the tradition of simply using the seas to navigate and fish. Not surprisingly, issues over who should hold rights to these rich resources began to emerge. Some states naturally had a wide shelf, while others had virtually no shelf at all. The topics of the border drawing processes for control over the continental shelves and the exploitation of the resources drew significant interest from the UNCLOS members. Truman's declaration that all resources around the United States' continental shelf belonged to the United States sparked the same types of claims from other nations. However, what the "continental shelf" included was ill-defined in these claims. Many states simply claimed rights to the seabed and subsoil near their territorial seas, obscuring who was claiming what. The first task facing the members was to clearly define a continental shelf. The first Convention settled on "the seabed and subsoil of the submarine areas adjacent to the coast but outside the area of the territorial sea, to a depth of 200 meters, or, beyond that limit to where the

[45] United Nations (1998).
[46] Ibid.

depth of the suprajacent waters admits of the exploitation of the natural resources of the said areas."[47]

With at least the definition settled, it was generally agreed that coastal states should have control over ocean resources out to 200 miles from their shore so that the outer limits coincide with the EEZ. However, the matter remained of those states who wanted control of the entire geographic shelf extending beyond 200 miles wide for broader economic jurisdiction. These geographically advantaged nations, including Argentina, Canada, India, Madagascar, Mexico, Sri Lanka, and France, were given the possibility of establishing a boundary extending out to 350 miles from their shores or even further depending on specific geographic criteria. The Convention resolved geographical disadvantages by setting the 200 miles EEZ limits as the boundary of the continental shelf for seabed and subsoil exploitation. Importantly, the rights to the continental shelf would not affect the legal status of the waters of the airways above the continental shelf.[48]

**Deep-Sea Mining**: Even with defined territorial limits and continental shelves claimed, there is still a sizeable part of the open ocean that is non-territorial with immense resources. It was discovered in 1874 that huge nodules containing metals such as nickel, copper, and cobalt existed in surplus, covering an area of 1.35 million square miles in the Pacific Ocean between Hawaii and Tahiti in an area known as the Clarion Clipperton Zone. During the UN general assembly in 1970, these resources were determined to be beyond the limits of national jurisdiction,[49] putting the zone and its resources under the jurisdiction of the International Seabed Authority. The International Seabed Authority has the power to grant mining rights to deep-sea mining contracts, currently covering 400,000 square miles of the zone, but has deemed nine areas as Areas of Particular Environmental Interest (APEIs) that are protected from mining activities.[50]

**Protection of the Marine Environment**: Pollution of the Earth's waters has been a growing concern, especially with offshore oil interests and the threat of oil tanker accidents and pollution. One of the most historical and devastating industrial disasters occurred in 2010 at an offshore oil rig in the Gulf of Mexico. An explosion at Deepwater Horizon caused 40 million barrels of oil to spill into the ocean for more than 87 days.[51] The oil spread at 1,100–1,300 mile depth, contaminating deepwater habitats. Long-term effects for fish species, deep-sea corals, sea turtles, and cetaceans are still not fully understood, while salt marshes and seabird populations were deeply impacted.[52] It is not just industrial disasters that cause harm to the oceans, however. Every time a tanker cleans its tanks at sea or a factory channels toxic residues to coastal waters, a city releases raw sewage into the sea, or the oil from an automobile gets poured into the sewers, the oceans become a little more polluted. The oceans' ability to heal and rejuvenate itself is not endless and will eventually lead to catastrophe if not properly addressed.

---

47 Ibid.

48 Ibid.

49 Ibid.

50 National Oceanic and Atmospheric Administration (n.d).

51 Environmental Protection Agency (n.d.).

52 Beyer, et al. (2016).

UNCLOS addressed six primary sources of pollution – land-based and coastal activities, continental-shelf drilling, potential seabed mining, ocean dumping, vessel-source pollution, and atmospheric pollution.[53] The Convention made it obligatory to strive for the protection and preservation of the marine environment. Additionally, all states were urged to cooperate on a global and regional basis in formulating rules and standards for marine environment protection. Every state was given jurisdiction to protect and preserve the marine environment of its EEZ. State jurisdiction allows coastal states to enforce anti-pollution measures to prevent and reduce marine pollution from dumping, land-based sources, or seabed activities. For marine vessels with the right of navigation, coastal states are allowed to exercise jurisdiction only to enforce laws and regulations outlined by the Convention or generally accepted rules and standards for international waters.

Furthermore, it is the duty of the "flag state," the state where the ship is registered and whose flag it flies under, to enforce rules adopted to control marine pollution. The "port state," which is the State where a ship is destined, has enforcement powers when dealing with shipping standards, marine safety, and pollution prevention. The International Seabed Authority is given discretionary powers to assess for potential environmental impacts of deep-sea mining and can recommend changes or formulate rules, regulations, or monitoring programs to ensure the marine environment is protected sufficiently.

**Binding Procedures for Settlement Disputes**: Regarding dispute settlements, the LOS is unique compared to other international treaties in that the mechanism for dispute settlements is incorporated into the document, rendering it mandatory for members to go through the settlement procedures for disputes. Any dispute process, whether through the International Tribunal for the LOS, the International Court of justice, international arbitration procedures, or special arbitration tribunals with expertise in specific disputes, would be subject to third-party settlements.[54] The mandatory dispute makes the LOS unique in international law.

### 1.2.1.2 UN Fish Stocks Agreement (UNFSA)

The UNFSA was signed in 1995 and came into force in 2001. The Agreement builds upon LOS and is considered one of the most important steps forward in implementing a comprehensive legal framework for the conservation and sustainability of straddling fish stocks[55] and highly migratory species.[56] The Agreement specifically addresses the issues of IUU fishing, over-capitalization and overcapacity of fishing vessels, and cooperation between countries in addressing inadequate fishing practices and poor fisheries management. The Agreement furthers the LOS by requiring that states cooperate beyond areas under national jurisdiction to ensure conservation and optimum utilization of resources on the high seas. RFMOs, under the Agreement, are the primary vehicles ensuring cooperation between coastal states and high seas

[53] United Nations (1998).
[54] Ibid.
[55] Those stocks that occur both within a country's EEZ and in adjacent high seas, such as cod, halibut, pollock, jack mackerel, and squid (United Nations Department of Information, 2020).
[56] Such as tuna, swordfish, and oceanic sharks (United Nations Department of Information, 2020).

fishing states in managing straddling and highly migratory fish stocks.[57] It also strengthens the responsibility of flag states over the fishing vessels flying their flags on the high seas.[58] Under the Agreement, fishery management must be based on a precautionary approach using the best available science.[59] There are currently 77 parties to the Agreement, including the European Union and most of the world's top fishing countries. Participating in the Agreement is an essential way for a nation to indicate its commitment to responsible fishing and marine ecosystem preservation.[60]

#### 1.2.1.3 FAO Compliance Agreement

The Agreement to Promote Compliance with International Conservation and Management Measure by Fishing Vessels on the High Seas (shortened to The Compliance Agreement) sought to improve the regulation of fishing vessels on the high seas. FAO approved the Compliance Agreement in 1993 at the 27th session and took effect in 2003. Like the UN Fish Stocks Agreement, the FAO Compliance Agreement was meant to supplement the LOS and fill in gaps not addressed by the LOS. Acceptance to the Agreement is open to any Member or Associate Member of the FAO and any United Nations Member State. Currently, there are 42 members of the Agreement.[61] The Agreement gives flag states greater responsibility over their fishing vessels to ensure compliance with international conservation and fishery management measures. Flag states were made responsible for ensuring no fishing vessels flying their flags fishes on the high seas unless authorized. The Compliance Agreement also deters re-flagging of fishing vessels on the high seas under flags of states that are less able to enforce international rules for conservation and management.[62]

#### 1.2.1.4 FAO Agreement on Port State Measures to Prevent, Deter, and Eliminate IUU Fishing

Following the approval of the FAO Compliance Agreement and the UN Fish Stock's Agreement, the Agreement on Port State Measures to Prevent, Deter, and Eliminate IUU Fishing (PSMA) was the first binding international agreement to target IUU fishing specifically. The original model for the PSMA was first introduced in 2001 but was non-binding, leaving room for gaps in successful implementation. After much work and difficult negotiations between the FAO and states, the PSMA was approved by the FAO in 2009 at its 36th session under paragraph 1 of Article XIV of the FAO Constitution, with 30 parties initially entering the Agreement. The PSMA entered into force in June of 2016.[63] Since its enforcement, 73 states have become participants in PSMA.[64]

---

57 United Nations Department of Public Information (2020).
58 Ibid.
59 International Seafood Sustainability Foundation (n.d.).
60 United Nationals Department of Public Information (2020).
61 Food and Agricultural Organization of the United Nations (2018a).
62 Fishery and Aquaculture Economics and Policy Division (1995).
63 Food and Agricultural Organization of the United Nations (2021a).
64 Food and Agricultural Organization of the United Nations (2021b).

The ultimate goal of the PSMA is to contribute to the long-term conservation and sustainable use of the oceans and their resources by increasing port state measures and individual states' authority over the enforcement of said measures.[65] The PSMA prevents fishing vessels engaged in IUU fishing from using ports and landing their catches, targeting vessels seeking entry into a port of a State different from the vessel's flag State.[66] The prevention includes not only denying access to ports but also to port services (such as refueling or repair). Additionally, many countries prohibit trade with countries that lack port state measures.[67] By reducing the economic incentives of IUU fishing, the PSMA deters IUU fishing while also blocking the products of IUU fishing from reaching national and international markets.[68] Cooperation between coastal states, flag states, and RFMOs has improved, and fisheries management and governance at all levels has been strengthened as a result of the binding agreement.[69]

#### 1.2.1.5 RFMOs

Thus far, the agreements described outline treaties that set the foundation for international efforts in managing global fisheries. RFMOs are the organizations responsible for ensuring the conservation and sustainable use of marine resources on the high seas to ensure efforts outlined in the agreements are correctly managed and maintained.[70] RFMOs are treaty-based multilateral bodies and organizations that oversee the management of shared fish stocks and other living marine organisms and are responsible for international efforts for sustainable marine fishing and the use of resources. There are 18 RFMOs globally[71] covering nearly all of the global high seas.[72] The UNFSA requires that states exploiting the resources contained in the high seas cooperate in managing those resources. Because fishery resources do not recognize the boundaries between domestic EEZs or the high seas, RFMOs must manage these transboundary resources through cooperation with each other and with coastal states containing EEZs.[73] The effectiveness of RFMOs has historically been questioned given the current state of the world's oceans and the level of fish stocks that are depleted or overexploited.[74] The complex nature of handling political and legal factors (issues of compliance and enforcement, sovereignty, and factors intrinsic to common pool resources) is more likely to be blamed than the management comprehensiveness of RFMOs.[75] More modernly, concerns over the organizations' ability to handle issues related to climate change have come into question.[76] Rather than reforming the RFMO management framework, progress on negotiating international

[65] Food and Agricultural Organization of the United Nations (2021a).
[66] Ibid.
[67] Ibid.
[68] Ibid.
[69] Ibid.
[70] Pentz, et al. (2018), p. 14.
[71] Ibid.
[72] Gonçalves (2021), p. 12.
[73] Pentz, et al. (2018), p. 14.
[74] Cullis-Suzuki & Pauly (2010), p. 1042.
[75] Pentz, et al. (2018), p. 18.
[76] Pentz, et al, (2018), p. 14.

biodiversity conservation measures between nations should be explored to address these issues and give RFMOs new tools to address climate change-related issues.[77]

#### 1.2.1.6 Soft Laws: FAO Code of Conduct for Responsible Fishing (CCRF) and International Plan of Action (IPOA)

Unlike the previously discussed agreements, the FAO CCRF is a voluntary measure adopted by the 1995 FAO Conference to promote long-term conservation and sustainable use of fisheries resources. The Code serves merely as a reference framework for national and international conservation efforts and is not legally binding.[78] The Code defines the principles and international standards of behaviors for effective management of living aquatic resources concerning the marine ecosystem and biodiversity[79] and addresses the responsible advancement of aquaculture development.[80] Despite its non-binding nature, the Code is considered significant instead of international food law. Unlike other Agreements, the Code recognizes explicitly the nutritional, economic, social, environmental, and cultural importance of fisheries and the lives and livelihoods supported by fisheries.[81]

The Code was developed simultaneously as the UNFSA and was intended to reinforce the UNFSA's provisions, which were supplemental to the UNCLOS. The Code contains vital principles for the Compliance Agreement and the Agreement on Port State Measures, binding agreements that can be thought of as tools related to the Code.[82] In addition, the Code establishes IPOA, and some of the leading programs for CCFR implementation. There are four key IPOAs. The first is the International Plan of Action for the Conservation and Management of Sharks[83] (IPOA-Sharks). Due to the slow growth rate, late maturity, and low fecundity, sharks and other related pelagic species are less able to withstand fishing pressures compared to other fish species. These pelagic species are more threatened by overexploitation and unsustainable harvest or loss of habitat.[84] Since 1970, oceanic sharks have declined by 71 percent, owing primarily to overfishing putting three-quarters of shark populations at risk of extinction.[85] The IPOA-Sharks aims to address these issues and preserve stocks. The second IPOA is the International Plan of Action to Prevent, Deter, and Eliminate Illegal, Unreported, and Unregulated Fishing (IPOA-IUU). As previously discussed, IUU fishing undermines the overall objective of sustainable fisheries, leaving both short- and long-term loss of social and economic opportunities. The IPOA-IUU provides measures to prevent, deter, and eliminate IUU fishing, building further upon CCRF.[86] The third is the International Plan of Action for the Management of Fishing Capacity (IPOA-Capacity). This IPOA addresses the issue

---

[77] Pentz, et al., (2018), p. 18.
[78] Food and Agricultural Organization of the United Nations (2021c).
[79] Ibid.
[80] Food and Agricultural Organization of the United Nations (2011), p. 53.
[81] Food and Agricultural Organization of the United Nations (2021c).
[82] Ibid, see Other Related Agreements.
[83] "Sharks" refers to all species of sharks, skates, rays, and chimaeras. Food and Agricultural Organization of the United Nations (2021c) – see Activities, Introduction.
[84] Department of Agriculture-Bureau of Fisheries and Aquatic Resources (2004), p. 112.
[85] Pacoureau et al. (2021).
[86] Food and Agricultural Organization of the United Nations (2001).

of overcapacity of fishing vessels, which is an increasing concern and contributes substantially to overfishing, the degradation of marine fisheries resource, and the decline of food production potential.[87] The last is the International Plan of Action for Reducing Incidental Catch of Seabirds in Longline Fisheries (IPOA-Seabirds). Seabirds are incidentally caught in various long line commercial fisheries across the globe. The primary long-line fisheries where incidental seabird catch occurs are tuna, swordfish, billfish, Patagonian toothfish, halibut, various cod species, and haddock. Albatrosses and petrels in the Southern Ocean, northern fulmars in the North Atlantic, and albatrosses, gulls, and fulmars in the North Pacific Ocean are the most frequent seabird species to be involved in incidental catch.[88]

The Code's Implementation is monitored by the Committee of Fisheries (COFI), which also has the authority to update the Code as necessary.[89] The COFI creates implementational reports, questionnaires to monitor the implementation and execution of the Code. Questionnaires are intended for Members, regional fishery bodies (RFBs), and non-governmental organizations (NGOs). In 2014, questionnaires focusing on Article 9 Aquaculture Development and Article 11 Post-Harvest Practices and Trade were provided to Members, RFBs, and NGOs. Eighty-seven percent of responding members reported some constraints in implementing the Code, including insufficient financial resources, human resources, incomplete policies or legal frameworks, inadequate scientific data and research, and lack of awareness and information about the Code.[90]

#### 1.2.1.7 The Agreement to Prevent Unregulated Commercial Fishing on the High Seas of the Central Arctic Ocean

In 2015, five states with EEZs surrounding the high-seas portion of the central Arctic Ocean signed the Declaration Concerning the Prevention of Unregulated High Seas Fishing in the Central Arctic Ocean. The United States, Canada, Norway, Russia, and Denmark signed this non-legally binding declaration and agreed not to authorize their fishing fleets to engage in commercial fishing in the area until there was an effective international mechanism to manage Arctic fishing. As others grew interested in Arctic fisheries, the original nations were joined by China, Iceland, Japan, the Republic of Korea, and the European Union in negotiating the legally binding Agreement to Prevent Unregulated Commercial Fishing on the High Seas of the Central Arctic Ocean, which was signed in 2018[91] and came into full force on June 25, 2021.[92] The Central Arctic Ocean Agreement is the first multilateral agreement of its kind to take a legally binding, precautionary approach to protect the oceans from fishery activities. The Agreement is effective for up to 16 years, with renewal every 5 years[93] and is bound by obligations under international law, the UNCLOS,

[87] Food and Agricultural Organization of the United Nations (2021c), see Activities.
[88] Food and Agricultural Organization of the United Nations (2021c) – see Activities, International Plan of Action for Reducing Incidental Catch of Seabirds in Longline Fisheries.
[89] Ibid, see Monitoring.
[90] Ibid.
[91] US Department of State (2018).
[92] US Department of State (2021).
[93] Arctic Council (2020).

and the Agreement for the Implementation of the Provisions of the United Nations Convention on the Law of the Sea of December 10, 1982, relating to the Conservation and Management of Straddling Fish Stocks and Highly Migratory Fish Stocks of August 4, 1995.[94]

Due to the melting of the Arctic Ocean, the previously unavailable sea is now a viable commercial fishing area[95] during the summer months.[96] The primary objectives of this agreement are to prevent unregulated fishing in the high seas portion of the Arctic Ocean and to facilitate joint scientific research and monitoring.[97] This preventative approach to protect the Ocean from IUU fishing before fishing begins to better understand the mostly undisturbed ecosystem. This approach is a step in the right direction for global fisheries management and biodiversity. However, there are many that point out issues with the Agreement (such as not addressing the potential of nations not a part of the agreement to "free-ride" on the agreement and exploit the Arctic, ultimately derailing the purpose of the Agreement) and suggest ratifying it as a new multilateral treaty to achieve full scrutiny of existing failures or gaps.[98]

### 1.2.2 National Efforts

Because states are obligated to care for marine resources in their jurisdiction and within the EEZ, many domestic measures were taken to help support the international sustainability efforts.

#### 1.2.2.1 United States – National Oceanic and Atmospheric Administration (NOAA) and the Magnuson-Stevens Act (MSA)

Established in 1976, the Magnuson-Stevens Fishery Conservation and Management Act (MSA) is the primary law that governs marine fisheries management in the United States.[99] The MSA expanded U.S. jurisdiction from 12 to 200 nautical miles beyond its coast and established national standards for the management and conservation of the nation's waters.[100] Specifically, the MSA addresses the decline of fish populations off the coast of the United States, the highly migratory species of the high seas that occupy the continental shelf of the United States, and the species that exist in US rivers or estuaries. The MSA recognized that the survival of these fish species was being threatened by increased fishing pressure, inadequate fishery resource conservation and management practices, and habitat losses that resulted in a diminished capacity to support existing fish stocks.[101] Fisheries play a sizeable role in the U.S. economy, and improper management of fish stocks contributes to economic loss for the nation. Thus, the MSA aimed to establish national measures to preserve

94 Agreement to Prevent Unregulated High Seas Fisheries in the Central Arctic Ocean (2019), p. 2.
95 US Department of State (2018).
96 US Department of State (2021).
97 Ibid.
98 Bromund (2021).
99 National Oceanic and Atmospheric Administration (NOAA) (2021).
100 National Oceanic and Atmospheric Administration, US Department of Commerce (n.d.a).
101 Department of Commerce (2007), 16 U.S.C. 1801.

and manage the nation's waters where international agreements were insufficient for local fishery management.

There are ten total national standards[102] originally formed by the MSA that set the principles for fishery management in the United States. Eight Regional Fishery Management Councils were founded under the MSA – New England Council, Mid-Atlantic Council, South Atlantic Council, Caribbean Council, Gulf Council, Pacific Council, North Pacific Council, and Western Pacific Council[103] – which carry out the national standards with representatives from coastal states and fishery stakeholders.[104] These councils are similar to the RFMOs established under the UN; they develop and amend fishery management plans and research priorities, hold advisory meetings, select fishery management options, set annual catch limits, and develop rebuilding plans.[105]

Since its initial creation, the MSA has been revised twice: in 1996 under the Sustainable Fisheries Act and in 2007 with the MSA Reauthorization Act. The former strengthened the requirements that prevented overfishing and focused on rebuilding overfished stocks and added three new national standards addressing fishing vessel safety, fishing communities, and bycatch. The MSA Reauthorization Act established annual catch limits and strengthened the role science plays in fisheries management and specifically addressed IUU fishing and further addressed bycatch.[106]

The NOAA under the U.S. Department of Commerce is the primary government agency responsible for overseeing fishery management in the United States. The agency was founded in its current form in 1970 but has a history dating back to the 1870s when U.S. President Ulysses S. Grant created the U.S. Commission of Fish and Fisheries. Today, the NOAA is the primary agency responsible for managing fisheries under the framework of the MSA, including the rebuilding of fish stocks, safe sourcing of seafood, and protection of the ecosystem. One of the notable accomplishments in recent years of the NOAA is the cleaning up of the Deepwater Horizon oil spill, and restoring coastal habitats and improving the water quality.[107] As of 2019, 91 percent of all U.S. managed seafood species were not experiencing overfishing, 82 percent of stocks were not overfished, and 45 percent of stocks were fully rebuilt.[108]

#### 1.2.2.2 European Union Common Fisheries Policy (CFP)

The CFP is a set of rules for managing European fishing fleets and conserving fish stocks, and setting catch limits. The CFP was first introduced in the 1970s to increase productivity, stabilize the market, and provide a source of healthy food

102 Standard 1 – Optimum Yield; Standard 2 – Scientific Information; Standard 3 – Management Units; Standard 4 – Allocations; Standard 5 – Efficiency; Standard 6 – Variations and Contingencies; Standard 7 – Costs and Benefit; Standard 8 – Communities; Standard 9 – Bycatch; Standard 10 – Safety of Life at Sea. NOAA. National Oceanic and Atmospheric Administration, US Department of Commerce (n.d.b).

103 Electronic Code of Federal Regulations (e-CFR) (2021) Title 50 Chapter VI Part 600. Magnuson-Stevens Act Provisions.

104 National Oceanic and Atmospheric Administration, US Department of Commerce (n.d.a).

105 Ibid.

106 Ibid.

107 National Oceanic and Atmospheric Administration, US Department of Commerce (n.d.b).

108 Ibid, see Our History.

for consumers at reasonable prices.[109] The CFP also aims to conserve fish stocks and reduce overfishing. Under the previous rule, each coastal state has sovereignty over resources in its EEZ, but under the CFP, the fishing area of EU Member States is considered one zone[110] and national quotas are divided based on historical data and fishing rights negotiated by ministers from each Member State.[111] This regulation sets the rules for cooperation between the EU Member States and international organizations to manage shared fisheries. Additionally, the CFP sets standards on seafood products sold within the EU to ensure fair market competition, and to protect consumers through means such as clear product labeling requirements. Council Regulation (EC) 1224/2009 established a system for ensuring compliance with the rules of the CFP.[112] In recent years, revisions to the CFP have been made to support the EU's move toward sustainability and a circular economy.[113] Regulation (EC) No 178/2002 sets rules on traceability and Regulation (EC) No 1005/2008 to deter and prevent IUU fishing.[114]

In 2013, the CFP was reformed to include provisions for fish stock management to reach maximum sustainable yields by 2020, introduced landing obligations, and continued application of multilateral plans to manage fishers in different sea basins, issued fleet capacity ceilings per EU Member States in combination with obligations for EU countries to ensure a stable and continuing balance between fishing capacity and fishing opportunities.[115] However, despite best efforts, the EU failed to achieve its goal of reaching maximum sustainable yields.[116] A 2018 revision sought to modernize, strengthen, and simply the EU fisheries control systems in order to ensure sustainability and increase the playing field for fisheries control. The changes are in line with the EU's REFIT program to minimize regulatory burdens and simplify.[117]

Of particular note is the role the CFP and fisheries management have played in the Brexit-trade talks. Fishing was one of the major sticking points in the trade negotiations between the United Kingdom (UK) and the EU in the years and months leading up to the formal departure of the UK from the EU. The UK wanted to regain control over UK waters and forgo its obligations and constraints under the CFP.[118] Though neither negotiating party got exactly what they wanted, the UK did indeed leave the CFP on December 31, 2020.[119] The EU will still have access to the UK waters, a point of contention between the UK Prime Minister and many British fishermen, but British fishing boats get a larger share of the fish from UK waters. This transition will start in 2021 and continue into 2026, after which negotiations on the rights to the UK

109 European Commission (n.d.a).
110 Jack (2020).
111 Morris and Barnes (2021).
112 Food and Agricultural Organization of the United Nations (2018b).
113 European Commission (n.d.b).
114 Food and Agricultural Organization of the United Nations (2018b).
115 European Commission (n.d.a).
116 Clayton (2021).
117 European Commission (n.d.a).
118 Morris and Barnes (2021).
119 Harper (2020).

EEZ will occur annually, with the UK gaining the right to completely exclude EU boats after 2026.[120]

#### 1.2.2.3 Fisheries Law of the People's Republic of China

China is one of the leading global exporters of fisheries. The country's role in global fish stock management is crucial, and so it is essential to understand how China has contributed to the goals of conservation and preservation of the world's fish stocks. The Fisheries Law of the People's Republic of China has been in effect since 1986 and remains the law for fishing in China. The law was created for "the purpose of enhancing the protection, increase, development and reasonable utilization of fishery resources, developing artificial cultivation, protecting fishery workers' lawful rights and interests and boosting fishery production, so as to meet the requirements of socialist construction and the needs of the people."[121] All fishery activities, including aquaculture, are covered by the law.

China's primary seas – the Bohai Sea, the Yellow Sea, the East China Sea, and the South China Sea – all belong to semi-closed margin seas that, due to hydrological characteristics, result in a lack of diverse species and pelagic migratory fish stocks.[122] Despite this, China's waters are home to a wide variety of fish species and have the largest populations of fish.[123] China has some of the largest numbers of marine fishing vessels (around 166.3 thousand[124]), using methods such as trawling, seine fishing, drift netting, stow netting, dip netting, cover netting, square netting, long-line fishing, pole fishing, cage fishing, and squid jigging.[125]

Under UNCLOS, China has entered into several bilateral agreements with neighboring countries – The Sino-Japanese Fisheries Agreement with Japan, the Sino-Korean Fisheries Agreement with South Korea, and the Sino-Vietnam Fisheries Agreement in Beibu Gulf with Vietnam. These measures were developed for the conservation, management, and sustainable use of shared fishing areas. Despite measures such as these, overfishing, deterioration of the marine environment, and the decline of fishery resources continue to threaten China's seas. China also faced issues of illegal drug use and water pollution.[126] The large-scale fishing industry, the complexity of fishery regulation, and the weak implementation of fisheries administration in China pose a problem for the proper implementation of fisheries law. The existing legal system for fisheries management in China lacks modernization and struggles to keep up with the current state of the world's oceans. Given the level of fishery activity in the country, extra focus and attention are needed in China's fishery governance.

---

[120] Morris and Barnes (2021).

[121] Fisheries Law of the People's Republic of China (1986), Translation from Ministry of Ecology and Environment of the People's Republic of China.

[122] Huang and He (2019), p. 174.

[123] Ibid, p. 177.

[124] Ibid, p. 174.

[125] Ibid.

[126] Ibid, p. 179.

#### 1.2.2.4 Other Efforts

Numerous other national efforts have been made under the obligation of international agreements like the UNCLOS to combat overfishing, IUU fishing, bycatch, and overcapacity. In Norway, for example, bycatch is required to be used as food.[127] Indonesia, one of the major fishing nations that relies heavily on its fisheries industry, has implemented some of the toughest policies to combat IUU fishing, going as far as to sink illegal fishers' vessels. One publication reported that as of April 2017, the Ministry seized and blew up around 317 foreign vessels operating illegally in Indonesian waters.[128] The Philippines, also having a large fishing sector, pledged to work under the Compliance Agreement and the Agreement on Port States Measures. In 2012, the Philippines Bureau of Fisheries and Aquatic Resources issued a closed season for sardine catch in major fishing grounds in order to address overfishing of sardines in its territories.[129] In the United States, as per the 2019 International Fisheries Management Report to Congress, three nations – Mexico, Ecuador, and Russia – were identified as having vessels in IUU fishing. As part of their efforts to protect fish stocks and comply with the UNCLOS, the United States held consultations with the identified nations to encourage them to take action against IUU activities.[130]

## 1.3 CONCLUSION

There are many working to rebuild and conserve the ocean and its vast resources. As concluded from the UN's 2020 State of the Worlds Fisheries report, there has been much improvement in recent years. However, there is still much work that needs to be done. Overfishing, IUU fishing, bycatch, and overcapacity of fishing vessels threaten the biodiversity of the Earth's oceans. Improved fishery management and more concrete and robust national efforts are needed to disincentivize and prevent IUU fishing and overfishing. Bycatch threatens pelagic species, better fishing methods are required to reduce and prevent the catch of unwanted species, and better regulations are needed to detail how to handle bycatch when it does happen. International efforts have been vigorous but alone are insufficient in rebuilding the Earth's marine resources. As demand for fish increases, so too does the world's population, and the pressures facing the world's ocean will continue to rise.

## REFERENCES

Afriansyah, A. (2018 November 05). Indonesia's practice in combatting illegal fishing: 2015–2016. In *Asian yearbook of International Law*, 22, pp. 283–300. https://doi.org/10.1163/9789004379633_015.

Agreement to prevent unregulated high seas fisheries in the central Arctic Ocean (2019). Government of Canada. Retrieved from https://www.dfo-mpo.gc.ca/international/agreement-accord-eng.htm.

---

127 Waldman (2014), p. 360.

128 Afriansyah (2018).

129 Wan (2018).

130 Oliver and Jacobs (2019).

Arctic Council (2020 October 28). Exploring the Arctic Ocean: The agreement that protects an unknown ecosystem. Retrieved from https://arctic-council.org/en/news/exploring-the-arctic-ocean-the-agreement-that-protects-an-unknown-ecosystem/.

Beyer, J., Trannum, H. C., Bakke, T., Hodson, P. V., Collier, T. K. (2016 Sep 15). Environmental effects of the Deepwater Horizon oil spill: A review. *Marine Pollution Bulletin*, 110(1), 28–51. doi: 10.1016/j.marpolbul.2016.06.027.

Bromund, T. (2021 April 16). The central Arctic Ocean fishing agreement: A challenge for U.S. diplomacy. Margaret Thatcher Center for Freedom. Retrieved from https://www.heritage.org/sites/default/files/2021-04/BG3606.pdf.

Cheney, J. (2020 July 9). Three experts talk about MSC's potential bluefin tuna certification. Retrieved from https://sustainablefisheries-uw.org/experts-talk-about-east-atlantic-bluefin-tuna/.

Clayton, A. (2021 March). Lessons from implementation of the EU's common fisheries policy. The Pew Charitable Trusts. Retrieved from https://www.pewtrusts.org/-/media/assets/2021/03/eone-lessonslearned.pdf.

Cullis-Suzuki, S., and Pauly, D. (2010). Failing the high seas: A global evaluation of regional fisheries management organizations. *Marine Policy*, 34(5), 2010, 1036–1042. ISSN 0308-597X, https://doi.org/10.1016/j.marpol.2010.03.002.

Department of Agriculture-Bureau of Fisheries and Aquatic Resources (2004). In turbulent seas: The status of Philippines marine fisheries. Coastal Resource Management Project of the Department of Environment and Natural Resources. Cebu City, Philippines.

Department of Commerce (2007 May). Magnuson-Stevens fishery conservation management act, Public Law 94–264, 16U.S.C. 1801. National Oceanic and Atmospheric Administration.

Electronic Code of Federal Regulations (e-CFR) (2021 July 15). Magnunson-Stevens act. Title 50C.F.R. Chapter IV Part 600.

Environmental Protection Agency (n.d.). Deepwater horizon–BP Gulf of Mexico oil spill. Retrieved from https://www.epa.gov/enforcement/deepwater-horizon-bp-gulf-mexico-oil-spill

European Commission (n.d.a). Common Fisheries Policy (CFP). Retrieved from https://ec.europa.eu/oceans-and-fisheries/policy/common-fisheries-policy-cfp_en.

European Commission (n.d.d). The Common Fisheries Policy (CFP). Retrieved from https://ec.europa.eu/fisheries/cfp/.

Fishery and Aquaculture Economics and Policy Division (1995). Agreement to promote compliance with international conservation and management measures by fishing vessels on the high seas. Food and Agricultural Organization of the United Nations. Rome, Italy. Retrieved from http://www.fao.org/documents/card/en/c/8cb30770-3145-55ed-a0db-315cbbb722a6.

Food and Agricultural Organization of the United Nations (2001). International plan of action to prevent, deter and eliminate illegal, unreported and unregulated fishing. Rome. Retrieved from http://www.fao.org/publications/card/en/c/71be21c9-8406-5f66-ac68-1e74604464e7.

Food and Agricultural Organization of the United Nations (2011). The code of conduct for responsible fisheries. Rome. Retrieved from http://www.fao.org/3/i1900e./i1900e.pdf.

Food and Agricultural Organization of the United Nations (2018a July 2). Agreement to promote compliance with international conservation and management measures by fishing vessels on the high seas. Retrieved from http://www.fao.org/fileadmin/user_upload/legal/docs/012s-e.pdf.

Food and Agricultural Organization of the United Nations (2018b). EU requirements for food safety and traceability of fish and fishery products. Shangahi, China. Retrieved from http://www.fao.org/3/CA2915EN/ca2915en.pdf.

Food and Agricultural Organization of the United Nations (2020a). The state of world fisheries and aquaculture. Food and Agriculture Organization of the United Nations. Retrieved from http://www.fao.org/3/ca9229en/CA9229EN.pdf.

Food and Agricultural Organization of the United Nations (2020b). The state of world fisheries and aquaculture 2020: Interactive STORY. Retrieved from http://www.fao.org/state-of-fisheries-aquaculture.

Food and Agricultural Organization of the United Nations (2020c). Overfishing of the world's major tuna stocks going down, bycatch and pollution reduced and 18 new areas protecting vulnerable marine ecosystems established. Retrieved from http://www.fao.org/news/story/en/item/1258859/icode/

Food and Agricultural Organization of the United Nations (2020d). Transshipment: A closer look. Rome. Retrieved from http://www.fao.org/3/cb2339en/CB2339EN.pdf

Food and Agricultural Organization of the United Nations (2021a). Agreement on Port State Measures (PSMA). Retrieved from http://www.fao.org/port-state-measures/en/.

Food and Agricultural Organization of the United Nations (2021b). FAO treaties database. Retrieved from http://www.fao.org/treaties/results/details/en/c/TRE-000003/.

Food and Agricultural Organization of the United Nations (2021c). Code of conduct for responsible fisheries. Retrieved from http://www.fao.org/fishery/code/en.

Food and Agricultural Organization of the United Nations (n.d.). Illegal, Unreported and Unregulated (IUU) fishing. Retrieved from http://www.fao.org/iuu-fishing/background/what-is-iuu-fishing/en/.

Food and Agricultural Organization of the United Nations Regional Office for Asia and the Pacific (2007). Managing fishing capacity and illegal, unreported, and unregulated fishing in Asia. Phuket, Thailand. Retrieved from http://www.fao.org/3/ah999e/ah999e00.pdf.

Garcia, S. M., & Rosenberg, A. A. (2010). Food security and marine capture fisheries: Characteristics, trends, drivers and future perspectives. *Philosophical Transactions of the Royal Society of London. Series B, Biological Sciences*, 365(1554), 2869–2880. https://doi.org/10.1098/rstb.2010.0171.

Gillett, R. (2008). Global study of shrimp fisheries. FAO technical paper. No 475. Food and Agricultural Organization of the United Nations, Rome 2008. Retrieved online from http://www.fao.org/3/i0300e/i0300e00.htm

Gilman, E. et al. (2020). Benchmarking global fisheries discards. *Scientific Reports*, 10, 14017. https://doi.org/10.1038/s41598-020-71021-x.

Gonçalves, L. R. (2021). *Regional Fisheries management Organizations. The Interplay between Governance and Science*. Springer, São Paulo.

Harper, J. (2020 December 31). UK fisheries unhappy with EU trade deal. Retrieved from https://www.dw.com/en/uk-fisheries-unhappy-with-eu-trade-deal/a-56104046.

Huang, S. and He, Y. (2019). Management of China's capture fisheries: Review and prospect. *Aquaculture and Fisheries*, 4(5), 173–182. https://doi.org/10.1016/j.aaf.2019.05.004.

Hunt, N. (2020 June 8). Overfishing on the rise as global consumption climbs: U.N. agency. Retrieved from Reuters https://www.reuters.com/article/us-food-fao-fish/overfishing-on-the-rise-as-global-consumption-climbs-u-n-agency-idUSKBN23F1CD.

International Labour Organization (2013 May 31). Caught at sea – Forced labour and trafficking in fisheries. Retrieved from https://www.ilo.org/global/topics/forced-labour/publications/WCMS_214472/lang--en/index.htm.

International Seafood Sustainability Foundation (n.d.). United Nations fish stocks agreement. Retrieved from https://iss-foundation.org/glossary/united-nations-fish-stocks-agreement/.

Jack, M. T. (2020 February 25). Common fisheries policy. Retrieved from https://www.instituteforgovernment.org.uk/explainers/common-fisheries-policy.

Mavrellis, C. (2017 March 17). Transnational crime and the developing world. Retrieved from https://gfintegrity.org/report/transnational-crime-and-the-developing-world/.

Ministry of Ecology and Environment, The People's Republic of China. (1986). Fisheries law of the people's Republic of China, translated. Retrieved from http://english.mee.gov.cn/Resources/laws/envir_elatedlaws/200710/t20071009_109918.shtml.

Morris, C. and Barnes, O. (2021 January 20). Brexit trade deal: What does it mean for fishing? Retrieved from https://www.bbc.com/news/46401558.
National Oceanic and Atmospheric Administration (2021). Laws & policies: Magnuson-Stevens act. Retrieved from https://www.fisheries.noaa.gov/topic/laws-policies.
National Oceanic and Atmospheric Administration, US Department of Commerce (n.d.a). Laws & policies. Retrieved from https://www.fisheries.noaa.gov/topic/laws-policies.
National Oceanic and Atmospheric Administration, US Department of Commerce (n.d.b). About us. Retrieved from https://www.fisheries.noaa.gov/about-us#overview.
NOAA (n.d.a) IUU fishing. Retrieved from National Ocean Council Committee on IUU Fishing and Seafood Fraud https://www.iuufishing.noaa.gov/FAQs/IUUFishingFAQs.aspx.
NOAA (n.d.b). Understanding bycatch. Retrieved from https://www.fisheries.noaa.gov/insight/understanding-bycatch#can-fishermen-use-or-sell-bycatch?.
Oliver, C. and Jacobs, N.A. (2019 September). Improving international fisheries management, 2019 report to congress. National Oceanic and Atmospheric Administration. Retrieved from https://media.fisheries.noaa.gov/dam-migration/improvingintlfisheriesmgmt_2019_report_final.pdf.
Pacoureau, N., Rigby, C. L., Kyne, P. M. et al. (2021). Half a century of global decline in oceanic sharks and rays. *Nature*, 589, 567–571. https://doi.org/10.1038/s41586-020-03173-9.
Pentz, B. et al. (2018). Can regional fisheries management organizations (RFMOs) manage resources effectively during climate change? *Marine Policy*, 92, 13–20. https://doi.org/10.1016/j.marpol.2018.01.011.
Pew Environment Group (2009 July). Overcapacity, what overcapacity? Retrieved from https://www.fishsec.org/app/uploads/2011/03/1247756350_75124.pdf.
Reeves, R. R., McClellan, K., and Werner, T. B. (2013). Marine mammal bycatch ain gillent and other etnagling net fisheris, 1990 to 2011. *Endangered Species Research*, 20(71–97). https://doi.org/10.3354/esr00481.
Statute of the International Tribunal for the Law of the Sea, Dec. 10, 1982, 1833 U.N.T.S. 561.
Sumaila, U. R. (2020 February 26). Illicit trade in marine fish catch and its effect on ecosystems and people worldwide. *Science Advances*, 6(9). Retrieved from https://advances.sciencemag.org/content/6/9/eaaz3801/tab-pdf.
U.S. Department of State (2018 October 1). U.S. signs agreement to prevent unregulated commercial fishing on the high seas of the central Arctic Ocean. Retrieved from https://2017-2021.state.gov/u-s-signs-agreement-to-prevent-unregulated-commercial-fishing-on-the-high-seas-of-the-central-arctic-ocean/index.html.
U.S. Department of State (2021 June 25). The agreement to prevent unregulated high seas fisheries in the central Arctic Ocean enters into force. Retrieved from https://www.state.gov/the-agreement-to-prevent-unregulated-high-seas-fisheries-in-the-central-arctic-ocean-enters-into-force/.
UN (1998). The United Nations convention on the law of the sea: A historical perspective. Retrieved from https://www.un.org/Depts/los/convention_agreements/convention_historical_perspective.htm#Exclusive%20Economic%20Zone.
UN (n.d.). Oceans and the law of the sea. Retrieved from https://www.un.org/en/sections/issues-depth/oceans-and-law-sea/.
United Nations Department of Public Information (2020 May). Fish Stocks Agreement: Overview of what the Agreement says and its impact. New York. Retrieved from https://www.un.org/depts/los/convention_agreements/reviewconf/FishStocks_EN_B.pdf
United Nations Office on Drugs and Crime (2011). Translational organized crime in the fishing industry. Retrieved from https://www.unodc.org/documents/human-trafficking/Issue_Paper_TOC_in_the_Fishing_Industry.pdf.
Waldman, J. (2014). How Norway and Russia made a cod fishery live and thrive. *Yale Environment*, 360, 13.

Wan, L. (2018 April 18). The Philippines agrees to address illegal fishing amid UN and scientists' pressure. Retrieved from https://www.foodnavigator-asia.com/Article/2018/04/18/The-Philippines-agrees-to-address-illegal-fishing-amid-UN-and-scientists-pressure.

World Wild Life Foundation (WWF) (2018 October 17). €12.5 million illegal bluefin tuna trade exposes threat to sustainable fisheries in Europe. Retrieved from https://www.wwf.eu/?336830/125-million-illegal-bluefin-tuna-trade-exposes-threat-to-sustainable-fisheries-in-Europe.

# 2 Perspectives and Predicaments on GE Salmon

*Audrey Amescua*
Northeastern University

## CONTENTS



## 2.1 INTRODUCTION

While many international environmental, political, and advocacy groups are focusing on international agreements and regulations to address the issues facing the world's oceans, others have turned to technology-based solutions and are even finding ways to capitalize on the issue. AquaBounty Technologies, Inc., a biotechnology company, is positioning itself as a "food security game-changer." The company has

DOI: 10.1201/9781003137733-4

become the first to produce a commercially feasible and federally approved genetically engineered (GE)[1] fish, known as AquAdvantage Salmon. The company gained approval for the commercial sale of the fish to humans from the U.S. Food and Drug Administration (FDA) in 2015 and the Canada Food Inspection Agency (CFIA) in 2016. Most notably, the fish is FDA approved as an animal drug! Though this is the first approved use of GE technology in an animal for commercial sale and consumption by humans, it is certainly not the only one being researched and investigated. Other private companies around the world are working to develop more species of GE fish, shellfish, and even land-based farm animals. The potential consequences for sustainable food systems could be devastating.

### 2.1.1 Technology

Though the approval of GE animals for human consumption is novel, the technology used to create the AquAdvantage Salmon is not new. GE crops have been around for decades, as has the research on the genetic engineering of mice. Based on these successful research studies from the mice experiments, a scientist named Garth Fletcher and his colleagues began to explore using the same principles in Atlantic salmon in the 1980s. It is Fletcher's work on which AquaBounty bases its technology.[2] Because salmon growth hormones are only active during certain times of the year, salmon experience slow growth. Ocean Pout, another type of fish species, on the other hand, have special proteins that enable it to grow and survive in frigid waters year-round. Fletcher took the DNA that keeps this protein on and running and connected it to a recombinant DNA (rDNA) growth hormone gene found in Chinook salmon,[3] effectively allowing the salmon to grow during periods it ordinarily would not. The AquAdvantage Salmon grows to market size in as little as 16–20 months rather than the usual 28–32 months for traditional non-GE farmed salmon.[4]

### 2.1.2 Overview of Genetically Engineered Animal Regulatory Framework

#### 2.1.2.1 United States

In the United States, genetically modified organisms are broadly regulated under general statutory authority of environmental, health, and safety laws and the Coordinated Framework for Regulation of Biotechnology, a policy statement originating in 1986 by the Executive Office of the President, Office of Science and Technology Policy (OSTP). In essence, the position taken in the OSTP was that regulatory jurisdiction and regulation over biotechnology products would be no different and conducted in the same manner as conventional products. GE regulations in the United States

---

[1] Transgenic and genetically modified are other terms used interchangeably to refer to genetically engineered animals.

[2] Thompson et al. (2019).

[3] Ibid; Conley (2016).

[4] Conley (2016).

focus on the product's contents rather than the fabrication process.[5] Furthermore, there is no federal legislation specific to the regulations of Genetically Modified Organisms (GMOs) or one federal agency with sole oversight. Rather, multiple jurisdictional agencies have roles in the management of GE products, and specific oversight is dependent upon the target organism or substance. For example, plant GMOs are under the purview of the U.S. Department of Agriculture's Animal and Plant Health Inspection Service (APHIS) through the Plant Protection Act; Genetically Modified (GM) foods, drugs, and other biological products are regulated by the Food and Drug Administration (FDA) under the Federal Food, Drug, and Cosmetic Act (FFDCA) and the Public Health Act; and GMO pesticides and microorganisms are subject to Environmental Protection Agency (EPA) oversight, pursuant to the Federal Insecticide, Fungicide, and Rodenticide Act and the Toxic Substance Control Act.[6]

Notably, the biotechnology industry plays a vital role in the U.S. economy, making the development of GE products relatively favorably regarded in the United States from both an economic and political perspective. The United States accounted for over 40 percent of all GE crops produced in 2017.[7] As of 2013, 93 percent of soybeans, 90 percent of cotton, and 90 percent of corn grown in the United States were genetically engineered.[8] Despite the general support for GE agriculture in the United States, some people would like to see more robust and concrete regulations around genetically modified materials. There have been two recent sets of federal standards that could have significant impacts on the treatment and perception of GE products. The most recent comes from the United States Department of Agriculture (USDA) offices of APHIS and the Food Safety and Inspection Service (FSIS). They have proposed a rule establishing regulations for the movement of animals modified or developed with genetic engineering. The comment period ended in May of 2021. If the rule passes, the APHIS, under the jurisdiction of the USDA, would be responsible for conducting safety assessments of animals that have been modified or developed using gene-editing technology. The agency is looking for issues that may increase the animal's susceptibility to pests or disease, including zoonotic disease, or the ability to transmit these diseases to other animals or humans.[9] Under the same rule, the FSIS would conduct pre-slaughter food safety assessments to ensure the slaughter and processing of certain animals modified or developed using genetic engineering would not result in a product that is adulterated or misbranded.[10] Because the statutory authority of these two organizations comes from the Animal Health Protection Act, the Federal Meat Inspection Act, and the Poultry Products Inspection Act, GE fish would not be subject to this new ruling as fish falls under the FFDCA, which gives jurisdiction of most fish and fish products to the FDA.[11]

The second major piece of new federal compliance standards comes from the National Bioengineered Food Disclosure Standard, passed in 2016 and published in

---

5 Library of Congress (2020a).
6 Ibid.
7 Library of Congress (2020a).
8 Ibid.
9 Federal Registrar (2020).
10 Federal Registrar (2021).
11 United States Marketing Service (2019).

2018. The new standard clarifies the definition of bioengineered foods as *those that contain detectable genetic material that has been modified through certain label techniques and cannot be created through convention breeding or found in nature*,[12] and makes the term "bioengineered" the standard terminology for GE products[13] (United States Marketing Service, 2020). By 2022, food makers, importers, and certain retailers must disclose to consumers specific food and food product ingredients that are bioengineered or have bioengineered ingredients (FDA, 2020). Four options for disclosure exist: on package text "Bioengineered food" or "Contains a bioengineered food ingredient"; USDA approved symbol for bioengineered food; electronic or digital link that must include instructions to "scan here for more food information" or similar language and the phrase "Call [1-000-000-0000] for more food information"; or text message disclosure must include the statement, "Text [command word] to [number] for bioengineered food information" (Agricultural Marketing Service, 2019). The implementation start date began on January 1, 2020, with a mandatory compliance date of January 1, 2022. Foods requiring disclosure include foods subject to the FDCA that are under the FDA's jurisdiction, foods subject to the Federal Meat Inspection Act, the Poultry Products Inspection Act, or the Egg Products Inspection Act under the USDA if the predominant ingredient would be independently subject to labeling under the FDCA, or the predominant ingredient is broth stock, water, or similar solution.[14] The USDA Agricultural Marketing Service (AMS) indicates the specific GE products that require disclosure, including AquAdvantage Salmon.

Importantly, restaurants, similar retail food establishments, and very small food manufactures are exempt from the National Bioengineered Food Disclosure Standard.[15] The result means that any GE fish, or any other GE food product that would ordinarily be subject to disclosure, would not need to be disclosed if sold in a restaurant or similar retail food establishment, leaving many to criticize the Act's ability to increase transparency for the consumer.

#### 2.1.2.2 European Union

The European Union (EU) has an established strict and comprehensive legal regime for regulating GM foods and products containing GM ingredients. Regulation (EC) No. 1829/2003 on Genetically Modified Food and Feed and Directive 2001/18/EC on the Deliberate Release into the Environment of Genetically Modified Organism are the bedrock pieces of EU regulation governing the use, sale, and entry of GMOs into the EU market. Additionally, Regulation (EC) No. 1830/2003, which amends Directive 2001/18/EC, governs the traceability and labeling requirements of GMOs. Directive 2009/41/EC covers contained the use of genetically modified microorganisms.[16]

Contrary to the United States – which bases regulation of GMOs on whether or not GE products are different from conventional, non-GE products – the EU's policy on GMOs is based on concerns from consumers, farmers, and environmentalists.

[12] Food and Drug Administration (2020a).
[13] Agricultural Marketing Service (n.d.).
[14] Agricultural Marketing Service (2019).
[15] Agricultural Marketing Service (n.d.).
[16] European Commission (n.d.).

The policy is designed to prevent adverse effects on the environment, as well as human and animal health, safety, and welfare. Additional key features are labeling, transparency, and traceability.[17] Any food entering the union that is GE or contains GE ingredients is stringently regulated and must be labeled; this includes consumer packaging or pre-packaged products. Products that may contain traces of GE material below 0.9 percent if technically unavoidable are exempt from the labeling requirement, as are medicinal products.[18] Though the cultivation of GMOs in each EU Member State's territory resides with each Member State, under Directive (EU) 2015/412, which amends Directive 2001/18/EC, GMOs for cultivation necessitate an individual risk assessment before being placed on the market. Once given authorization, as of April 3, 2017, Member States have the right to exclude the cultivation of GMOs in their territory.[19] Nineteen of the 27 EU Member States have voted to partially or fully ban GMOs.[20]

#### 2.1.2.3 International

The diverging approaches to GE legislation and regulation of GE products taken by the United States and EU have led to disparate impacts on consumers within each orbit. The United States governs the development and use of the technology leveraging these products to boost its agricultural economy, but gives little information to consumers, ultimately hindering transparency; while the EU has strict regulations in place for the protection of the environment, farmers, and consumers and gives much thought into the health and wellbeing of consumers. Given their prominent stature in the global economy and globalized food trade, many other countries and regulatory bodies follow their lead, resulting in unnecessary additional trade barriers. National GMO policies and regulations vary significantly, from some countries having loosely enforced policies; to some countries, like Austria, completely restricting the cultivation of GMOs in their area; while other countries (e.g., Israel) allow the development and growth of GMOs for research purposes only; to some such as the United States and Argentina having loose regulations around the cultivation and sale of GMOs; and still others have yet to adopt any policy around GE products.[21]

Despite this patchwork of GMO regulations around the globe, two major international protocols attempt to bridge the gaps: the Cartagena Protocol of 2000 and the Nagoya-Kuala Lumpur Supplementary Protocol of 2010, both of which are attached to the Convention on Biological Diversity of 1993. The Cartagena Biosafety Protocol (the Protocol) is an international agreement between 168 countries and the EU that aims to ensure the safe transfer, handling, and use of GMOs between countries to protect and conserve the sustainable use of biological diversity. The Protocol supplements the U.N. Convention on Biological Diversity, signed in 1992 at the U.N. Rio Conference. Contained within are measures to protect genetic resources from potential risks that the release of genetically modified organisms might pose.

---

[17] Ibid.
[18] EUR-Lex (2016).
[19] EUR-Lex (2015).
[20] Europa (n.d.).
[21] Library of Congress (2020b).

The agreement guarantees States the control over GMO movement, such as the GE salmon, within their territory, if such organisms are introduced into the environment.

The Protocol has three divisions for Genetically Modified Organism: those voluntarily introduced into the environment, those destined for contained use, and GMOs intended for direct use as food, food ingredients, or processing. The result means that States have full autonomy to ban or regulate the aquaculture of GE fish. However, many nations have refused to ratify the Protocol, particularly those such as the United States, Argentina, Canada, and Australia which rely on high levels of agricultural exports. Ironically, because salmon do not recognize national borders, GE salmon may implicate the Protocol, which gives States the authority to ban or regulate the aquaculture of GE fish. The Protocol does not, however, guarantee States the power to control the import of such organisms for the use as food.[22]

The 2010 Nagoya-Kuala Lumpur Supplementary Protocol (Supplement) was adopted to augment the Cartagena Protocol and aims to bolster the conservation and sustainable use of biodiversity by providing rules and procedures in liability and redress related to living modified organisms (LMOs). This addresses damages that may result from LMOs which find their origin in a transboundary movement, and what response measures are to be taken in the event of damage caused by LMOs.[23]

## 2.2 UNITED STATES AND CANADA APPROVAL AND POTENTIAL IMPLICATION FOR INTERNATIONAL REGULATIONS

### 2.2.1 Overview of Approval

In 2015, the U.S. FDA approved AquaBounty's technology to produce the AquAdvantage salmon for human consumption. Eighty companies vowed not to sell it. Canada followed with approval of the salmon in 2016, where it hit the market with 20,000 pounds of product sold. Thus far, the salmon has only been sold to distributors, meaning that most of the salmon on the market is in restaurants or food service locations rather than supermarkets. There is currently no requirement that restaurants or foodservice locations label the salmon as genetically modified on menus or disclose it to consumers in any way.[24] In opposition, some U.S. senators, like Lisa Murkowsi, fought hard against the approval of the GE salmon and delayed its sale in the United States for years. None-the-less, AquaBounty successfully demonstrated that the GE salmon were not only near equivalents to their non-GE counterparts and therefore safe for consumers, but also that GE salmon was mostly sterile and, therefore, could not breed with wild salmon, reducing the environmental risk of the product. In addition, proof showed that the farmed, modified salmon would not survive for very long in the wild themselves, and had low risk of escaping to the wild, resulting in the FDA ultimately approving the fish to be sold.[25]

Since the approval of the technology, however, the U.S. District Court for the Northern District of California ruled in November 2020 that the FDA violated core

[22] Bratspies (2008).
[23] Convention on Biodiversity (2018).
[24] Thompson et al. (2019).
[25] Ibid.

environmental laws when they approved the GE salmon.[26] The Court found that the FDA ignored the potential environmental consequences, particularly because AquaBounty plans to cultivate the sale and commercialization of the salmon not just in the United States and Canada, but around the world. The Court ruled this was a violation of the National Environmental Policy Act and the Endangered Species Act.[27] The FDA was ordered to go back and thoroughly analyze the environmental consequences of an escape of the genetically engineered salmon into the wild.[28] The ruling leaves the issue of whether the GE salmon could escape and threaten wild populations. The ruling, however, has not deterred AquaBounty, as the FDA will likely maintain its approved status of the technology, with AquaBounty seeing its first harvest of the AquAdvantage Salmon and samples sent to customers in the United States.[29]

### 2.2.2 Regulation in the United States

The FDA regulates GE animals under the New Animal Drug Application (NADA) provisions of the Federal Food, Drug, and Cosmetic Act (FFDCA).[30] GE animals are regulated as animal drugs because the FDA considers the transfer of genetic information to be a means of delivering a drug (hormone, protein, etc.) into the animal's tissue. Regarding the AquAdvantage Salmon, the promoter that delivers the growth hormone to the fish's tissue is considered a drug. Because GE animals are regulated in this way, producers must complete a NADA to gain FDA approval for GE animals, including fish. AquaBounty first approached the FDA for approval of its technology to be used commercially in salmon in 1993, filing for formal approval in 1995. AquaBounty submitted its last required study of AquAdvantage Atlantic salmon in 2009 for their NADA, at which point the FDA issued a final guidance to the industry explaining its regulation of GE foods animals using the NADA.

### 2.2.3 AquaBounty Operation

AquaBounty currently has three facilities, one for research and development purposes and two serving as farms. AquaBounty had its first farm approved in Canada on Prince Edward Island. This location now serves primarily as an R&D facility. AquaBounty's 89,570 square foot Rollo Bay facility on Prince Edward Island expects its first harvest in early 2021.[31] In 2018, the FDA granted approval for an AquaBounty facility in Albany, Indiana,[32] measuring 122,000 square feet and capable of raising 1,200 metric tons of salmon each year. The first harvest of conventional Atlantic salmon concluded at the end of June 2020, and the first harvest of

[26] Add case number.
[27] Center for Food Safety (2020).
[28] Ibid.
[29] Hill (2021).
[30] 21 U.S.C.A. §§ 360(a)(1), 360(a)(3).
[31] AquaBounty (2021a).
[32] FDA (2020).

GE salmon, originally expected in Q4 of 2020,[33] wrapped up March–April 2021 due to the global COVID-19 pandemic. AquaBounty is looking into potential sites for a third fish farm to serve as a large-scale commercial facility and is expected to be able to have a 10,000-metric ton capacity. It is aiming to have construction of this facility began in 2021 with production in 2023.[34] The company previously had operations in Panama after Canada granted approval for the company to export its eggs to the nation. However, after the FDA lifted import restrictions on the GE salmon eggs in 2019, the company moved operations to the Indiana facility, closing operation in Panama.[35]

### 2.2.4 Potential Human Health and Safety Concerns

The FDA's safety assessment of the GE salmon concluded that the AquAdvantage Salmon are identical to conventional salmon. Many, however, argue there could be unintended consequences of altering the DNA of the salmon and that the presence or absence of added or deleted substances could cause unforeseen compositional changes.[36] According to an evaluation by the National Research Council on GE plants, animals, and microorganisms, there is evidence to suggest that unintended compositional changes do arise from genetic engineering. The health effects of those unintended changes depend on the nature of the substances altered and the biological consequences of the compounds.[37] The potential that those unintended consequences can pass to humans greatly concerns to many opponents of the GE fish. One of these opponents' primary focus is on the rDNA construct of the GE salmon and the use of the growth hormone from the Chinook salmon inserted into the receptor from the ocean pout, which is then used to modify the Atlantic salmon. Levels of the insulin-like growth factor 1 (IGF-1) hormone in the GE salmon may be elevated as a result, which could create unintended health consequences in both the animal itself and the humans consuming it.[38] That said, there is currently no documented adverse health effect in humans attributed to genetic engineering.[39] According to the FDA's review of the public commentary on the concerns of the IGF-1 hormone, the hormone is produced naturally by humans and is a normal component of food derived from animals and has found no demonstrable causative association between IGF-1 levels and long-term health consequences. The FDA also found that levels of IGF-1 in AquAdvantage salmon were minute, below any detectable limit.[40]

Another major concern in the GE Salmon is the potential for increased allergenicity, or individuals developing an allergic response, to the fish.[41] Worry over the increased predominance of food allergens in general throughout industrialized

[33] AquaBounty (2021a).
[34] Hill (2021).
[35] AquaBounty (2019).
[36] Upton and Cowan (2015).
[37] Ibid.
[38] Food and Water Watch, Consumers Union, and the Center for Food Safety (2012).
[39] Upton and Cowan (2015).
[40] Food and Drug Administration (2018).
[41] Upton and Cowan (2015).

nations has increased over the past two decades. As many as 8 percent of Americans may be affected by immunoglobulin E (IgE)[42] food allergies.[43] The reason for the increased allergenicity is unclear. However, because proteins are the primary cause of most allergic reactions to foods, some point to proteins produced in genetically modified foods as the culprit behind the rise in allergens throughout heavily industrialized nations.[44] Specifically, in the GE salmon, the engineering of the salmon gave the fish three sets of chromosomes rather than the usual two, making it a triploid rather than a diploid organism. The extra chromosome may increase the potency of allergenic compounds present in the fish. However, in its assessment, the FDA found that the triploid salmon pose no additional allergenic risk compared to non-GE Atlantic salmon. Furthermore, fish is considered one of the major food allergens, and a person who would be allergic to conventional salmon would likely avoid all salmon, including the GE salmon.[45]

### 2.2.5 Environmental Concerns

On November 5, 2020, the U.S. District Court of Northern California ruled that the FDA's approval of the GE salmon violated the Endangered Species Act and the National Environmental Policy Act. The Court's major findings revolve around the fact that the FDA did not thoroughly evaluate the consequences of an escape of the GE salmon into the wild and the impacts it could have on the wild populations.[46]

Many scientists and environmental groups are greatly concerned about the harm that might be wrought by GE fish on the environment and believe that there is uncertainty in the identification of environmental problems early on and doubt of solving problems once identified, leaving many unanswered questions around the approval of the GE fish. In the FDA's original environmental assessment, the agency concluded that the engineered salmon were highly unlikely to escape from the two original facilities owned by AquaBounty, one in Prince Edward Island in Canada and the other facility in Panama, and even if the transgenic salmon were to escape, they were unlikely to survive or establish themselves as a population in the wild. The issue, however, is that the FDA did not meaningfully analyze what would happen to wild salmon if the transgenic salmon did survive and establish themselves in the wild.[47]

The two primary concerns regarding the environmental impact of the AquAdvantage Salmon are: first, the potential for AquAdvantage Salmon to interbreed with wild Atlantic salmon; and second, that the AquAdvantage salmon will compete with native fish for food, habitat, mates, and other resources.[48] Given that the FDA's original analysis focuses on only the two facilities in operation by

---

[42] Immunoglobulin E is one of the five classes of immunoglobulins and is responsible for the body's immune system antibody response to perceived threats, including specific proteins found in foods (Godwin et al. 2020).

[43] Dunn et al. (2017).

[44] Ibid.

[45] Food and Drug Administration (2018).

[46] Case 3:16-cv-01574-VC http://earthjustice.org/documents/legal-document/court-ruling-on-genetically-engineered-salmon.

[47] United States District Court of Northern California (2020).

[48] Wodder et al. (2010).

AquaBounty at the time, many opponents feel that the farmed GE salmon will eventually escape from future land-based aquaculture systems and interbreed with wild Atlantic salmon populations. Part of the Court's order is to go back and thoroughly analyze the potential environmental consequences in the event that the GE salmon were to escape the facilities, keeping the new locations in mind.[49] However, the FDA did stipulate in the approval of the AquAdvantage Salmon that the salmon must be produced and farmed in landlocked facilities and may not be connected to the ocean by way of "net pens."

### 2.2.6 Lack of Transparency in the Approval Process

Another major concern from opponents of the GE salmon and GE animals is the lack of general transparency in the approval of GE animals.[50] As described in section 4.4.2, the U.S. FDA regulates GE animals using the NADA application. The rDNA construct in the GE animal meets the FDA's definition of a new animal drug as it affects the structure and function of the salmon.[51] A new animal drug is considered unsafe under section 512(a)(1) of the FD&C Act unless it is given approval by the FDA by way of the NADA application. Companies may submit a NADA to the FDA for review, providing all appropriate information, risk assessments, and studies demonstrating the new animal drug is safe. However, due to the Trade Secrets Act,[52] the agency is prohibited from revealing any information it obtains during the review to the public for comment.[53] In fact, the FDA does not have the authority to disclose whether an application even exists until after publication of the approval in the Federal Register.[54] Companies submitting NADA may disclose the application, as AquaBounty Technologies did, but still lack data or information for public review.

### 2.2.7 Public Perception and Consumer Acceptance

Acceptance of GMOs around the world is mixed. In the United States, the percentage in favor of GMOs hovers around 26–27 percent of the population, while the percentage of opponents ranges from 46 to 58 percent of the population.[55] Most consumers want labeling and transparency for foods that have been produced through modern biotechnology, as many consumers express anxiety over the safety of GMOs in foods. Polls conducted indicated that nearly half of consumers would not eat genetically modified crops while another three-quarters stated they would not eat genetically modified fish, and two-thirds stated they would not eat genetically modified meat.[56]

---

49 United States District Court of Northern California (2020).

50 Consumers Union (2008), p. 1; McEvilly (2013), p. 421.

51 Food and Drug Administration (2017).

52 Trade Secrets Act, 18 U.S. Code Chapter 90 – Protection of Trade Secrets gives people the right to file a civil law suit at the federal level in the event that a trade secret related to interstate or foreign commerce has been misappropriated (American Bar Association, 2016).

53 See 21 U.S.C. § 331(j) (2006) (prohibiting information acquired under the authority of several sections related to the Animal Drug approval process from being revealed).

54 21 C.F.R. § 514.11 (2012).

55 Library of Congress (2020a).

56 Ibid.

Major medical groups in the United States, such as The National Research Council, the American Association of the Advancement of Science, and the American Medical Association, have presented research and statements stating that there is no evidence that GMOs present a unique safety risk compared to conventionally bred products. Organizations that oppose GMOs are typically environmental, farming, and consumer organizations.[57] The latter does bring to question the cost and impact GMO organisms have on trade and impact on agricultural farmers, with farmers having to rely on new seeds every year rather than being allowed to plant seeds from the previous crop year.[58]

### 2.2.8 Labeling Issues

Federal food labeling policy is regulated under the 1938 Federal Food, Drug, and Cosmetic Act and the 1966 Fairy Packaging and Labeling Act. Under Section 403(a)(1) of the FD&C, a food is misbranded if its labeling is false or misleading. Section 201(n) describes misleading as failing to reveal material facts in light of any representations that are made or suggested in the labeling. When the FDA initially approved the GE salmon, there was no labeling requirement, as there was no material difference detected between non-GE salmon and GE salmon, and the GE food is "substantially equivalent" to the non-GE food. Because of this immaterial difference between the food, the FDA does not have the authority to require labeling on the basis of consumer interest alone.[59]

One of the greatest concerns from consumers and advocacy groups has been the lack of labeling requirements for the engineered salmon that address the consumer's right-to-know. Polls show that consumers increasingly support the special labeling of GE products.[60] Proponents of the GE salmon argue that because the GE salmon is considered "substantially equivalent," requiring special labeling for the GE salmon would imply the product is somehow different and cast the product in a negative light.

However, the National Bioengineered Food Disclosure Act now requires that certain GE foods be labeled. Though this Act has been criticized as being incoherent, a list of GE foods, including the AquAdvantage Salmon, can be found as part of the List of Bioengineered Foods. As a result, the product must be labeled as bioengineered once on the market in at least one of the four ways specified in the act – text, symbol, electronic/digital link, or text message. AquaBounty's CEO Sylvia Wolf told reporters that the company is indeed committed to transparency and will label the salmon as genetically engineered.[61] Of importance, due to the exemptions for restaurants and certain retail establishments, any AquAdvantage salmon distributed and sold in restaurants or similar retail settings would not need to be disclosed to the consumer as bioengineered.[62]

---

57 Ibid.
58 Singh & Kaur (2020).
59 Food and Drug Administration (2010).
60 Kopicki (2013).
61 English & Entine (2021).
62 Agricultural Marketing Service (2019).

## 2.3 FUTURE GE FISH AND ANIMAL APPLICATIONS

### 2.3.1 Future of GE Fish

AquaBounty currently ships eggs from Canada to Indiana for farming and growing to market size. Notably, this fishing occurs on land. During its original risk assessment and review of the AquaBounty's application, the FDA considered the environmental impact and risk of escape from the two facilities that were intended for production at the time: the facility in PEI, Canada, and the facility in Albany, Indiana, in the United States. However, AquaBounty has plans to expand its operations aggressively. The first commercial harvest of the salmon at the Indiana farm concluded at the end of May 2021 with the harvest of 5 metric tons of the GE Atlantic salmon.[63] The company plans to continue ramping up production to full capacity at the Indiana facility and intends to open another farm in the United States. The company announced in late October 2020 that Mayfield, Kentucky, was a potential location for a 10,000 metric ton salmon farm,[64] but has since reconsidered that proposal. The company does intend to open a third farm to begin production in 2023.[65] With AquaBounty planning such an ambitious harvest in the future, many are concerned about the environmental risks associated with the increased potential for the GE salmon to escape given the increasing production. Though a federal judge has ordered the FDA to re-analyze the potential impacts on the wild species if the GE salmon were to escape, the company has no plans on slowing down its production, as the FDA only needs to conduct a more thorough assessment and does not pose a significant threat to AquaBounty's plan to sell the salmon.[66]

In addition, AquaBounty has received regulatory approval from Brazil's National Biosafety Technical Commission (CTNBio) for the sale of the GE Atlantic salmon in Brazil. The approval provides a third country in which AquaBounty can sell its salmon and creates an opening in the South American market.[67] Brazil is the largest and most populous country in South America, with a booming agricultural industry and demand for salmon. CTNBio conducted a standard and regulatory review and concluded that the AquAdvantage salmon is safe for environmental and human health. It also based its review on previous approvals by Canada and the United States. Brazil regulates genetically modified organisms under the Biosafety Law No. 11, 105 (2005), which is the law that created the CTNBio. A GE product must go through five approval steps before it can be sold. After these steps have been completed and approved, the product goes through a political assessment prior to marketing to decide if it is advantageous for the country to bring the new GE product to market.[68]

---

[63] AquaBounty (2021b).
[64] AquaBounty (2020).
[65] Hill (2021).
[66] English (2021).
[67] AquaBounty (2021c).
[68] Library of Congress (2020b).

### 2.3.2 Other Animal Applications

In December of 2020, the FDA issued approval of another genetically engineered animal for human consumption. The GalSafe pigs are a line of domestic pigs intended to be safe for consumption for people who have Alpha-gal syndrome (AGS),[69] which causes mild to severe allergic reactions to alpha-gal sugar found in red meat.[70] The pigs, licensed to the United Therapeutics subsidiary – Revivicor Inc., have also been approved for human therapeutics in addition to use as food, making it the first of its kind and unlike the GE salmon. Simply put, the GalSafe pigs have been engineered to remove alpha-gal sugar from their genome. The pigs contain intentional genomic alteration (IGA) accomplished by inserting rDNA into their genome. The pPL657 rDNA construct disrupts the glycoprotein galactosyltransferase alpha-1,3 (GGTA1) gene, the gene normally responsible for producing alpha-gal sugar.[71] Studies examining the effectiveness and safety of the genetic modification are still being conducted, but the technology's sponsors believe in its high potential to be a both safe source of human food and human therapeutics for those with AGS. The GalSafe pigs are undergoing the NADA application, but this time under the FDA Center for Veterinary Medicine's Veterinary Innovation Program (VIP), which was created to facilitate advancements in the development of innovative animal products by providing greater certainty in the regulatory process.[72]

The VIP program is intended to help sponsors navigate the regulatory framework for approving GE products and support innovation. The primary goal is to ensure the safety of biotechnology plant and animal products while avoiding regulatory barriers to future innovation.[73] As many predicted after the FDA approved the AquAdvantage Salmon, there is growing interest in the application of other GE animals, either for human consumption or drug production. Based on the FDA's implementation of the VIP program, it is attempting to make the regulatory approval less confusing for those that would produce GE products. It appears that the interest in this sector will continue, and we will likely see more new animal drug applications for GE animals.

## 2.4 CONCLUSION

It is a political and economic decision to permit any GE fish to enter the food chain. The risks for the environment and the integrity for the food system are out of proportion. Simply put, there is cost-benefit analysis here from the perspective of planetary stewardship on behalf of future generations. Human diets and the environment stand to suffer irreparable harm as GE fish are increasingly pushed onto food chains.

[69] Alpha-gal syndrome often begins after a person has been bitten by a Lone Star tick. The tick transmits alpha-gal sugar into the person's body, creating an immune response that later produces mild to severe reactions to alpha-gal sugar, a sugar molecule found in most mammals. As a result, many medical products that have been derived from animal origin, including drugs and other medications, contain alpha-gal. See Centers for Disease Control and Prevention (2020).

[70] Food and Drug Administration (2020b).

[71] Ibid.

[72] Food and Drug Administration (2020c).

[73] Food and Drug Administration (2018).

## REFERENCES


Agricultural Marketing Service (AMS) (2019 July). National bioengineered food disclosure standard fact sheet. Retrieved from https://www.ams.usda.gov/sites/default/files/media/BEFactSheet.pdf.

Agricultural Marketing Service (n.d.). BE disclosure. Retrieved from https://www.ams.usda.gov/rules-regulations/be.

American Bar Association (2016 September). Explaining the defend trade secrets act. Retrieved from https://www.americanbar.org/groups/business_law/publications/blt/2016/09/03_cohen/.

AquaBounty (2019 March 16). AquAdvantage, the first GMO salmon, is coming to America. Retrieved from https://aquabounty.com/investors/media-relations/aquadvantage-the-first-gmo-salmon-is-coming-to-america.

AquaBounty (2020 October 29). AquaBounty Technologies, Inc. Announces mayfield, Kentucky as potential location for planned 10,000 metric ton salmon farm. Retrieved from https://investors.aquabounty.com/news-releases/news-release-details/aquabounty-technologies-inc-announces-mayfield-kentucky

AquaBounty (2021a). Our farms. Retrieved from https://aquabounty.com/our-farms.

AquaBounty (2021b May 10). AquaBounty sells out the first commercial-scale harvest of genetically engineered Atlantic salmon from its Indiana farm. Retrieved from https://investors.aquabounty.com/news-releases/news-release-details/aquabounty-sells-out-first-commercial-scale-harvest-genetically.

AquaBounty (2021c June 1). AquaBounty announces first harvest of GE Atlantic salmon; Receives approval for the sale of GE Atlantic salmon in Brazil. Retrieved from https://investors.aquabounty.com/news-releases/news-release-details/aquabounty-announces-first-harvest-ge-atlantic-salmon-receives.

Bratspies, R. (2008). Farming the genetically modified seas—the perils and promise of transgenic salmon.

CDC (2020 October 6). Alpha-gal syndrome. Retrieved from https://www.cdc.gov/ticks/alpha-gal/index.html

Center for Food Safety (2020 November 05). Federal court declares genetically engineered salmon unlawful. Retrieved from https://www.centerforfoodsafety.org/press-releases/6186/federal-court-declares-genetically-engineered-salmon-unlawful

Conley, D. (2016). AquAdvantage Salmon: Climate-smart aquaculture. Retrieved from https://aquabounty.com/wp-content/uploads/2016/10/FFE_2016_EAS-Aquabounty.pdf.

Consumers Union (2008). Comments on Docket No. FDA-2008-D-0394, "Draft Guidance for Industry #187: Regulation of Genetically Engineered Animals Containing Heritable rDNA Constructs via the Docket's Division.

Convention on Biodiversity (2018 October 26). About the Nagoya - Kuala Lumpur supplementary protocol on liability and redress. Retrieved from https://bch.cbd.int/protocol/supplementary/about/.

Dunn, S. E., Vicini, J. L., Glenn, K. C., Fleischer, D. M., and Greenhawt, M. J. (2017 September 01). The allergenicity of genetically modified foods from genetically engineered crops. *Annals of Allergy, Asthma, and Immunology*, 119(3), pp. 214–222. Retrieved from https://doi.org/10.1016/j.anai.2017.07.010.

English, C. (2021 March 22). AquaBounty's GE salmon is headed for U.S grocery stores. Can Anti-GMO groups block its commercialization? American Council on Science and Health. Retrieved from https://www.acsh.org/news/2021/03/22/aquabountys-ge-salmon-headed-U.S-grocery-stores-can-anti-gmo-groups-block-its-commercialization-15416.

EUR-Lex (2015 March 11). Directive (EU) 2015/412 of the European parliament and of the council of 11 March 2015. *Official Journal of the European Union*. Retrieved from https://eur-lex.europa.eu/legal-content/EN/TXT/?uri=CELEX:32015L0412.

EUR-Lex (2016 April 18). Genetically modified organisms – Traceability and labeling. Retrieved from https://eur-lex.europa.eu/legal-content/EN/TXT/?uri=LEGISSUM:l21170.

Europa (n.d.). Several European countries move to rule out GMOs. Retrieved from https://ec.europa.eu/environment/europeangreencapital/countriesruleoutgmos/.

European Commission (n.d.). GMO legislation. Retrieved from https://ec.europa.eu/food/plant/gmo/legislation_en.

Food and Drug Administration (2018 January 16). AquAdvantage Salmon - FDA response to public comments to the veterinary medicine advisory committee. Retrieved from https://www.fda.gov/animal-veterinary/animals-intentional-genomic-alterations/aquadvantage-salmon-fda-response-public-comments-veterinary-medicine-advisory-committee.

Food and Drug Administration (2020a April 22). How GMOs are regulated for food and plant safety in the United States. Retrieved from https://www.fda.gov/food/agricultural-biotechnology/how-gmos-are-regulated-food-and-plant-safety-united-states.

Food and Drug Administration (2020b December 14). FDA approves first-of-its-kind intentional genomic alteration in line of domestic pigs for both human food, potential therapeutic uses. Retrieved from https://www.fda.gov/news-events/press-announcements/fda-approves-first-its-kind-intentional-genomic-alteration-line-domestic-pigs-both-human-food.

Food and Drug Administration (2020c February 15). VIP: Veterinary innovation program. Retrieved from https://www.fda.gov/animal-veterinary/animals-intentional-genomic-alterations/vip-veterinary-innovation-program.

Federal Registrar (2020 December 28). Regulation of the movement of animals modified or developed by genetic engineering. Retrieved from https://www.federalregister.gov/documents/2020/12/28/2020-28534/regulation-of-the-movement-of-animals-modified-or-developed-by-genetic-engineering

Federal Registrar (2021 March 08). Regulation of the movement of animals modified or developed by genetic engineering. Retrieved from https://www.federalregister.gov/documents/2021/03/08/2021-04716/regulation-of-the-movement-of-animals-modified-or-developed-by-genetic-engineering.

Food & Water Watch, Consumers Union, and the Center for Food Safety (2012 February 7). Petition to deem ABT technologies' genetically engineered AquAdvantage Salmon an unsafe food additive. Retrieved from http://www.centerforfoodsafety.org/files/fdafoodadditivepetitiongesalmon.pdf.

Food and Drug Administration (2010). Background document: Public hearing on the labeling of food made from the AquAdvantage Salmon. Retrieved from http://www.fda.gov/Food/GuidanceRegulation/.

Food and Drug Administration Center for Veterinary Medicine (2017). Guidance 187, guidance for industry regulation of genetically engineered animals containing heritable recombinant DNA constructs. Retrieved from https://www.fda.gov/regulatory-information/search-fda-guidance-documents/cvm-gfi-187-regulation-intentionally-altered-genomic-dna-animals

Godwin, L., Sinawe, H., Crane, J. S. (2021 January). Biochemistry, Immunoglobulin E. [Updated 2020 Sep 8]. In: StatPearls [Internet]. Treasure Island (FL): StatPearls Publishing. Retrieved from https://www.ncbi.nlm.nih.gov/books/NBK541058/.

Hill, S. (2021 February 24). AquaBounty sends GE salmon samples to customers ahead of first commercial harvest. Retrieved from https://www.seafoodsource.com/news/aquaculture/aquabounty-sends-ge-salmon-samples-to-customers-ahead-of-first-commercial-harvest

FDA. (2015). Industry: Regulation of engineered animals containing heritable rDNA constructs. Retrieved from https://www.fda.gov/media/135115/download

Kopicki, A. (2013). Strong support for labeling modified foods. NY Times. Retrieved from http://www.nytimes.

Library of Congress (2020a December 30). Restrictions on genetically modified organisms: United States. Retrieved from https://www.loc.gov/law/help/restrictions-on-gmos/usa.php.

Library of Congress (2020b December 30). Restrictions on genetically modified organisms: Brazil. Retrieved from https://www.loc.gov/law/help/restrictions-on-gmos/brazil.php.

McEvilly, M. (2013). Lack of transparency in the premarket approval process for Aquadvantage Salmon. *Duke Law & Technology Review*, 11, pp. 413–433. https://www.researchgate.net/publication/228311271_Farming_the_Genetically_Modified_Seas_The_Perils_and_Promise_of_Transgenic_Salmon.

Singh, S. J. and Kaur, G. (2020 December 11). Indian farmers are right to be outraged. Retrieved from https://www.cnn.com/2020/12/11/opinions/indian-farmer-protests-simran-singh-gunisha-kaur/index.html.

Thompson, M., Saltzman, M., and Bittman, M. (2019 June 22). Future of food: This genetically engineered salmon may hit U.S. markets as early as 2020. *PBS News Hour*. Retrieved from https://www.pbs.org/newshour/show/future-of-food-this-genetically-engineered-salmon-may-hit-u-s-markets-as-early-as-2020.

United States District Court Northern District of California (2020). Institute for Fisheries Resources, et al., v. United States Food and Drug Administration et al. Case No. 16-cv-01574-VC.

United States Marketing Service (2019). National Bioengineered Food Disclosure Standard Fact Sheet. Retrieved from https://www.ams.usda.gov/sites/default/files/media/BEFactSheet.pdf

United States Marketing Service (2020). BE Disclosure. Retrieved from https://www.ams.usda.gov/rules-regulations/be

Upton, H. and Cowan, T. (2015 December 8). Genetically Engineered Salmon. Congressional Research Service. Retrieved from https://sgp.fas.org/crs/misc/R43518.pdf

Wodder, R. et al. (2010). Letter to Margaret Hamburg, M.D., Commissioner, FDA. Retrieved from http://www.foe.

# *Editor Suggestions for Future Discussion*

## *JULIANA v. UNITED STATES*

In light of the trailblazing lawsuit of 11 children against governments that are failing at protecting the global environment, GE fish could be a notable friction point in the creation of new case law. Consider the following materials in determining where GE fish and overfishing play into the rights of future generations to protect what is left of the integrity of the food system and the environment:

> In 2015, 21 youth and organizational plaintiff Earth Guardians, filed their constitutional climate lawsuit, *Juliana v. United States*, against the U.S. government. Their complaint asserts that, through the government's affirmative actions that cause climate change, it has violated the youngest generation's constitutional rights to life, liberty, and property, as well as failed to protect essential public trust resources.
>
> After a denial of the youth's petition to rehear their landmark constitutional climate lawsuit *en banc*, in March 2021, the youth plaintiffs in *Juliana v. United States*, including eleven Black, Brown, and Indigenous youth, requested to amend their complaint in the U.S. District Court to revise their requested relief to address the concern raised in the Ninth Circuit's opinion. The youth plaintiffs currently seek a declaration of their constitutional rights and a declaration that the U.S. national energy system is unconstitutional.
>
> Since President Biden took office, the *Juliana* plaintiffs' attorneys made multiple requests to the Department of Justice (DOJ) for a meeting of the parties to discuss the Biden administration's position in the case and to discuss potential resolutions of the case, but they did not receive a response. At a scheduling conference on May 13, 2021, the District Court judge stressed that all three branches of government need to work together to address the climate crisis and ordered the parties to engage in settlement discussions.
>
> Oral argument on the motion to amend was held on June 25, 2021. In June, 18 Republican Attorneys General sought intervention to obstruct any potential settlement and to oppose the youth plaintiffs' motion for leave to amend their complaint.

DOI: 10.1201/9781003137733-5

On July 6, 2021, the Attorneys General from New York, Delaware, Hawai'i, Minnesota, Oregon, and Vermont, filed an amicus brief in support of the case.

On November 1, 2021, settlement talks between the youth plaintiffs and the DOJ ended without resolution. Despite good faith efforts on the part of the youth plaintiffs, they see no reason to continue to pursue settlement discussions until the decision-makers for the federal defendants come to the settlement table. Currently, the *Juliana* 21 anticipate a favorable ruling on their motion to amend their complaint in the District Court. Unfortunately, despite the new administration, the DOJ seems to be set on closing the courthouse doors on the youth plaintiffs.

*Juliana* became one of the most significant targets of the Trump administration's "shadow docket," seeking to have cases decided by the Supreme Court before trial and without factual records. The Biden administration is now the third administration to be defendants in the case. Instead of fighting the youth and the public interest, the Biden administration should change the DOJ's previous scorched earth approach to *Juliana*, stand for the constitutional rights of children, and work with the youth to come to a sensible resolution based on technically and economically feasible solutions to the climate crisis.

The Biden administration and DOJ can remedy this injustice through one of two available pathways:

1. Immediately reverse the unsupported legal position of Trump's DOJ and recognize that a declaration of constitutional rights is well within the Article III authority of federal courts and the 1934 Declaratory Judgment Act passed by Congress and that such a declaration suffices to establish redressability.
2. Engage in meaningful settlement discussions with the *Juliana* plaintiffs, their attorneys, and their experts to commit to science-based, durable climate remedies.

The DOJ has broad discretion when settling litigation.

President Biden declared that climate change is the "number one issue facing humanity" and that tackling the climate crisis is his top priority. He told children their government will now work for them. Until the courts declare that youth have fundamental rights that the government is infringing by causing climate change, youth and all humanity will face ever more threatening climate chaos. The futures of these youth plaintiffs and of children around the nation and across the globe depend on the President following through on the recent executive orders and policies promising to address climate injustice and the climate crisis. The Biden administration has an urgent and timely opportunity to resolve Plaintiffs' claims and grant them the justice they seek.[1]

[1] https://www.ourchildrenstrust.org/biden-doj.

# Part 2

## Bovine Animals

# Editor's Note

Bovine animals have long played an essential role in human existence, supplying humans with high-value nutrition for sustenance, manure for fertilization, leather for warmth and protection, and labor for cultivation. In many ways, bovines have shaped current society in both developed and developing worlds. Unfortunately, these animals have not always been treated with their deserved respect and appreciation. Frequent reports of mistreatment and abuse have led many to shun the animal agricultural industry. Additionally, these animals find themselves at the forefront of the climate crisis and are largely blamed for contributing consequential amounts of greenhouse gas (GHS) to the environment. Indeed, dairy and cattle operations are responsible for more GHS emissions than any other food source. With a growing population and the threat of climate change worsening, it is vitally important to future food security that these animals are protected and that we increase the efficiency by which humans work with bovines; both to lessen current emission rates and as feeding a larger population will necessitate that the entire agriculture sector provides more food than ever in the coming years. This starts with crafting better animal welfare laws, improving animal husbandry practices, and protecting against infectious and zoonotic diseases.

Because humans rely so heavily on these animals for valuable resources, one might assume they would be well protected. Disappointingly, the opposite has proven true. There currently is stringent legislation defining food safety and human worker safety, but there is a distinct lack of thorough animal safety rules and enforcement. In the U.S. for example, the United States Department of Agriculture's Food Safety Inspection Service (USDA-FSIS) enforces food safety laws for meat production activities, and the Occupational Safety and Health Administration (OSHA) oversees worker safety. However, there is no one agency that is responsible for enforcing animal safety and welfare. These animals have been exploited – injected with hormones, oxytocin, antibiotics, and other unnatural substances; impregnated by artificial insemination; and in some cases, even smuggled in horrific transboundary mafia operations. Though this is certainly not the case in every operation, the number of operations that do not concern themselves with animal welfare because they know that enforcement agencies' focus is limited primarily to rules for food safety

DOI: 10.1201/9781003137733-7

and human safety (i.e. not animal safety) is enough to harm to the industry at large. Though there are many examples of responsible bovine operations that treat animals with humanity and respect, as one author will present in this section with a case study of the Kisan Dairy in India,[1] there are many other examples of mistreatment – even of bulls and buffalo being used as a form of entertainment in certain communities.[2]

There has even been legislation in the U.S. rendering it illegal for so-called "whistleblowers" to report on egregious activities happening in agricultural facilities. Dubbed "ag-gag", these laws were passed with the intent to criminalize whistleblowers attempting to document animal welfare violations.[3] This, coupled with the lack of regulatory oversight, should be alarming for anyone who relies on bovine animal resources.

In regard to climate, because of the domestication of these animals and current agricultural practices, the dairy industry and beef cattle industry have been largely blamed for exacerbating the effects of climate change. Undoubtedly, livestock products are responsible for more greenhouse gas emissions than most other food sources.[4] The dairy sector alone produces 3.1 gigatons of $CO_2$ every year[5]; livestock supply chains account for 7.1 gigatons (14.5 percent of global anthropogenic greenhouse gas emissions).[6] Cattle are responsible for two-thirds of that total.[7]

Despite this, much can be done to combat greenhouse gas emissions resulting from livestock activities. Reducing the demand for meat would seem an obvious choice. However, with the growing human population, higher incomes, and urbanization, demand for livestock products in low- and middle-income countries is ever on the rise, and the livestock sector is rapidly growing. Meat and milk constitute a substantial portion of the protein consumed globally, along with many essential micronutrients,[8] and are crucial to ensuring a food secure future in developing nations. Accordingly, the products cannot simply be removed from the diets of all without negatively impacting this potentially bright future.

The Food and Agricultural Organization of the United Nations (FAO) suggests the following: "productivity improvements that reduce emission intensities; carbon sequestration through improved pasture management; better livestock integration in the circular bioeconomy."[9] Productivity improvements include better feed and nutrition and improved animal health and husbandry. Interestingly, these concepts go hand-in-glove with improved animal welfare practices. Improving feed quality no doubt has an impact on the overall health of the ruminant, and by improving grassland management, the ruminant has better access to optimum nutrition. Better animal health and husbandry practices focusing on the health of the animal contribute to reduced incidents of zoonotic disease. By increasing the performance of each

---

1 Verkest, A., Dairy Cows and Goats: The US Regulatory and Protective Challenges.
2 Ramdas, S., The Disintegration of Bovine Animal Protection in India.
3 Johnson, A., AG-GAG: Agriculture, Whistleblowers, and the 1st Amendment.
4 https://www.fao.org/3/i8098e/i8098e.pdf.
5 Verkest, A., Dairy Cows and Goats: The US Regulatory and Protective Challenges.
6 https://www.fao.org/3/i8098e/i8098e.pdf, pg 3.
7 Id.
8 Id.
9 Id.

animal through these improved practices, FAO estimates that GHG emissions can be reduced by 20–30 percent.[10] One-quarter of the Earth's surface is dominated by permanent pastures. Unfortunately, growth in the livestock sector and poor grazing management have led to the overgrazing of these pastures. Restoring this degraded pastureland using new technologies and practices can significantly reduce carbon emissions.[11]

If humans are to continue to rely on bovine animal protein and dairy products, the current global regulatory framework must be improved to promote animal welfare. Thus, we will not only be increasing human, animal, and environmental health, but we will also help secure a safer, more resilient future in the face of rising populations and climate change.

This section will discuss ag-gag laws in detail, providing case studies that demonstrate how these rules impinge upon first amendment rights and contribute negatively toward animal welfare. Next, the section will examine the limited scope of animal law in India and explain the failures of that law in protecting and promoting animal welfare. In the same vein, an overview of the global dairy industry and the regulations which govern it will be given with a focus on animal welfare, human health, and food safety. The section will close with an explanation of the perceived relationship between animal disease and human health. All ideas presented represent the current regulatory framework and industry practices. Each author provides evidence of the lacking laws and regulations for the protection of animal welfare. It will be clear after reading this section that protecting animal welfare means ensuring health and safety for animals, humans, and the environment.

10 Ibid, pg. 4.
11 Ibid, pg. 5.

# *Textbox: Zoonotic Diseases*

## *When Animals Are Sick, People Get Sick*[1]

*Christian Mulgrew de Laire*

## INTRODUCTION

Simply put, when animals get sick, people get sick. On the one hand, animals live close to people, whether they are insects carrying malaria or geese that are consumed, people can get a multitude of ailments from animals. On the other hand, people make animals sick, such as by confining them in Concentrated Animal Feeding Operations (CAFOS), or battery cages, transporting them in trucks, or slowly and brutally killing them. While there are uncountable questions of ethics and conscience, there is limited space in this volume to address them all. This textbox illuminates some of the connections that cause infections to hop species.

In the era of globalization, internationalized representations of infectious disease threats have profound implications for understanding contagious disease problems and their management in developing countries. By examining public policy around public health, animal health, and trade, it becomes clear that resource-constrained countries at the nexus of public-animal health and trade are perceived as the source and victim of infectious disease. It is essential to think about how governments can approach risk management concerning pandemic scale threats that cross the species barriers to humans. The feasibility of the risk is flexible in how the risk is framed across animal health, public health, and trade.

[1] This chapter is based on an earlier chapter by Kennedy Mwacalimba (Associate Director, Outcomes Research at Zoetis), and it was updated and adapted for the purpose of this volume. Internal citations are omitted.

DOI: 10.1201/9781003137733-8

## TRADE, AGRICULTURE, AND HEALTH

There is a significant threat that comes from industrial animal agriculture, aquacultures, wildlife exploitation, and also gaming, all of which decrease risk resilience of the species that are consumed by humans and pass diseases on to humans – with a ripple effect across economies, policies, trade, environmental integrity, and food safety. A combination document published by the Food and Agricultural Organization (FAO), World Organization for Animal Health (OIE), and World Health Organization (WHO) sets forth the criteria of the One Health approach, which is combined with the Sustainable Development Goals (SDG):

> addresses the interconnectedness of health and its social and economic determinants aligns with the SDG framework. Health is a critical consideration in achieving the 17 goals and taking a One Health approach in health activities will support making progress in achieving the SDGs. The SDGs themselves reflect a One Health approach, ensuring that healthy people and animals live on a healthy planet.[2]

Specifically, the WHO explains that

> new pathogens from animals particularly viruses remain unpredictable and continue to emerge and spread across the countries. The diseases are also a concern to global health owing to their epidemic potential, high case fatality ratio and the absence of specific treatment and vaccines available to control the spread of most of these zoonotic diseases (with the exception of the yellow fever vaccine) As the world is increasingly inter-connected, emerging zoonoses in one country can potentially constitute a threat to global health security[3].

The WHO emphasizes that

> zoonoses matter not just because they are so common, but because they cause morbidity and mortality, high burden on health systems but more importantly, it causes significant economic loses to the countries by way of losing animal trade, travel as well as loss of economic opportunities for the people through loss of livestock.[4]

### The Human-Animal Interface and Biological Influencers

Only a few pathogens are human-specific; the rest are spread from other forms of life. "The human-animal interface" is the sharing of environments; thus, it encompasses the ecological and biological influences of disease transmission, which allow for human-animal disease management. While the spread of pathogens is not new, it is magnified by the globalized world. The risks of the human-animal interface that spread pathogens, including changes in land use, livestock production, chosen routes for economic growth, and trade promotion, provide an essential conceptual for examining public health through health policy and their relationships with risk enabling anthropogenic activities.

---

[2] https://www.fao.org/3/ca2942en/CA2942EN.pdf.

[3] http://www.emro.who.int/about-who/rc61/zoonotic-diseases.html.

[4] ibid.

### Relationship between Trade and Health: The Microbial Perfect Storm

As a comprehensive, multi-faceted phenomenon rapidly transforming society, globalization is central in shaping the debate around trade and health. There are different understandings of what globalization means.

According to a National Academies of Science report,[5] globalization has snowballed on infectious disease emergence, creating a sort of microbial "perfect storm" that will be a recurring event.[6] Changes in land use, livestock production, climate change, routes of commerce, etc., influence the "perfect microbial storm." Livestock, for example, is a way for the zoonotic disease to incubate and spread as the animals are moved. Therefore, globalization is a conduit for the spread of infectious disease since increased contact allows for the spread of disease.

## TRADE EFFECTS ON PUBLIC HEALTH

Trade and its effects on public health, through disease spread, is a matter of both historical and contemporary policy significance. The spread of the disease has been linked to traded products moved through carriage vehicles, such as ships, as seen during the "Black Death."[7] It was this realization that led to the first International Sanitary Conferences in 1851.[8] At several points in history, trade has been restricted in order to preserve health and protect trade interests. However, the economy has been prioritized over health recently.[9] Similar arguments have been made about global health policy,[10] access to medicines,[11] food safety,[12] and infectious disease spread.[13]

The view of public health officials that health concerns are secondary to global commerce becomes essential when considering the role that trade policy plays in disease spread and control. For example, Lipson's[14] review of the World Trade

---

[5] Navarro, V., 1998, Comment: Whose globalization?: *American Journal of Public Health*, v. 88, pp. 742–3.

[6] Mann, J., 1990, The global lesson of AIDS: *New Scientist*, v. 126, p. 30.

[7] Bettcher, D. W., Yach, D., and Guindon, G. E., 2000, Global trade and health: key linkages and future challenges: *Bull World Health Organ*, v. 78, pp. 521–34.

[8] ibid; Aginam, O., 2002, International law and communicable diseases: *Bull World Health Organ*, v. 80, pp. 946–51; Hoffman, S. J., 2010, The evolution, etiology and eventualities of the global health security regime: *Health Policy Plan*, v. 25, pp. 510–22.

[9] Lee, K., and Koivusalo, M., 2005, Trade and health: is the health community ready for action?: *PLoS Med*, v. 2, p. e8.
Lang, T., 1999, Diet, health and globalization: five key questions: *Proceedings of the Nutrition Society*, v. 58, pp. 335–43.

[10] Lee, K., Fustukian, S., and Buse, K., 2002, A introduction to global health policy. In: Lee, K., Buse, K., and Fustukian, S. (eds) *Health policy in a globalizing world*. Cambridge University Press, Cambridge.

[11] Kerry, V. B., and Lee, K., 2007, TRIPS, the Doha declaration and paragraph 6 decision: what are the remaining steps for protecting access to medicines?: *Global Health*, v. 3, p. 3.

[12] Rowell, A., 2003, *Don't Worry, It's Safe to Eat: The True Story of GM Food, BSE, and Foot and Mouth*: Earthscan Publications, London.

[13] Kimball, A. M., 2006, *Risky trade: infectious diseases in the era of global trade*. Ashgate Publishing, London.

[14] Lipson, D. J., 2001, The World Trade Organization's health agenda: *BMJ*, v. 323, pp. 1139–40.

Organization (WTO) health agenda and the study done by Shaffer et al.[15] on ethics in public health research both suggest that trade agreements limit a country's ability to police its food safety and health restrictions leading to increasing interconnectedness of infectious disease spread through trade.[16]

The tension between trade promotion and health protection has existed, but globalization has increased the tension. Some diseases that spread through trade are Monkeypox, SARS, avian Influenza, and African Swine Fever. Monkeypox infected 71 people in six US states when prairie dogs acquired the disease from the African rodents hosed with them.[17] SARS supposedly spread through bat and human interaction.[18] Avian Influenza came from human exposure at live animal markets in Southeast Asia.[19] African Swine Fever, an animal health problem, spread rapidly along with the Atlantic dynamic coastal trading network of West Africa.[20]

Looking at the trade and health problem from another perspective, perhaps the issue is not simply about the priorities of WHO, WTO, FAO, and OIE, since the "legally binding" and "soft-law" of international organizations, such as, the WHO, FAO, and OIE have been reshaping trade agreements, approaches to disease control, and international health guidelines to fit international standards.

## THE ROLE OF AGRICULTURE IN DISEASE SPREADS

Agriculture, which includes animal health, plays a role in the interface between trade and public health. It has been suggested that the OIE create a global perspective on health similar to that of the IHRs.[21] The perspective would help combat transboundary diseases and guide the global approach to preventing zoonosis disease outbreaks.[22] To date, no overhaul has been made, but progress has been made to achieve these goals. Simultaneously, the international animal health community has taken a pro-trade stance while maintaining the guidelines of the Sanitary and Phytosanitary (SPS) Agreement, that is created under the WTO's Terrestrial Animal

---

[15] Shaffer, E. R., Waitzkin, H., Brenner, J., and Jasso-Aguilar, R., 2005, Global trade and public health: *American Journal of Public Health*, v. 95, pp. 23–34.

[16] ibid.

[17] Morse, S. S., 2004, Factors and determinants of disease emergence: *Revue Scientifique et technique*, v. 23, pp. 443–51; Kahn, L. H., 2006, Confronting zoonoses, linking human and veterinary medicine: *Emerging Infectious Diseases*, v. 12, pp. 556–61

[18] Fevre, E. M., Bronsvoort, B. M., Hamilton, K. A., and Cleaveland, S., 2006, Animal movements and the spread of infectious diseases: *Trends in Microbiology*, v. 14, pp. 125–31.

[19] Cowen P., and Morales, R., 2002, Economic and trade implications of zoonotic diseases. In: Burroughs, T., Knobler, S., and Lederberg, J. (eds) *The emergence of zoonotic diseases: understanding the impact on animal and human health*. National Academies Press, Washington; Morse, S. S., 2004, Factors and determinants of disease emergences. *Revue Scientifique et Technique*, v. 23, pp. 443–51; Karesh, W. B., Cook, R. A., Bennett, E. L., and Newcomb, J., 2005, Wildlife trade and global disease emergence: *Emerging Infectious Diseases*, v. 11, pp. 1000–2.

[20] ALive, 2006, Avian influenza prevention and control and human influenza pandemic preparedness in Africa: assessment of financial needs and gaps: *Fourth International Conference on Avian Influenza*, Bamako, Mali, Dec. 2006.

[21] ibid.

[22] Blancou, J., Chomel, B. B., Belotto, A., and Meslin, F. X., 2005, Emerging or re-emerging bacterial zoonoses: factors of emergence, surveillance and control: *Veterinary Research*, v. 36, pp. 507–22.

Health Code.[23] Thus, countries that are involved in livestock and livestock product trade are expected to comply with the SPS Agreement in order to reap the full benefits of international commerce.[24] Zepeda et al.[25] uphold that public health measures should be based as much as possible on international standards to avoid risk analysis of local policy. The SPS Agreement states that measures to protect public health, animal health, and plant health should only minimally interfere with trade. This clause is problematic nationally as it is also found in the IHRs.[26] The clause does demonstrate the increasing interactions of different areas, like health and trade, within international policy.

Similarly, widely spread infectious diseases, such as SARS and pandemic avian Influenza became a global issue. Given their unique situations, while a global approach is beneficial, the assumptions of how infectious disease should be prioritized, the risk, and how to manage the risk are not easy for developing countries to follow.

## SOLUTIONS

Shortened supply chains and small, closely held farms, integrated into information tracking systems, such as blockchain, have the potential to provide smaller foci for infection. The smaller the foci, the lower the risks for epidemic or even pandemic spreads. Considering the CAFOs and other industrial forms of animal farming do the exact opposite, i.e. increase the risk and climate for infection, speaks against these forms. Thus, in addition to an animal welfare favoring approach, attention to detail rather than mass animal production can help to reduce the risk for disease spread.

23 Bruckner, G. K., 2009, The role of the World Organisation for Animal Health (OIE) to facilitate the international trade in animals and animal products: *Onderstepoort Journal of Veterinary Research*, v. 76, pp. 141–6; OIE Terrestrial Animal Health Code 2010 available at http://web.oie.int/eng/normes/mcode/en_index.htm.

24 Thiermann, A. B., 2005, Globalization, international trade and animal health: the new roles of OIE: *Preventive Veterinary Medicine*, v. 67, pp. 101–8.

25 Zepeda, C., Salman, M., Thiermann, A., Kellar, J., Rojas, H., and Willeberg, P., 2005, The role of veterinary epidemiology and veterinary services in complying with the World Trade Organization SPS agreement: *Preventive Veterinary Medicine*, v. 67, pp. 125–40.

26 Merianos A., and Peiris M., 2005, International health regulations. *Lancet*, v. 366, pp. 1249–52. doi:10.1016/S0140-673(05)67508-3.

# 3 Global Regulatory Overview of Farm Animal Welfare[1]

*Christian Mulgrew de Laire*

## CONTENTS



[1] This chapter is based on an earlier chapter by Lewis Bollard (Farm Animal Welfare Senior Program Officer at Open Philanthropy), and it was updated and adapted for the purpose of this volume. Internal citations are omitted.

DOI: 10.1201/9781003137733-9

## 3.1 INTRODUCTION

In the last half century, globalization, a growing world population, and increased prosperity have caused the proliferation of global laws, institutions, and industrialized animal agriculture. But it was not until recently that the three were closely linked. Industrialized animal agriculture had largely escaped even domestic regulation, which led to global institutions, driven by domestic actors, to ignore regulation as well.

### 3.1.1 The Globalization of "Factory Farming"

In the last half century, growth in poultry consumption has driven the dramatic expansion of animal agriculture. The FAO[2] compared the growth of ducks, cattle, pigs, chickens, turkeys over a period from 1961 to 2013, finding that only pigs grew proportionally to human population growth. Ducks, cattle, chickens, and turkeys all grew proportionally more than the human population.

At the same time, animal agriculture has industrialized. Until the mid-twentieth century, the majority of farmers were families in low-density areas raising a small herd of diverse animals.[3] In the early twentieth century, farmers realized they could be more efficient by focusing on and more intensely rearing one animal. These farmers were aided by the development of artificial insemination, antibiotics, and medicated feeds that allowed the animals to be kept in unnatural conditions like confined environments, and produce at a higher level.[4] Artificial insemination, for example, went from being uncommon to involving almost every animal within a century.[5]

The Green Revolution of the 1950s and 1960s brought industrialized agriculture to the developing world. The Revolution's focus on irrigation, chemical fertilizer, and crop genetic selection more than doubled crop production after 1960. These efficiencies increased the price of agricultural land and the opportunity cost of grazing livestock outside. On the other hand, the price of grains decreased, reducing costs for "factory farms" or places that confined animals indoors. As a result, the industry focused more on pig and poultry since they eat grains, unlike cows and sheep that require roughage.[6]

Breaking from the tradition of small-scale farming,[7] China has driven a huge portion of the growth in factory farming. When Deng Xiaoping began opening China's economy in 1978, the nation's economy boomed, and meat consumption quadrupled driven by huge increases in pork and chicken consumption.[8] Much of the increased consumption is of industrially raised meat, which is viewed as safer than backyard-raised meat.[9] Since 2006, when China experienced an outbreak of porcine reproductive and respiratory syndrome, the government encouraged the consolidation and

---

[2] http://www.fao.org/3/y4252e/y4252e07.htm.
[3] Cochrane (1993), pp. 122–150.
[4] ibid.
[5] Wilmot (2007).
[6] Sherman (2002), p. 79.
[7] Schneider and Sharma (2014), p. 12.
[8] FAO (2014a).
[9] Schneider and Sharma (2014).

industrialization of the nation's pork industry. This consolidation has resulted in less than one-third of pigs being raised in a backyard with fewer than 50 other pigs.[10]

The globalization of industrial animal agriculture has been a boon for multinational agribusinesses, which now control the global supply chain for pork, beef, and poultry production. Brazilian-based JBS, with estimated revenues of $50 billion,[11] became the world's largest meat company by acquiring companies in Australia, Argentina, and the United States, including Swift & Company. Chinese-based Shuanghui/WH became the largest pork producer when it acquired the largest American producer, Smithfield Foods. Tyson Foods and Cargill, with revenues of $34 billion and $20 billion, respectively, produce much of the rest of the world's beef, pork, and poultry.[12] Even many distinctive brands, including Pilgrim's Pride Chicken, Jimmy Dean's sausages, Swift beef and pork, Hillshire Farm pork, Wright bacon, Shady Brook turkey, and Amour pork, are owned by these four huge multinational meat companies.

### 3.1.2 The Animal Welfare Effects of the Globalization of "Factory Farming"

The spread of industrial animal agriculture created an atmosphere of inhumane treatment of animals. The animals are raised in indoor confinement systems without sunshine, clean air, or space, only to be mutilated, separated prematurely from their mothers, and slaughtered in inhumane manners.

The "Brambell Report," written in 1964 when the British government wanted to investigate the spreading intensive farming on animal welfare, found that the modern factory farm was detrimental to animal welfare.[13] The report wrote that animals should be given enough room to live, and, if required, surgeries should be conducted with pain relief. In short, the report called for a system efficient for the animals, not just the balance sheet. Unfortunately, the government ignored the report and their unique opportunity to change the farming system.

Almost a half century later, the Pew Commission on Industrial Farm Animal Production, consisting of 15 experts including a former US Secretary of Agriculture, concluded that factory farming was harming animal welfare.[14] It suggested reforms such as the phasing out of confinement systems, such as veal crates, gestation and farrowing crates, and battery cages, ending painful mutilations, and improved flooring and housing systems.[15] The Commission noted that, "Recently, the [factory farming] model has begun to spread to all corners of the world, especially the developing world."[16]

To this day, most animal factory farms have yet to implement the conditions suggested by the Brambell report or the Pew Commission. The poultry industry forces

---

10 Schneider and Sharma (2014), p. 19.
11 Samora (2014).
12 Clyma (2014).
13 Brambell (1965).
14 Pew Commission (2008).
15 ibid.
16 ibid.

chickens to live in crowded sheds deprived of sunlight, fresh air, or space. Chickens are selectively bred to grow so fast that many can no longer walk and experience chronic pain.[17] Prior to slaughter, birds are often roughly handled while being placed into transport trucks due to the high (1,000–1,500 birds per hour) rate at which they are caught. Unreliable stun baths and cutting lines at the slaughterhouse result in millions of birds being boiled alive each year.[18]

Chickens in the egg industry have it even worse. Hens are stored with 3–9 other birds in wire-meshed cages called batteries. Birds receive only 48–72 square inches of space, and the batteries are stacked up to six levels high.[19] The American median shed stocking density is 30,000–70,000 hens.[20] At birth, male chicks are typically ground up alive while female chicks are "de-beaked" to stop hens from cannibalizing each other in extreme confinement.[21]

Indeed, most farm animals are subjected to intensive confinement and acts that inflict acute pain. Almost all pigs have their tails and genitalia mutilated at birth without the use of painkillers. Most breeding pigs spend their lives immobilized in gestation (pregnancy) crates and farrowing (birthing) crates.[22] Fish are raised in aquaculture conditions that are essentially water-based factory farms, with dirty water and parasites eating at their faces.[23] Dairy cows are forced to carry excess amounts of milk in their utters, while unwanted male calves are sold to veal farms that usually use "veal crates" that prevent calves from even socializing.[24] Beef cattle have the best conditions with open pastures; however, they are still castrated, branded, and endure a period of feedlot confinement.[25]

### 3.1.3 The Failure of Domestic Laws at Addressing Globalized Factory Farming

Domestic laws have failed to keep up with the globalization of factory farming practices. Most developed countries do not regulate factory farming domestically. In the United States, there are no federal laws preventing animal cruelty on farms or routine agricultural practices, and in many states, cruelty laws that apply to cats and dogs exclude farm animals.[26] The European Union (EU) has better protection for domestic farm animals but these protections largely apply only to animals within their borders. New Zealand with its strong animal welfare laws cannot stop the nation's largest dairy cooperative, Fonterra, from establishing low-welfare factory farms in China.[27]

---

17 HSUS (2014a).
18 HSUS (2014a); Kindy (2013).
19 HSUS (2008).
20 ibid.
21 ibid.
22 HSUS (2010).
23 Yue (2013).
24 HSUS (2014b).
25 HSUS (2014c).
26 Wolfson and Sullivan (2004).
27 Radio New Zealand (2010).

Developing nations have varying levels of protection. According to the group World Animal Protection, most African nations still lack basic animal protection laws.[28] Russia has anti-cruelty laws which should cover farm animals, but never specifically names them.[29] Brazil has adopted more comprehensive laws protecting animals during raising, transport, and slaughter, but this protection is still minimal.[30] Mexico adopted farm animal welfare provisions, but enforcement remains inconsistent.[31]

Animal protection laws in developing nations are important because they consist of 9 out of the 10 largest poultry producers,[32] 8 of the world's 10 largest egg producers, and 5 of the world's 10 pork-producing nations.[33] This is a trend that will continue as multinational producers expand into developing nations to decrease costs. As developing nations get richer, the rising domestic demand for animal protein will further augment this trend. The rise of factory farming in China has the potential to undermine much of the progress on farm animal welfare in developing nations. China produces 43 percent of the world's pork, almost twice as much as the second largest producer, the EU.[34] Additionally, China has a more diversified poultry production than the United States and Brazil, who have similar production statistics.[35] Both Cargill and Tyson are investing in fully integrated chicken production facilities in China to produce safe chickens, and China is a major importer of pork in America, especially the less popular parts.[36] Unfortunately, most of China's farm animals are on factory farms.

China's legal framework has been slow to adapt, given China's historical lack of meaningful animal protection and welfare laws. A 2005 law passed by the Tenth People's Congress adopted a law increasing agricultural productivity and food safety while failing to improve animal welfare.[37] In fact, the Law's 73 articles never mention animal welfare, instead promoting practices like increased selective breeding and intensification of production systems, which are detrimental to the welfare of animals.

However, there are signs of change. A 2011 survey of Chinese citizens found that while two-thirds of respondents had never heard of animal welfare, and were more concerned with food safety and quality, they would nonetheless support a national law protecting farm animals.[38] A majority also disapproved of routine farming practices such as killing chickens in front of chickens and keeping pigs on concrete floors.[39] In August 2008, the Chinese Commerce Ministry added humane slaughter requirements for pork, but compliance with these requirements remains unknown.[40]

---

28 Animal Protection Index (2014).
29 ibid.
30 ibid.
31 ibid.
32 HSI (2012a).
33 HSI (2012b); HSI (2012c).
34 HSI (2012c); HSI (2012b); https://apps.fas.usda.gov/psdonline/circulars/livestock_poultry.pdf.
35 HSI (2012a).
36 Everett (2014).
37 Chinese National People's Congress (2005).
38 You (2014).
39 ibid.
40 Xinhua (2009).

## 3.2 EUROPE'S FARM ANIMAL WELFARE RETURNS

Europe pioneered the first transnational approach to farm animal welfare and is spreading the approach globally. Because of Europe's history of progressively improving animal welfare standards, European standards are today more stringent than those of any other government or international institution. As a result, Europe could provide an example for the rest of the world.

### 3.2.1 The Council of Europe's Standards

Europe has long led the world in transnational animal welfare standards. In 1976, the Council of Europe began working on pan-European standards for animal welfare and economics. The standards continued when the EU was founded and included members that would never join the EU, like Turkey and Switzerland.

The Council's "Convention for the Protection of Animals Kept for Farming Purposes" bound the members to a set of animal welfare principles.[41] The convention stated that animals must be housed and cared for in a manner "appropriate to their physiological and ethological needs,"[42] as well as, providing "[t]he freedom of movement appropriate to an animal…and shall not be confined to cause it unnecessary suffering or injury."[43] However, the convention did not outline any penalties meaning the European Court ruled the convention indicative only.[44] As a result, many European countries that had ratified the convention continued using the likes of battery cages and gestation crates.

### 3.2.2 The EU's Standards

In the 1990s the Council of the European Communities (later the Council of the EU, hereby called "Council") developed these general principles into specific directives. These directives only apply to the members of the EU, which has a more limited membership than the Council of Europe, but still created the first enforceable transnational farm animal welfare protection. In 1991, the Council adopted minimal standards for veal calves and strong enforcement standards.[45] While the Council did not include veal crates and tethering, these practices were abolished by the European Commission in 1997.[46] The directive also required member states to adopt these standards and enforce compliance among their producers.

The directives had an even larger impact on egg and pork production by creating the first transnational ban on battery cages and gestation crates, as well as protection for laying hens.[47] The directive phased out barren battery cages in 2012.

---

[41] Council of Europe (1976).
[42] ibid.
[43] ibid.
[44] Judgment of the Court of 19 March 1998 – The Queen v Minister of Agriculture, Fisheries and Food, ex parte Compassion in World Farming Ltd.
[45] Council of Europe (1991).
[46] European Commission (1997).
[47] Council of Europe (1999).

The Commission adopted parallel base standards for pig's flooring, noise, lighting, and other housing factors.[48]

These directives act as both a sword and a shield: coercing European nations to adopt certain animal welfare standards, while simultaneously shielding against trade restrictions from nations with higher standards. This dual-purpose became apparent when the European Court of Justice ruled on a case about British veal exports. Because British animal welfare standards were more stringent than those in the rest of Europe, British dairy producers would export calves to countries with more lenient standards. In 1995, Compassion in World Farming, a British animal welfare advocacy group, sued in British court to try to stop this trade. The European Court of Justice found that Britain was not only allowed but required to continue its export of live calves, reasoning that European free trade required harmony of export and import standards. Nations were thus barred from adopting stricter animal welfare measures than the present domestic standards.[49, 50]

Europe's reforms were driven by public opinion and a pro-regulatory legislative system claiming concern over animal welfare and possible benefits from improved animal welfare like better quality, safety, and taste of products.[51] Matheny and Leahy argue that Europe's reforms are made possible by a long tradition of animal protection, a vibrant field of academic animal welfare science, smaller agriculture market, and less powerful agriculture lobbyists than in the United States.[52] Though these factors are lacking in the rest of the world, European advocacy is providing the boost to improve global animal welfare.

### 3.2.3 European Advocacy for Global Farm Animal Welfare Regulation

Europe is now pushing other countries to increase their farm animal welfare standards through free trade agreements (FTAs) that traditionally ignore animal welfare. For example, the North American Free Trade Agreement (NAFTA) had separate agreements for labor and the environment, but none for animal welfare. This treatment of animal welfare restrictions on trade as invalid reduces the countries' ability to regulate animal welfare.

The EU has begun adding animal welfare provisions to their bilateral FTAs, forcing trading partners to keep up with European standards. The EU found higher standards blocks import countries with lower standards from undercutting their own products and meets its citizens' desire for more humane treatment of animals.

The first inclusion of an animal welfare clause by Europe was in 2003 in an FTA with Chile. While the FTA did not outline enforceable standards, it did express a common understanding of animal welfare between parties, and the long-term goal of

---

[48] European Commission (2001).

[49] Judgment of the Court of 19 March 1998 – The Queen v Minister of Agriculture, Fisheries and Food, ex parte Compassion in World Farming Ltd.

[50] ibid.

[51] Harper, Henson (2001).

[52] Matheny, Leahy (2007).

harmonizing European and Chilean animal welfare standards upwards.[53] Cabanne argues that the EU-Chile FTA "played a positive role in the institutionalization of animal welfare in Chile, in particular for livestock production."[54] Since signing the FTA, Chile has passed an animal welfare law that includes farm animals and created a sub-department to cover animal welfare issues. Furthermore, Chile and Uruguay with OIE (World Organization for Animal Health) support established Collaborating Centre on Animal Welfare. The EU has since included animal welfare language in bilateral FTAs with South Korea, Japan, and United States (only to have the United States rebuff them).[55]

## 3.3 OTHER GLOBAL APPROACHES TO FARM ANIMAL WELFARE

Beyond the EU, other global institutions have been slow to adopt animal welfare policies. There are two main reasons for this slow adoption. The first is that animal welfare is outside their main goals of international harmony and economic growth. The second is their reluctance to accept animal welfare as important. This trend seems to be changing as global institutions have begun recognizing a need to be proactive and to keep up with global trends.

### 3.3.1 The World Bank

International financial institutions have not only failed to address the globalization of factory farming but have often unintentionally abetted it. The World Bank, a United Nations agency, provided loans to developing countries to reduce poverty, but never took a stance on animal welfare or factory farming. Furthermore, the World's Bank International Finance Corporation (IFC) which finances private sector projects, invested millions of dollars in factory farms that use battery cages and gestation crates.[56] Other farming projects in which they invest typically spread American style factory farms.

The IFC has belatedly published a "Good Practice Note on Animal Welfare in Livestock Operations," a set of animal welfare guidelines for agribusinesses that receive an investment in 2006.[57] The Note recommended giving animals space to stand, turn around, stretch, sit, and lie down comfortably without banning battery cages and gestation crates.[58] Since the Note, the IFC has invested in projects that use both extreme confinement systems.[59] The Note referenced minimal animal welfare audit standards of the U.S. National Council of Chain Restaurants and the Food

---

[53] EU-Chile agreement on sanitary and phytosanitary measures applicable to trade in animals and animal products, plants, plant products, and other goods and animal welfare (2003), Article 1 of the Annex.

[54] Cabanne (2013).

[55] Farming UK (2013); European Commission (2014); Kopperud (2014); https://www.google.com/url?q=https://www.eurogroupforanimals.org/news/missed-opportunity-eu-japan-free-trade-agreement&sa=D&source=docs&ust=1637554100849000&usg=AOvVaw1URt19E7gY-rivSdQpPPrI.

[56] HSI, Compassion in World Farming, Four Paws (2013).

[57] International Finance Corporation (2006).

[58] ibid.

[59] HSI, Compassion in World Farming, Four Paws (2013).

Marketing Institute, apparently endorsing these standards.[60] However, the IFC standards are voluntary with no clear punishment for disobeying.[61]

In 2014, the IFC updated the Note largely without strengthening its language. The new guidelines replace the reference to the U.S. chain restaurant animal welfare standards with the stricter standards of the Global Animal Partnership, a non profit that set the standards for Whole Foods Supermarkets.[62] The guidelines also acknowledged an "international trend to move from the use of sow stalls to group housing systems."[63] But the 2014 Note maintained the voluntary nature of the guidelines, encouraged the adoption of weak OIE animal welfare standards, referenced the abysmal standards of the U.S. Chicken Council, and removed the 2006 language that animals need space to turn around.[64]

### 3.3.2 The World Organization for Animal Health (OIE)

The World Organization for Animal Health (OIE) is the intergovernmental organization tasked with improving animal welfare. They created the first internationally applicable animal welfare standards, which carry sway with the member states despite remaining voluntary. The OIE identified animal welfare as a priority in its 2001–2005 strategic plan. In that spirit, the OIE created the Animal Welfare Working Group in 2002 whose suggestion was adopted by the OIE a year later.[65] Later, the OIE integrated animal welfare guiding principles and 12 other global animal welfare standards into its Terrestrial Animal Health Code, which consisted of guidelines for transportation and handling concerning animal health and zoonotic disease.[66]

The OIE standards are minimal, largely enshrining current industry practices. For example, the "broiler chicken production system" standards only provide vague descriptions of best practices and do not address overcrowding, or require precise space requirements per bird.[67] The poultry slaughter standards, while stronger than American standards, do not provide goals to prevent cruelties like the immersion of live birds in the scalding tanks.[68] Furthermore, pork and poultry standards do not mention gestation crates or battery cages, probably since new standards need the approval of all 180 members.

Nonetheless, OIE standards provide the best route to improving animal welfare standards globally for three reasons. First, the OIE standards are the only standards accepted by the majority of the countries. Its 180 members consist of the largest meat-producing nations and developing nations without domestic laws. The standards require the members to commit to implementation, though the results are unknown, and involvement in standard preparation through a consensus-based approach.[69]

---

[60] International Finance Corporation (2006).

[61] ibid.

[62] Compare International Finance Corporation (2006) with International Finance Corporation (2014).

[63] International Finance Corporation (2014).

[64] ibid.

[65] OIE (2014a).

[66] OIE (2014b), Chapter 7: Animal Welfare.

[67] OIE (2014b), Article 7.10.4.

[68] OIE (2014b), Article 7.5.7.3(b).

[69] OIE (2013).

The OIE has developed a global animal welfare infrastructure through workshops and conferences aimed at building national regulator's capacity to adopt and enforce standards, developing a strategic plan for implementation throughout the OIE's five regions, and appointing implementation liaisons to each member state.[70] This animal welfare infrastructure is unrivaled by any other organization.

Second, under the World Trade Organization (WTO), the OIE standards are a presumptively valid basis for regulating trade and the international reference to animal health through the Application of Sanitary and Phytosanitary Measures (SPS Agreement).[71] This means high-welfare nations could refuse animal imports from a nation if the animals were not treated in compliance with OIE standards. Contrarily, if a nation meets the OIE standards, a high-welfare nation has no ground to impose additional standards. This explains why the EU has taken an active role in increasing global animal welfare.

Third, acknowledging pro-animal welfare principles forms the basis for stricter standards in the future. In particular, the "Guiding Principles" include the "five freedoms" from the Brambell Report (freedom from hunger, thirst, and malnutrition; freedom from pain, injury, and disease; freedom from physical and thermal discomfort; freedom from fear and distress; and freedom to express normal patterns of behavior). When applied consistently, these standards would eliminate battery cages, gestation crates, and other routine factory farm practices restricting normal animal behavior.[72] The standards also acknowledge the inspirational principle, "That the use of animals carries with it an ethical responsibility to ensure the welfare of such animals to the greatest extent practicable," leading to upward evolving animal welfare standards.[73]

### 3.3.3 Other International Institutions

Among the other international institutions that play a more hands-on role in farm animal welfare, The European Bank of Development and Reconstruction and the Food and Agriculture Organization of the United Nations (FAO) stand out. The European Bank of Development and Reconstruction, owned by 64 countries and two EU institutions, had invested $10 billion in agribusiness without concern for animal welfare.[74] However, in May 2014 the bank announced that it would only invest in future projects that meet or surpass European welfare standards.[75] The bank also created a report comparing welfare standards in beef, pork, and poultry industries.[76] This hands-on approach has the potential to significantly improve animal welfare conditions at new factory farms in developing nations.

70 *See, e.g.*, OIE (2012); IFC (2014).

71 *See* Agreement on Sanitary and Pytosanitary Measures: Preamble, Article 3, and Annex A, paragraph 3(b).

72 OIE (2014b).

73 ibid.

74 HSI, Compassion in World Farming, Four Paws (2013).

75 Humane Society International, EBRD Sets Milestone for Animal Welfare: Adopts Rules to Stop Financing Extreme Confinement of Farm Animals.

76 FAO (2014b).

The FAO, with 192 members, is tasked with combating hunger and promoting agricultural development. Despite considering environmental issues over animal welfare, the FAO hosted a meeting of experts who created a report with vague suggestions such as "animal welfare should not be treated as a stand-alone issue," and nations should "search for improvements that will be practical in the given situation."[77] The FAO also runs an online "Gateway to Farm Animal Welfare" that publishes animal welfare resources and events.[78] Furthermore, their legal experts help countries draft animal welfare legislation and build capacities to monitor welfare. In 2014, the UN Committee on World Food Security mentioned farm animal welfare, showing the influential force of the FAO.[79]

## 3.4 INTERNATIONAL TRADE LAW AND FARM ANIMAL WELFARE

Traditionally viewed by the FTAs and the WTO as impeding free trade, animal welfare is becoming incorporated into FTAs in part due to the work by the EU and loosening restrictions by the WTO.

### 3.4.1 Trade and Farm Animal Welfare

International trade regulations are crucial to farm animal welfare since in 2007 roughly 9 percent of the world's meat was traded internationally.[80] The European Commission explained that higher domestic animal welfare standards could be avoided by importing meat from a country with lower standards.[81] An example is when Great Britain adopted a ban on the tethering and crating of mother pigs in 1999. European nations undermined this law by exporting their pork raised with lower animal welfare standards to Great Britain and putting British producers at a comparative disadvantage.[82] In 2005, the majority of pork sold in the UK was imported and produced using confinement systems illegal in Great Britain.[83]

The EU has overcome this predicament by imposing animal welfare standards on a transnational basis. International trade rules do not stop countries from mutually agreed animal welfare standards, but even EU standards are threatened by imports from countries not following EU regulations. As a result, nations committed to high animal welfare are seeking ways to regulate imported products, a practice that trade law has traditionally viewed as an unfair barrier to trade.

### 3.4.2 WTO Regulation of Animal Welfare-Based Trade Restrictions

The WTO, which governs most disputes through the General Agreement on Tariffs and Trade (GATT), has historically been hostile to animal welfare and the barriers to

---

77 FAO (2008).

78 *See* http://www.fao.org/ag/againfo/themes/animal-welfare/aw-awhome/en/?no_cache=1.

79 UN Committee on World Food Security (2014), p 16.

80 Matheny, Leahy (2007).

81 European Commission (2000).

82 Matheny, Leahy (2007).

83 ibid.

trade that it creates. Sometimes, however, dissolving barriers has a positive effect on animal welfare, as seen by the reduced subsidies on grains and live animal exports. Unfortunately, the spread of factory farming was another outcome of reduced trade barriers. Schneider and Sharma argue that China's reduction of soybean prices to join the WTO led to a rise in pig factory farming.[84]

Furthermore, GATT appears to explicitly preclude imposing of animal welfare standards on imports and sales through Article 1 and Article 3. Article 1 establishes the "most favored nation status" principle, requiring all nations to treat other members equally, and on a non-discriminatory basis.[85] As a result, nations cannot favor animal products from nations with higher animal welfare laws. Article 3 establishes the "national treatment" principle, which means countries cannot favor domestic products over imports, limiting the impact of any national animal welfare law by introducing international competition.[86]

The WTO's Agreement on Technical Barriers to Trade (TBT) further entrenches the traditional view that animal welfare standards are not a legitimate basis to discriminate against foreign trade.[87] The TBT distinguishes between process and production methods, allowing for discrimination based on processes that impact the final product but not based on processes that only impact production.[88] The agreement classifies animal production methods in the latter category, thus largely barring nations from imposing animal welfare standards on imported products.[89]

But three potential exceptions do exist to the seeming prohibition on sales or import restrictions based on animal welfare. First, animal welfare measures can be justified as pursuit of a "legitimate objective" permissible under the TBT agreement.[90] One of the "legitimate objectives" recognized by the Agreement is the protection of "animal or plant life or health. But the TBT agreement only applies to measures that are considered "technical" regulations—and these measures still have to survive review under the GATT.[91]

Second, animal welfare measures may be permissible under the WTO Agreement on the Application of Sanitary and Phytosanitary Measures (SPS). The agreement states that laws regarding animal treatment are permissible if they are enacted to prevent the spread and risk of disease in a territory.[92] To qualify, nations must prove the measures are scientifically justifiable, necessary to stop the disease, and are no more trade-restrictive than necessary.[93] So far, no WTO member has tried to make the difficult showing for an animal welfare regulation.

Third, animal welfare measures may be permissible under Article XX of GATT, which provides ten trade restriction exceptions to Articles I and II. Two exceptions allow measures "necessary to protect public morals" and "necessary to protect

---

[84] Schneider and Sharma (2014), p. 13.
[85] *See* General Agreement on Tariffs and Trade (1994) 1867 U.N.T.S. 187.
[86] ibid.
[87] Thiermann, Babcock (2005).
[88] Agreement on Technical Barriers to Trade (1994) 1868 U.N.T.S. 120.
[89] ibid.
[90] ibid.
[91] ibid.
[92] Thiermann, Babcock (2005).
[93] ibid.

animal life or health."[94] However, the necessity standards are tough because GATT defines a necessity as an act to which "an alternative measure which [the member nation] could not reasonably expect to employ and which is not inconsistent with other GATT provisions is available to it."[95] Furthermore, the provision must be enacted to protect public morals or animal health within a territory. This means it would not apply to foreign animals, making it hard to restrict imports based on the treatment of animals.[96]

### 3.4.3 The WTO EC-Seal Products Decision

Conventional wisdom suggests that the three exceptions outlined above do not justify animal welfare regulations. Thiermann and Babcock argued similarly that "attempting to resolve animal welfare concerns under the WTO may not be the best place to address these issues."[97]

But conventional wisdom was called into question by the 2009 *EC-Seal Products* decision by the WTO dispute settlement body. In that case, Canada and Norway challenged the EU's 2009 ban on the import and marketing of seal products. The countries argued that the EU violated GATT Articles I and II and the TBT for imposing an unnecessary restriction on trade. Canada and Norway also argued that the EU discriminated against them by banning their products while accepting seal imports from subsistence hunting by indigenous communities and from marine resource management culls.

On November 25, 2013, the WTO Dispute Settlement Panel laid an important precedent for farm animal welfare trade restrictions by largely rejecting Canada and Norway's arguments.[98] The Panel determined that the EU had a legitimate basis to ban imports after finding that seal hunts cause the inhumane treatment of seals. However, the panel also found that the exceptions for seal products from subsistence indigenous hunting were not applied evenly and the exception for seal products from marine resource culls lacked a legitimate basis.[99] Thus, the panel rejected the ban under TBT and Article XX, which requires justified trade restrictions be applied without discrimination.[100] The WTO's appellate body largely affirmed the opinion but determined that the TBT did not apply since the European ban was not a technical regulation.[101]

The rulings represent a long-term victory, despite its short-term effects of striking down Europe's ban. Both bodies upheld the core purpose of the ban. Crucially, the Panel rejected Canada and Norway's proposed alternative to the ban of conditioning market access for seal products on labeling and certified compliance with animal

94 ibid.

95 General Agreement on Tariffs and Trade (1994) 1867 U.N.T.S. 187.

96 ibid.

97 Thiermann, Babcock (2005).

98 WTO Panel Reports (2013) European Communities – measures prohibiting the importation and marketing of seal products, WT/DS400/R and WT/DS401/R.

99 Ibid.

100 ibid.

101 WTO Appellate Body Report (2014) European Communities – measures prohibiting the importation and marketing of seal products, WT/DS400/AB/R and WT/DS401/AB/R.

welfare standards, finding this would not achieve the same benefits for seal welfare as with the ban.[102] Furthermore, the Appellate Body rejected claims that Europe could only ban seals if they banned slaughterhouses and other forms of animal abuse.[103]

Most importantly, the WTO Panel and Appellate Body concluded that animal welfare was a legitimate reason to restrict trade under Article XX of the GATT. The Panel's ruling that the ban on seal imports was permissible under Article XX to protect the public's morals was upheld by the Appellate Body.[104] The Body furthered the ruling by adding the importance of seal welfare in Europe.[105] Unfortunately, the ruling focused on the moral concerns of Europeans, but the cruelty of seal clubbing occurs largely in Canada.

## 3.5 RECOMMENDATIONS FOR FUTURE GLOBAL PROGRESS ON FARM ANIMAL WELFARE

As illustrated, the current framework of global rules and institutions fails to protect farm animal welfare. While some scholars call for a new global framework, for example through a UN Declaration on Animal Welfare, the passage of such framework is unlikely. More likely is incremental progress through the liberal interpretation of the GATT's Article XX and the SPS Agreement, the development of stronger voluntary practices at the OIE and FAO, and the implementation of binding standards through FTAs.

### 3.5.1 Interpret WTO Agreements Liberally

The WTO should liberally interpret governing agreements in order to allow animal welfare to restrain trade. In particular, the WTO dispute resolution bodies could expand their *EC-Seals* decision and announce that the SPS Agreement and GATT Article XX include farm animal welfare restrictions. However, the bodies do not render advisory opinions and would need one member to challenge another in order to render an opinion.

A more liberal interpretation would allow countries to expand their animal welfare standards and make EU standards more effective by allowing welfare import bans. In turn, countries will need to increase standards to gain access to the customers in Europe and other high animal welfare markets. Additionally, there could be reduced pressure from farmers who are at a disadvantage against cheaper imports from low-welfare countries. The result would be a change from a "race to the bottom" to a "race to the top" mentality in farming.

102 WTO Panel Reports (2013) European Communities – measures prohibiting the importation and marketing of seal products, WT/DS400/R and WT/DS401/R.

103 WTO Appellate Body Report (2014) European Communities – measures prohibiting the importation and marketing of seal products, WT/DS400/AB/R and WT/DS401/AB/R.

104 WTO Panel Reports (2013) European Communities – measures prohibiting the importation and marketing of seal products, WT/DS400/R and WT/DS401/R.

105 WTO Appellate Body Report (2014) European Communities – measures prohibiting the importation and marketing of seal products, WT/DS400/AB/R and WT/DS401/AB/R.

### 3.5.2 Develop Best Practices at International Institutions

While the current OIE standards are a good start, the OIE needs to strengthen the voluntary standards and do more to ensure their adoption. The standards recognize the "five freedoms" and the need to comprehensively address farm animal welfare from birth until slaughter, but also enshrine current practices. This is likely the result of the consensus-based process, where animal industries are given significant influence through their national governments. Fortunately, by requiring all standards to conform to the "five freedoms," the OIE could eliminate pain in slaughter and grant freedom to laying hens and pigs. Furthermore, this would set a vision for developing nations to follow.

The FAO should declare its support for higher welfare systems and its opposition to low-welfare practices. The FAO has the ability to influence animal welfare in developing countries and to bring together industry groups, national governments, and animal protection organizations. While the FAO is using its influence, it has remained agnostic to outcomes and should instead adopt a clear position in favor of constant improvement and opposed to harmful animal practices.

### 3.5.3 Implementing Binding Standards through FTAs

Nations with high farm animal welfare should nudge other nations to adopt higher standards using FTAs. The EU has led the way with its agreements with Chile and South Korea and insistence on the inclusion of animal welfare provisions in future agreements. Such provisions ensure a level-playing field for high-welfare producers and reduce WTO scrutiny, since each nation must sign to the provisions of its FTA.

The EU should continue demanding inclusion of animal welfare provisions in FTAs, and countries like the United States should adopt more progressive policies promoting welfare in trade negotiations. Other nations that have championed animal welfare, from India and Israel to New Zealand, should insist on nations signing FTAs with animal welfare provisions.

The inclusion of animal welfare provisions in a growing number of bilateral FTAs will facilitate the eventual inclusion of the provisions in multilateral FTAs, such as the Asia Pacific Economic Cooperation (APEC). Then the world will finally have a form of global farm animal welfare regulation.

## 3.6 CONCLUSION

Global institutions need to adopt stronger farm animal welfare policies, and nations need to establish a basis for farm animal welfare laws. The failure to adopt policies and laws as factory farming has grown has allowed for severe animal abuse. Though global policies and laws are starting to address this gap, reform must move faster. The moral credibility of the global order depends on it.

## REFERENCES

Animal Protection Index (2014) World animal protection. Http://api.worldanimalprotection.org/indicators. Accessed 3 Jan 2015.

Brambell FWR (1965) Report of the technical committee to enquire into the welfare of animals kept under intensive livestock husbandry systems. Her Majesty's Stationery Office, London.

Cabanne C (2013) The EU-Chile free trade agreement—a boost for animal welfare. Eurogroup for Animals, Brussels.

Chinese National People's Congress (2005) Animal husbandry law of the People's Republic of China. Http://www.npc.gov.cn/englishnpc/Law/2007-12/13/content_1384134.htm. Accessed 3 Jan 2015.

Clyma K (2014) Slideshow: ranking the meat and poultry industry, Meat+Poultry, April 4, 2014. http://www.meatpoultry.com/articles/news_home/Business/2014/04/Slideshow_Ranking_the_meat_and.aspx?ID={4EB1C430-7CF8-41B3-8457-9FB76625F9F8}. Accessed 24 Jan 2015.

Cochrane W (1993) *The Development of American Agriculture: A Historical Analysis.* University of Minnesota Press, Minneapolis.

Council of Europe (1976) European convention for the protection of animals kept for farming purposes, Strasbourg, 10.III.1976. Http://conventions.coe.int/Treaty/en/Treaties/html/087.htm. Accessed 3 Jan 2015.

Council of Europe (1991) Council Directive of 19 November 1991 laying down minimum standards for the protection of calves (91/629/EEC). Http://ec.europa.eu/food/fs/aw/aw_legislation/calves/91-629-eec_en.pdf. Accessed 3 Jan 2015.

Council of Europe (1999) Council Directive 1999/74/EC of 19 July 1999 laying down minimum standards for the protection of laying hens. Http://eur-lex.europa.eu/LexUriServ/LexUriServ.do?uri=OJ:L:1999:203:0053:0057:EN:PDF. Accessed 3 Jan 2015

European Commission (1997) Commission Decision of 24 February 1997 amending the Annex to Directive 91/629/EEC laying down minimum standards for the protection of calves (97/182/EC). Http://ec.europa.eu/food/fs/aw/aw_legislation/calves/97-182-ec_en.pdf. Accessed 3 Jan 2015.

European Commission (2000) European Communities proposal: animal welfare and trade in agriculture. Http://www.wto.org/english/tratop_e/agric_e/ngw19_e.doc. Accessed 3 Jan 2015.

European Commission (2001) Commission Directive 2001/93/EC of 9 November 2001 amending Directive 91/630/EEC laying down minimum standards for the protection of pigs. Http://eur-lex.europa.eu/legal-content/EN/TXT/PDF/?uri=CELEX:32001L0093&from=en. Accessed 3 Jan 2015.

European Commission (2014) EU-Japan trade talks continue, Europa, January 31, 2014. http://trade.ec.europa.eu/doclib/press/index.cfm?id=1024. Accessed 3 Jan 2015.

Everett (2014) China's rising meat consumption draws U.S. companies, Midwest Center for Investigative Reporting, May 15, 2014. http://investigatemidwest.org/2014/05/15/chinas-increasing-chicken-consumption-draws-u-s-companies/.

FAO (2008) Capacity building to implement good animal welfare practices. http://www.fao.org/fileadmin/user_upload/animalwelfare/AW_Exp-meeting_EN_1.pdf. Accessed 24 Jan 2015.

FAO (2014a) International livestock database. Http://faostat.fao.org/site/573/DesktopDefault.aspx?PageID=573#ancor. Accessed 3 Jan 2015.

FAO (2014b) Review of animal welfare legislation in the beef, pork, and poultry industries. http://www.fao.org/3/a-i4002e.pdf. Accessed 24 Jan 2015.

Farming UK (2013) EU presses for animal welfare standards in trade agreements, Farming UK, May 25, 2013. Http://www.farminguk.com/news/EU-presses-for-animal-welfare-standards-in-trade-agreements_25596.html.

Harper G, Henson S (2001) Consumer concerns about animal welfare and the impact on food choice, EU FAIR CT98-3678. Http://ec.europa.eu/food/animal/welfare/eu_fair_project_en.pdf. Accessed 3 Jan 2015.

HSI (2012a) An HSI fact sheet: top 10 global producers: chickens & chicken meat. http://www.hsi.org/assets/pdfs/hsi-fa-white-papers/top_10_chicken_meat_producers.pdf. Accessed 24 Jan 2015.

HSI (2012b) An HSI fact sheet: top 10 global producers: hen eggs. http://www.hsi.org/assets/pdfs/hsi-fa-white-papers/top_10_producers_of_eggs.pdf. Accessed 24 Jan 2015.

HSI (2012c) An HSI fact sheet: top 10 global producers: pigs and pig meat. http://www.hsi.org/assets/pdfs/hsi-fa-white-papers/top_10_producers_of_pigs.pdf. Accessed 24 Jan 2015.

HSI, Compassion in World Farming, Four Paws (2013) International finance institutions, export credit agencies and farm animal welfare. http://www.hsi.org/assets/pdfs/hsi_ifi_report_june_2013.pdf. Accessed 24 Jan 2015.

HSUS (2008) An HSUS report: the welfare of animals in the egg industry. Http://animalstudiesrepository.org/cgi/viewcontent.cgi?article=1016&context=hsus_reps_impacts_on_animals. Accessed 3 Jan 2015.

HSUS (2010) An HSUS report: the welfare of animals in the pig industry. Http://animalstudiesrepository.org/hsus_reps_impacts_on_animals/28/. Accessed 3 Jan 2015.

HSUS (2014a) An HSUS report: the welfare of animals in the chicken industry. Http://files.givewell.org/files/shallow/industrial-agriculture/HSUS%20Chickens.pdf. Accessed 3 Jan 2015.

HSUS (2014b) An HSUS report: the welfare of cows in the dairy industry. Http://www.humanesociety.org/assets/pdfs/farm/hsus-the-welfare-of-cows-in-the-dairy-industry.pdf. Accessed 3 Jan 2015

HSUS (2014c) An HSUS report: the welfare of animals in the meat, egg, and dairy industries. Https://www.petfinder.com/docs/library-articles/animal-welfare-overview.pdf. Accessed 3 Jan 2015.

International Finance Corporation (2006) Good practice note on animal welfare in livestock operations. Http://www.ifc.org/wps/wcm/connect/7ce6d2804885589a80bcd26a6515bb18/AnimalWelfare_GPN.pdf?MOD=AJPERES. Accessed 3 Jan 2015.

International Finance Corporation (2014) Good practice note: improving animal welfare in livestock operations. Http://www.ifc.org/wps/wcm/connect/c60a6680439041ae9a25ba869243d457/IFC-GPN-AnimalWelfare_2006vs2014.pdf?MOD=AJPERES. Accessed 3 Jan 2015.

Kindy K (2013) USDA plan to speed up poultry-processing lines could increase risk of bird abuse, Washington Post, October 29, 2013. Http://www.washingtonpost.com/politics/usda-plan-to-speed-up-poultry-processing-lines-could-increase-risk-of-bird-abuse/2013/10/29/aeeffe1e-3b2e-11e3-b6a9-da62c264f40e_story.html.

Kopperud S (2014) 'Global animal welfare' — who's on first? Brownfield Ag News, October 3, 2014. Http://brownfieldagnews.com/2014/10/03/global-animal-welfare-whos-first/.

Matheny G, Leahy C (2007) Farm-animal welfare, legislation, and trade. *Law Contemp Probl* 70:325–358.

OIE (2012) Third global conference on animal welfare: implementing the OIE standards – addressing regional expectations. Http://www.oie.int/eng/AW2012/intro.htm. Accessed 3 Jan 2015

OIE (2013) International trade: rights and obligations of OIE member countries. Http://www.oie.int/fileadmin/Home/eng/Internationa_Standard_Setting/docs/pdf/Legal_rights_and_obligations/A_Rights_and_obligations_April_2013.pdf. Accessed 3 Jan 2015.

OIE (2014a) OIE's achievements in animal welfare. Http://www.oie.int/animal-welfare/animal-welfare-key-themes/. Accessed 3 Jan 2015.

OIE (2014b) Terrestrial animal health code. Http://www.oie.int/index.php?id=169&L=0&htmfile=titre_1.7.htm. Accessed 3 Jan 2015.

Pew Commission on Industrial Farm Animal Production (2008) *Putting Meat on the Table: Industrial Farm Animal Production in America.* Pew Charitable Trusts, Washington.

Radio New Zealand (2010) Fonterra intensive dairy farming in China criticized, February 3, 2010. Http://www.radionz.co.nz/news/national/2116/fonterra-intensive-dairy-farming-in-china-criticised.

Samora R (2014) UPDATE 1-CEO of Brazil's JBS meat company sees 2014 revenues up 22 pct, Reuters, November 13, 2014. Http://www.reuters.com/article/2014/11/13/brazil-jbs-china-idUSL2N0T319B20141113.

Schneider M and Sharma S (2014) China's pork miracle? Agribusiness and development in China's pork industry. Institute for Agriculture and Trade Policy. Http://www.iatp.org/files/2014_03_26_PorkReport_f_web.pdf. Accessed 24 Jan 2015.

Sherman D (2002) *Tending Animals in the Global Village: A Guide to International Veterinary Medicine*. Lippincott Williams & Wilkins, Philadelphia.

SPS Agreement: Preamble, Article 3, and Annex A, paragraph 3(b). Http://www.wto.org/english/tratop_e/sps_e/spsagr_e.htm. Accessed 24 Jan 2015.

Thiermann AB, Babcock S (2005) Animal welfare and international trade. *Rev Sci Tech Off Int Epiz* 24(2):747–755.

UN Committee on World Food Security (2014) Principles for responsible agricultural investments. Http://www.fao.org/fileadmin/templates/cfs/Docs1314/rai/CFS_Principles_Oct_2014_EN.pdf. Accessed 24 Jan 2015

Wilmot S (2007) From 'public service' to artificial insemination: animal breeding science and reproductive research in early twentieth-century Britain. *Stud Hist Philos Biol Biomed Sci* 38(2):411–441.

Wolfson D, Sullivan M (2004) Foxes in the hen house: animals, agribusiness, and the law: a modern American fable. In: Sunstein C, Nussbaum M (eds) *Animal Rights: Current Debates and New Directions*. Oxford University Press, Oxford.

Xinhua (2009) Humane slaughtering spearheads China's drive to promote animal welfare, September 5, 2009. Http://china.org.cn/environment/2009-09/05/content_18519951.htm.

You X (2014) A survey of Chinese citizens' perceptions on farm animal welfare. *PLoS ONE* 9(10):1371. Http://journals.plos.org/plosone/article?id=10.1371/journal.pone.0109177. Accessed 24 Jan 2015.

Yue S (2013) An HSUS report: fish and pain perception. Http://www.humanesociety.org/assets/pdfs/farm/hsus-fish-and-pain-perception.pdf. Accessed 3 Jan 2015.

WTO Panel Reports (2013) European Communities – measures prohibiting the importation and marketing of seal products, WT/DS400/R and WT/DS401/R.

WTO Appellate Body Report (2014) European Communities – measures prohibiting the importation and marketing of seal products, WT/DS400/AB/R and WT/DS401/AB/R.

# 4 AG-GAG
## *Agriculture, Whistleblowers, and the 1st Amendment*

*Autumn T. Johnson*
Tierra Strategy

## CONTENTS



## 4.1 INTRODUCTION: WHAT IS AG-GAG?

Laws intended to criminalize whistleblowers attempting to document animal or environmental abuses have been passed all over the United States. These laws penalize the act of documenting activities related to agriculture, including video, photos, or scientific samples. Commonly referred to as "ag-gag," they are being challenged in the courts, often being overturned in whole or in part due to freedom of speech violations. To date, five federal courts have struck down their state's ag-gag laws. While this is a developing area of the law, this chapter attempts to explain the types of ag-gag laws, where they have been passed, and how federal courts have ruled on them thus far. The Animal Legal and Historical Center at Michigan State University offers the following summary:

Ag-gag laws generally fall into one of three categories or a combination of two or all types:

1. The first modern-day ag-gag laws are known as "agricultural interference" laws, and they ban recording images or sounds at industrialized farming operations without the owner's consent.

DOI: 10.1201/9781003137733-10

2. "Agricultural fraud" laws were also introduced early on; this kind of ag-gag law bans entering or applying for employment at industrialized farming operations under false pretenses.
3. "Rapid-reporting" legislation mandates that anyone who records an image or sound at an industrialized farming operation turn the recording over to authorities within a specified amount of time, usually 24–48 hours.[1]

While the third type, regarding the time to report, may seem innocuous, prosecutions of animal welfare violations typically require documentation over a substantial period of time. Having to report violations immediately prohibits collecting enough evidence to document a pattern of behavior. According to The Humane Society of the United States, "requiring that any violation be immediately reported [is] intended to prevent whistleblowers from establishing a pattern of abuse. Because of the power of the factory farming lobby, prosecutors will rarely file charges for farm animal cruelty unless a pattern has been established and the evidence is overwhelming."[2]

"Ag-gag" laws are also sometimes referred to as "farm protection laws" or "anti-whistleblower laws" or "anti-sunshine laws" or "farm interference laws" (hereinafter, "ag-gag"). Other related statutes are sometimes called "eco-terrorism" or "animal enterprise interference" or "agroterrorism" laws. Michigan State University keeps a table of all such laws by state.[3] Ag-gag laws vary from state to state.

The PEW Charitable Trusts, a nonprofit advocacy group, attributes the rise of ag-gag laws with animal welfare advocates successes in reforming the industry. "When animal welfare groups started about a decade ago to pay their employees to take jobs on farms to expose practices, the industry responded by pushing for what animal welfare advocates call ag-gag laws. Some of the laws made it a crime to take photos or videos of private farm property without the owner's permission, while others made it a crime for an employee of an animal welfare organization to lie about where they worked when they applied for a job on a farm."[4]

## 4.2 STATE STATUTES

Eleven states in the United States have passed ag-gag laws. They are Arkansas, Idaho, Iowa, Kansas, Missouri, Montana, North Carolina, North Dakota, Alabama, Wyoming, and Utah. These laws were generally passed between 2010 and 2015. The ASPCA keeps a full list of states that have introduced ag-gag legislation.[5] The list includes 29 states. While many of these statutes did not pass and others have

---

[1] Prygoski, A. (2015). A Brief Summary of Ag-Gag Laws. *Michigan State University, Animal Legal & Historical Center.* https://www.animallaw.info/intro/ag-gag-laws.

[2] The Humane Society of the United States. *Anti-Whistleblower Ag-Gag Bills Hide Factory Farming Abuses from the Public.* https://www.humanesociety.org/resources/anti-whistleblower-ag-gag-bills-hide-factory-farming-abuses-public (last visited Jan. 2021).

[3] Michigan State University. *Animal Legal & Historical Center.* https://www.animallaw.info/statutes/topic/ecoterrorism-or-agroterrorism?order=title&sort=desc (last visited Jan. 2021).

[4] PEW. (Nov. 29, 2016). *Farmers Push Back Against Animal Welfare Laws.* https://www.pewtrusts.org/en/research-and-analysis/blogs/stateline/2016/11/29/farmers-push-back-against-animal-welfare-laws.

[5] ASPCA. *What is Ag-Gag Legislation?* https://www.aspca.org/animal-protection/public-policy/what-ag-gag-legislation (last visited Jan. 2021).

already been struck down by the courts, others remain or are currently in the midst of litigation (described below).

## 4.3 CASE LAW

Many organizations track ag-gag laws around the country. Animal Legal Defense Fund (ALDF) is one such organization. Their map of ag-gag laws is seen in Figure 4.1. ALDF has been involved in litigation in an effort to prevent enforcement of state enacted ag-gag laws. Their first lawsuit was filed against the State of Utah in 2013. In 2014, they filed suit against the State of Idaho. In August of 2015, the U.S. District Court for the District of Idaho ruled that Idaho's law was unconstitutional. In 2016, ALDF joined others in suing the State of North Carolina over its "Anti-Sunshine Law." In July of 2017, the U.S. District Court of Utah found its ag-gag law to be unconstitutional. In 2017, the ALDF and others filed suit in Iowa against that state's ag-gag law. In January of 2018, the U.S. Court of Appeals for the Ninth Circuit issued

**FIGURE 4.1** Animal legal defense fund Ag-Gag map. (Animal Legal Defense Fund. *Ag-Gag Issues*. https://aldf.org/issue/ag-gag/ (last visited Jan. 2021).)

its ruling on the appeal from Idaho. That decision is excerpted below. In 2018, ALDF and others filed suit against the State of Kansas. In 2019, the Federal District Court for the Southern District of Iowa ruled that Iowa's ag-gag law was unconstitutional. Iowa then enacted a revised ag-gag law, which is currently being challenged in court. In 2019, ALDF filed suit against the State of Arkansas. In 2019, ALDF also filed suit against the State of North Carolina.[6]

There have been two federal appellate court decisions on the merits of ag-gag laws. They are particularly relevant for understanding the types of ag-gag laws enacted and the reasons they are often struck down. In 2017, the 10th Circuit Court of Appeals issued a decision regarding Wyoming's ag-gag law. It was different from many of the other ag-gag laws enacted because it specifically targeted "collecting resource data" (i.e., at environmental advocates collecting samples of pollution or other environmental degradation due to agriculture). In 2018, the 9th Circuit Court of Appeals issued a decision regarding Idaho's ag-gag law. This law was more similar to the type of ag-gag laws enacted in other states, which specifically penalized documentation of animal welfare within agricultural facilities. Both statutes were challenged on constitutional grounds. Both decisions are summarized below.

In addition to the two cases mentioned above, three federal trial courts have found ag-gag laws unconstitutional. Utah was the first to be struck down. In 2017, the District Court of Utah struck down its ag-gag statute. Most recently, in 2020, these statutes were struck down in Kansas and North Carolina. The District Court for the District of Kansas struck down that state's ag-gag law in January of 2020. The District Court for the Middle District of North Carolina struck down portions of its ag-gag law in June of 2020. All three cases are discussed below.

## 4.3.1 Circuit Court Cases

### 4.3.1.1 Western Watersheds Project v. Michael[7]

"Criminal trespass occurs when an individual enters or remains on the property of another with knowledge or subsequent notification that he is not authorized to do so." Trespass was already illegal in Wyoming when in 2015, it passed two statutes that prohibited people from entering "open land for the purpose of collecting resource data" without permission. "Resource data" is "data relating to land or land use," including anything pertaining to "air, water, soil, conservation, habitat, vegetation or animal species." "Collect" means: "(1) taking a 'sample of material' or a 'photograph,' or 'otherwise preserving information in any form' that is (2) 'submitted or intended to be submitted to any agency of the state or federal government.'" Data collected in any way violating this statute could not be used and the government was required to delete any information collected unlawfully and could not rely on it in making any official decisions.

The plaintiffs argued that the statutes were contrary to Constitution's 1st and 14th Amendments, Free Speech and Equal Protection (among others). The trial

[6] Animal Legal Defense Fund. *Ag-Gag Issues*. https://aldf.org/issue/ag-gag/ (last visited Jan. 2021).

[7] *Western Watersheds Project v. Michael*, 869 F.3d 1189 – Court of Appeals, 10th Circuit (2017).

court determined that free speech, petition, and equal protection claims were valid. Wyoming then modified the statutes.

The new statutes eliminated the term "open lands" and prohibited people from (1) entering private land "for the purpose of collecting resource data"; (2) entering private land to "collect resource data", or (3) "crosses private land to access adjacent or proximate land where he collects resource data." Collect was redefined as (1) "to take a sample of material" or "acquire, gather, photograph or otherwise preserve information in any form"; and (2) "recording… a legal description or geographical coordinates of the location of the collection."

The trial court then ruled for Wyoming and the plaintiffs filed an appeal. Though some of the statutes apply to conduct that occurs on private land and is, therefore, not subject to constitutional protection, the appellate court determined it was mistaken as to the provision about crossing private land to collect data on adjacent land, which could apply to activity on public property.

Because trespass was already illegal in Wyoming, this statute has the effect of increasing "a pre-existing penalty for trespassing if an individual subsequently collects resource data from public land." This results in differential treatment. The appellate court said, "the question is not whether trespassing is protected conduct, but whether the act of collecting resource data on public lands qualifies as protected speech."

The statutes penalized activity on public land, "so long as an individual also records where such data was gathered." These prohibited activities could include "collecting water samples, taking handwritten notes about habitat conditions, making an audio recording of one's observation of vegetation, or photographing animals."

The appellate court held that collecting "resource data" on public land is protected speech in that it impacts the creation of speech common in public policy advocacy and development. It stated that the United States "Supreme Court has explained that 'the creation and dissemination of information are speech within the meaning of the First Amendment.'"[8]

#### 4.3.1.2 Animal Legal Defense Fund v. Wasden[9]

In 2012, an undercover activist obtained employment at a farm and surreptitiously filmed inhumane treatment of the animals. In 2014, Idaho passed a statute prohibiting "interference with agricultural production." The law made it a crime to (1) enter an "agricultural facility by force, threat, misrepresentation or trespass"; (2) obtain records by the same means; (3) obtain employment through misrepresentation with the intent to harm the farm, including economically; or (4) record conduct at a farm.[10]

The appellate court considered the underlying motivations behind the State's enactment of the law. "The bill was drafted by the Idaho Dairymen's Association, a trade organization representing Idaho's dairy industry." One of the stated goals

[8] All quotes within this summary are *ibid.*

[9] *Animal Legal Defense Fund v. Wasden*, 878 F. 3d 1184 – Court of Appeals, 9th Circuit (2018).

[10] Idaho Code § 18-7042(1)(a)-(d).

was to protect the farming industry from "undercover investigators who expose the industry to the 'court of public opinion,' which destroys farmers' reputations, results in death threats, and causes loss of customers."

Plaintiffs asserted the statute violated the 1st and 14th Amendments of the U.S. Constitution. The appellate court explained the issue as whether or not "the misrepresentations prohibited in the Idaho statute constitute[d] speech protected by the First Amendment." It then went on to analyze the issue based on the United States Supreme Court case of Alvarez. In that case, the Supreme Court determined false statements can be protected speech. The appellate court determined that the misrepresentation clause regulated protected speech and must be constrained to "achieve a compelling government interest."

The appellate court also stated that the law was not the "least restrictive means among available, effective alternatives." It went on to state "we do not ignore that a vocal number of supporters were less concerned with the protection of property than they were about protecting a target group from critical speech."

The appellate court concluded that these concerns could be remedied by the removal of the word "misrepresentation" from the statute. "Because the proscription on misrepresentations is neither integral nor indispensable to the subsection's goal of protecting property rights, the offending term 'misrepresentation' should be stricken, leaving the remainder of the subsection intact."

The section on obtaining records by misrepresentation "criminalizes obtaining records of an agricultural production facility by misrepresentation." The appellate court reasoned that this portion of the law dealt with a specific harm and, therefore, was not prohibited under Alvarez. "Alvarez highlights that a false statement made in association with a legally cognizable harm or for the purpose of material gain is not protected. Unlike false statements made to enter a property, false statements made to actually acquire agricultural production facility records inflict property harm upon the owner and may also bestow a material gain on the acquirer."

The appellate court concluded that this statute impacted lawful property rights. It also determined that Idaho lawmakers had sufficiently demonstrated how the conduct implicated in this section of the law can harm farmers. "Although some legislators wanted to silence investigative journalists reporting on the agricultural industry, the full legislative history shows that a legitimate purpose for enacting the subsection was to prevent harm from damaged or stolen records."

The appellate court also determined that this section of the statute did not violate the Equal Protection Clause. "Legislation is generally presumed to be valid and will be sustained under the Equal Protection Clause 'if the classification drawn by the statute is rationally related to a legitimate state interest.'" The appellate court found that the "overall purpose" of the law was to protect farms from harm.

The employment section of the law prohibited "obtaining employment with an agricultural production facility by… misrepresentation with the intent to cause economic or other injury." The appellate court stated that this section of the law strictly adhered to the Alvarez decision. The appellate court thought it was very significant that the Supreme Court specifically included offers of employment.

The appellate court also found that this section of the law did not violate the Equal Protection Clause, because it serves a "legitimate governmental purpose."

This is because employees have different levels of access to the facility and its records, which may be confidential.

Finally, the appellate court considered the provision that prohibited video or audio recordings of animal abuse or other inappropriate conduct. "The Recordings Clause regulates speech protected by the First Amendment and is a classic example of a content-based restriction that cannot survive strict scrutiny."

The appellate court confirms that recordings are protected speech. This section of the law also targets specific content, which makes is prohibited regulation of "content-based speech." It is intended to prohibit speech targets at a particular subject matter. "A regulation is content-based when it draws a distinction 'on its face' regarding the message the speaker conveys or 'when the purpose and justification for the law are content based.'" The section does both. Therefore, it "prohibited public discussion of an entire topic." Further, because the section prohibits filming of the operations, it is specifically targeted at the content of the recording.

Because the law is content-based, it must be "necessary to serve a compelling state interest" and "is narrowly drawn to achieve that end." The appellate court did not think it could meet the latter part of the test, no matter if it could meet the former. "We are left to conclude that Idaho is singling out for suppression one mode of speech — audio and video recordings of agricultural operations — to keep controversy and suspect practices out of the public eye."

"For these reasons, the Recordings Clause cannot survive First Amendment scrutiny and is therefore unconstitutional."[11]

### 4.3.2 District Court Cases

#### 4.3.2.1 Animal Legal Defense Fund v. Herbert[12]

Utah's law was similar to Idaho's in that it criminalized misrepresentation to enter an agricultural facility and also obtaining video from inside. It was challenged on First Amendment grounds, as a violation of Freedom of Speech. The Court recites the history of government support for the agriculture sector and then states,

> In short, governmental protection of the American agricultural industry is not new and has taken a variety of forms over the last two hundred years. What is new, however, is the recent spate of state laws that have assumed an altogether novel approach: restricting speech related to agricultural operations. These so-called "ag-gag" laws have their genesis in the 1990s. Around that time, animal rights advocates had begun conducting undercover investigations to expose animal abuse at various facilities. After these initial investigations became public, Kansas, Montana, and North Dakota all enacted ag-gag laws.[13]

There was little uptake of these laws until the mid-to-late 2000s, after a series of egregious animal abuse videos were made public. Instead of updating animal welfare laws, within just a few years, 16 states proposed ag-gag laws. Of these, Iowa's was the first to be implemented, followed almost immediately by Utah's in 2012.

---

[11] All quotes within this summary are from the *Wasden* case, cited above.

[12] *Animal Legal Defense Fund v. Herbert*, 263 F. Supp. 3d 1193 - Dist. Court, D. Utah (2017).

[13] All quotes within this summary are from the *Herbert* case, cited above.

As explained above, the U.S. Supreme Court in Alvarez held that lies are not categorically unprotectable speech. To be unprotectable, they must cause "legally cognizable harm." The Court determined that was not the case, as lies vary in their significance and many would not cause harm, thereby not necessitating categorical exclusion.

The Court also declined to categorically characterize misrepresentation to enter property a trespass. It is a trespass only if the misrepresenting party is intending to interfere with ownership or possession of the property. The Court found the Act was overly broad and criminalized many trivial lies that cause no harm. The Court also declined to accept the State's assertion that the act was criminalizing lying to obtain employment because the act was much broader. The Court found that films, photographs, and sound recording were protected speech.

Finally, the State attempted to argue there were no First Amendment rights on private property. "[T]he State… confuses two related but distinct concepts: a landowner's ability to exclude from her property someone who wishes to speak, and the government's ability to jail the person for that speech." The First Amendment is generally not a defense to tortious conduct. "But that is not the question before the court. The question here is whether the State (not a private landowner) can prosecute (not sue for damages) a person based on her speech on private property." This is in direct conflict with the First Amendment.

After determining both provisions on the statute (the lying provision and the recording provision) were both matters of Free Speech under the First Amendment, the Court applied strict scrutiny to the lying provision because it was content based. "The speech in question is the lie itself, and the only way to know whether a lie is to review what was said. This is perhaps the quintessential example of a content-based restriction." Similarly, the Court found strict scrutiny to also apply to the recording provision because the statute criminalizes images "of" not "at" an agricultural facility. "[I]f a person walks off an agricultural facility with a recording, the only way to know whether she is criminally liable under the Act is to view the recording. That makes the provision content-based, and subject to strict scrutiny." The presumption under strict scrutiny is that the law is not constitutional. The State has to prove (1) the statute furthers a compelling government interest and (2) that it does so with the least restrictive means possible.

> [T]he State confined the court's analysis to four discrete government interests it contends support the Act, arguing: (1) the Act protects animals from diseases brought into the facility by workers; (2) it protects animals from injury resulting from unqualified or inattentive workers; (3) it protects workers from exposure to zoonotic diseases; and (4) it protects workers from injury resulting from unqualified or inattentive workers.[14]

The State did not assert any property or privacy interests in its briefs. And the Court notes that the legislative history of the statute did not bear out any of the four above stated bases for the law. In fact, the legislative discussion focused on vegetarian propaganda seeking to put the agricultural sector out of business. However, there is

---

[14] All quotes within this summary are from the *Kelly* case, cited above.

no evidence that lying to gain access or recording the operations spread disease or endangered workers. "[T]he harm targeted by the Act is entirely speculative. And harm that is 'mere[ly] speculat[ive]... does not constitute a compelling state interest.'" And even if it is was not, it would still be too broad to survive strict scrutiny.

"What the Act appears perfectly tailored toward is preventing undercover investigators from exposing abuses at agricultural facilities." Therefore, the Court found the statutes unconstitutional under the First Amendment.

#### 4.3.2.2 Animal Legal Defense Fund v. Kelly[15]

Kansas' Farm Animal and Field Crop and Research Facilities Protect Act has been on the books since 1990.

> [T]he Act makes it a crime to commit the following acts without the effective consent of the owner and with the intent to damage an enterprise conducted at the animal facility: (1) damage or destroy an animal facility or an animal or property at an animal facility; (2) exercise control over an animal facility, an animal from an animal facility or animal facility property with the intent to deprive the owner of it; (3) enter an animal facility that is not open to the public to take photographs or recordings; and (4) remain at an animal facility against the owner's wishes.

In 2012, the statute was amended to change the definition of consent to preclude garnering consent by "force, fraud, deception, duress or threat." The State asserted the plaintiffs (nonprofit organizations) did not have standing to sue. After a long and nuanced analysis, the Court found the plaintiffs had standing to sue regarding a majority of the issues at stake and proceeded with an analysis of the merits.

When determining the standard of review, the Court found the statute was intended to regulate speech, not just conduct (i.e., entry). Because the prohibited speech requires deception, the speech is content-based. "The law does not prohibit such conduct if the person has the intent to benefit the enterprise conducted at the animal facility, and in this respect, it impermissibly discriminates based on the speaker's views about animal facilities." Therefore, the Court applied strict scrutiny. The State did not attempt to prove they were attempting to protect a compelling interest and, even if they did, it was not narrowly tailored enough. The statute does "not prevent everyone from violating the property and privacy rights of animal facility owners — only those who violate said rights with intent to damage the enterprise conducted at animal facilities." Therefore, it is unconstitutional.

#### 4.3.3.3 People for the Ethical Treatment of Animals v. Stein[16]

In 2015, the North Carolina legislature passed the Property Protection Act, over the governor's veto. "The Act amended current law that provides a civil remedy for interference with certain property rights by creating a civil cause of action for the owner or operator of a premises."[17] Like the others discussed above, this law also prohibits the "record[ing] images or sound occurring within an employer's premises

---

[15] *Animal Legal Defense Fund v. Kelly*, 434 F. Supp. 3d 974 - Dist. Court, D. Kansas (2020).

[16] *People for the Ethical Treatment of Animals, Inc. v. Stein*, Dist. Court, MD North Carolina (2020).

[17] All quotes within this summary are from the *Stein* case, cited above.

and uses the recording to breach the person's duty of loyalty to the employer" or "[k]nowingly or intentionally placing on the employer's premises an unattended camera or electronic surveillance device and using that device to record images or data." Though civil, the violations were significant, including $5,000 a day in damages and an award of attorney's fees. In addition to agricultural facilities, this case also focused on an intent by the plaintiffs to investigate research conducted on animals at the University of North Carolina. In addition to First Amendment claims, the plaintiffs also asserted the statute violated the Fourteenth Amendment (Due Process and Equal Protection). The plaintiffs had previously lost this suit on standing grounds but won on appeal. Other claims were previously resolved in a Motion to Dismiss in this District Court.

> As far as the court can discern, nearly all other similar laws impose criminal liability while the Property Protection Act provides a civil cause of action for damages. But while the Act operates in the private sphere, it is state action to the extent the State has identified speech it wishes to allow to be proscribed and has empowered private parties to enforce the prohibition. "Calling a speech restriction a 'property right'. . . doesn't make it any less a speech restriction, and it doesn't make it constitutionally permissible."

Defendants attempted to argue that Free Speech was not a defense to this statute, because it was a law of "general applicability," such as trespass. However, the Court found that to be false given that the law itself "proscribes a form of expression" and speech is an element of proof or has more than incidental effect upon it. Unlike the courts above, the Court applied strict scrutiny to some provisions of the law and intermediate scrutiny to others because they were determined to be "content neutral."

Subsections (b)(1) and (b)(2) were found unconstitutional as the defendants offered no compelling interest for their enactment. Sections (b)(3) and (b)(5) were reviewed under intermediate scrutiny, which requires the government to prove the statute is "narrowly tailored to serve a significant government interest and leave[s] open ample alternative channels of communication." The Court determined that protecting private property is a legitimate government interest, but the State failed to prove it was a narrowly tailored way to address that interest. The Court struck down two sections as applied and two on their face. "Where a statute is unconstitutional in every scenario, the appropriate remedy is to strike down the law on its face." Sections (b)(2) and (b)(3) were entirely stricken from the law.

On the 14th Amendment challenge, the Court found the statute was not void for vagueness. On the Equal Protection claim, a ration basis review was applied. The Court did not find animus on the part of lawmakers towards groups such as the plaintiffs (i.e., animal rights advocates).

## 4.4 CONCLUSION

These cases make it clear that ag-gag laws have significant constitutional concerns. While predominantly subject to strict scrutiny, they rarely are found to serve a compelling government interest in the narrowest way possible, regardless of whether the speech is criminalized or on private property.

# 5 The Disintegration of Bovine Animal Protection
## *Fundamental Animal Rights vs. Speciesism in Indian Law*

*Sahana Ramdas*

**CONTENTS**



A cursory look at cow protection laws gives the impression that the paramount goal is to protect the well-being of cows. However, a deep dive into the cow slaughter ban laws reveals that the rules are not intended for the betterment of the living conditions of cattle, instead enacted to place the religious sentiments and political interests of the dominant religion on a higher pedestal. On one end, the cow slaughter trade is criminalized by the beef ban laws, which purport to protect cows, scientifically promote animal husbandry, and thus, stringently enforced against the targeted population of the Indian society. On the other end, dairy farming practices lead to prolonged

DOI: 10.1201/9781003137733-11

cruelty to all bovines and, yet, is an unfettered industry that garners national support and less public scrutiny.[1]

## 5.1 INTRODUCTION: FUNDAMENTAL ANIMAL RIGHTS

In the Animal Law movement, India is often recognized for its progressive animal law judgments and the wide variety of animal welfare legislation.[2] In *People for Animals v. MD Mohazzim*, the Delhi High Court in 2015 unclipped the wings of the caged birds of Gujarat by reaffirming their natural right to fly, holding that birds have a fundamental right to fly.[3] In the first of such pathbreaking judgments, the Uttarakhand High Court in 2018 interpreted the Fundamental Right to Life under Article 21 of the Indian Constitution to include animals' Right to Life.[4] This Judgment in *Narayan Dutt Bhatt v. Union of India* assigned legal personhood to non-human animal species.[5] In a recent 2019 judgment, *Karnail Singh and others v. the State of Haryana*, the High Court of Punjab and Haryana recognized the entire animal kingdom as separate legal entities.[6] Under the doctrine of *Parens Patriae*, in which the government protects those who cannot protect themselves, the Court noted humans' moral and legal obligation to act as legal guardians of all the non-human animals found within Haryana.[7] The second of such legal personhood judgments reasoned that legal personhood is evolving. The concept needs to keep up with the scientific and moral progress the human-made society is transitioning toward.[8] All animals, including avian and aquatic species, were declared to possess legal rights in both judgments. Both courts declared citizens of the states as Loco Parentis (in the position of parental protectors) of the animal kingdom, having the legal authority to represent and protect animals' interests.[9] In what appeared to be the most promising of such judgments, *AWBI v. Nagaraj of 2014*, the Supreme Court recognized animal exploitation as inherently cruel practice and held that animals have an inherent right to live a life with dignity, honor, and devoid of unlawful and barbaric suffering.[10] This watershed judgment in the Animal Law movement holistically addressed

---

[1] Sahana Ramdas is an Animal Lawyer from India. She has a Master's in Animal Law, Environmental Law, and General Law. She is a member of the New York State Bar Association and Bar Council of India. She is now working on legal projects in the three disciplines. She is a member of the TIPS Animal Law Committee and can be reached at saanaramdas@gmail.com.

[2] See: A compilation of Animal Law of India, by the Animal Welfare Board of India and Ministry of Fisheries, Animal Husbandry and Dairying (Department of Animal Husbandry and Dairying). http://www.awbi.in/policy_acts_rules.html.

[3] https://www.nonhumanrights.org/content/uploads/2015-May-DelhiHC-People-for-Animals-v.-Md.-Mohazzim.pdf.

[4] Id; https://www.downtoearth.org.in/blog/wildlife-biodiversity/decoding-uttarakhand-hc-s-judgment-on-animal-rights-61287.

[5] http://www.nluassam.ac.in/docs/lex%20terra/Lex%20Terra%20Issue%2032_7.pdf.

[6] https://www.livelaw.in/pdf_upload/pdf_upload-361239.pdf.

[7] Id.

[8] Narayan Dutt Bhatt vs/ UOI & Ors. https://www.nonhumanrights.org/blog/punjab-haryana-animal-rights/.

[9] Id.

[10] Animal Welfare Board of India vs. A. Nagaraja and Ors. (07.05.2014 - S.C.) Equivalent citations: 2014(4)ABR556, 2014(4)BomCR26, 2014(4)FLT618, JT2014(7)SC1, 2014(2)KLT717(SC), (2014)4MLJ89, 2014(6)SCALE468, (2014)7SCC547, 2014 (5) SCJ 614. https://www.nonhumanrights.org/content/uploads/Animal-Welfare-Board-v-A.-Nagaraja-7.5.2014.pdf.

animals' legal and moral rights. Even so, it has had no deterrent effect against animal cruelty incidents in India.[11] Furthermore, due to vote bank politics and the anthropocentric attitude of the partakers, legal personhood rights to all animals are yet to be conferred on a national level.

The above mentioned Public Interest Litigation Writ petitions in animal legal personhood matters were filed against the inhumane transport conditions in which domestic and livestock animals were moved in and around India.[12] In the background, anthropocentric laws deprioritize protecting the most persecuted animals—"livestock" animals.[13]

[11] In past 10 years India saw nearly 5 lakh cases of animal cruelty, unreported could be much higher. *India Times*. https://www.indiatimes.com/news/india/animal-cruelty-cases-india-overall-534574.html; Humans losing humanity: Rising instances of animal cruelty in India, land where they are worshipped. *Indian Express*. https://www.newindianexpress.com/nation/2020/jun/19/humans-losing-humanity-rising-instances-of-animal-cruelty-in-india-land-where-they-are-worshipped-2158506.html.

[12] https://www.nonhumanrights.org/blog/punjab-haryana-animal-rights/.

[13] Section 11(3) of the Prevention of Cruelty to Animals Act, 1960.

### 5.1.1 The Prevention of Cruelty to Animals Act: A Toothless Piece of Legislation

The Prevention of Cruelty to Animals Act, 1960 (the PCA), is the national legislation that addresses animal welfare issues and forbids animal cruelty.[14] While the Central Act criminalizes many types of animal cruelty, it has created utilitarian exceptions for permitting large-scale animal slaughter for food, religious sacrifices, and animal experimentation.[15] The Act is intended to prevent "*unnecessary pain and suffering*" in animals, implying that humans' demand for animal meat and animal products justifies the infliction of pain and suffering over livestock animals and further legitimizes the exploitation of animals' anatomy for science, and recreational purposes. [16]

The PCA, when enacted, was a progressive animal welfare legislation, casting a net on many unlawful acts of cruelty.[17] However, the 1960 Act has not been amended to keep up with inflation cycles or needs for criminal justice reform in Animal Law.[18] Due to the meager penalties the Central Act imposes, it has become a toothless piece of legislation where the penalty provisions have no punitive or deterrent effect on animal offenders.[19] The highest fine for the severest of animal crimes draws a paltry penalty of 50 Rupees, equivalent to 0.61 USD.[20] Adding to the loopholes is the lack of implementation of animal welfare standards and inadequate enforcement of the existing penal provisions by regulatory authorities.[21] Predictably, there has been a rise in crimes against animals where repeat animal abusers and other animal offenders often get away with gruesome animal cruelty crimes.[22] Despite this growing number of animal crimes, there has been improper maintenance of animal cruelty

---

14 The Prevention of Cruelty to Animals Act, 1960 (59 of 1960) as amended by Central Act 26 of 1982. https://www.iitk.ac.in/animalwelfare/data/PCA_act_brief.pdf.

15 Section 11(3) (a) - (e), Section 14 of Chapter IV, Section 28 of Chapter VI, Chapter V Performing Animals.

16 See Id.

17 The Prevention of Cruelty to Animals Act, 1960 (59 of 1960).

18 See: https://www.thehindu.com/opinion/op-ed/for-the-welfare-of-animals/article32633152.ece.

19 Paltry fine of Rs 50 allows abusers to get away with animal. *The Times of India*, March 18th 2016. https://timesofindia.indiatimes.com/india/paltry-fine-of-rs-50-allows-abusers-to-get-away-with-animal-cruelty/articleshow/51450261.cms; Badri Chatterjee, Cruelty to animals: Low conviction numbers, meagre fines to blame. *Hindustan Times*, February 19th 2018; https://www.hindustantimes.com/mumbai-news/cruelty-to-animals-low-conviction-numbers-meagre-fines-to-blame/story-zg5uVrw8MPHKIEtyjdluiO.html; Also see: https://thelogicalindian.com/inclusivity/animal-cruelty-29330; https://www.planetcustodian.com/animal-cruelty-in-india/14897/.

20 Section 11(o), Section 11(2), Section 12 of CHAPTER III-Cruelty to Animals Generally, Section 20, Chapter IV Experimentation of Animals, the PCA https://www.iitk.ac.in/animalwelfare/data/PCA_act_brief.pdf; Also see: Why is an animal's life still worth only 50 rupees? *Times of India*, https://timesofindia.indiatimes.com/city/bengaluru/why-is-an-animals-life-still-worth-only-rs-50/articleshow/57972960.cms?frmapp=yes&from=mdr.

21 See: https://www.hindustantimes.com/mumbai-news/cruelty-to-animals-low-conviction-numbers-meagre-fines-to-blame/story-zg5uVrw8MPHKIEtyjdluiO.html;https://factly.in/the-story-of-cruelty-to-animals-cases-in-india-lack-of-data-minimal-penalties/.

22 https://www.hindustantimes.com/mumbai-news/cruelty-to-animals-low-conviction-numbers-meagre-fines-to-blame/story-zg5uVrw8MPHKIEtyjdluiO.html; https://www.indiatimes.com/news/india/animal-cruelty-cases-india-overall-534574.html;https://www.indiatoday.in/india/story/almost-5-lakh-animals-became-victims-of-crimes-in-last-10-years-in-india-report-1770190-2021-02-17.

complaints where governments have failed to maintain up-to-date public records of animal cruelty data.[23] To date, various calls for the amendment have been unheeded, because of which the PCA largely fails the intended purpose of preventing animal cruelty.[24] A Private Member Bill by a Member of Parliament, and the Union Ministry of Fisheries, Animal Husbandry and Dairying had proposed a few amendments to the 1960 Act.[25] However, no progress has been made in addressing the need for stringent punishments against animal offenders.[26]

#### 5.1.1.1 Compromised Enforcement of Animal Protection Laws and the Demotion of Cattle Welfare through the Disharmonious Beef Bans

Protecting cows has historically been a political matter and a bone of contention between two religious groups in India.[27] In the Hindu religion, cows are regarded as sacred animals, and hence cow slaughter is religiously shunned.[28] Conversely, the Islam religion has no such religious bans. Trade in cattle skin and beef is a part of religious minorities' and tribal communities' (marginalized classes) dietary habits and is a source of livelihood.[29]

Article 51A(g) makes it a Fundamental Duty of every citizen to protect and improve the natural environment, including forests, lakes, rivers, and wildlife, and to have compassion for living creatures[30]; similarly, the Fundamental Duty under Article 48A lays down the directive principle for protecting and improving the

---

23 See: https://www.hindustantimes.com/mumbai-news/cruelty-to-animals-low-conviction-numbers-meagre-fines-to-blame/story-zg5uVrw8MPHKIEtyjdluiO.html;https://scroll.in/article/987579/despite-a-history-of-violence-against-animals-india-does-not-have-a-formal-record-of-this-cruelty; https://factly.in/the-story-of-cruelty-to-animals-cases-in-india-lack-of-data-minimal-penalties/.

24 Aarefa Johari, #NoMore50: Activists are rising to demand harsher punishments for cruelty to animals. Scroll.in, May 16th 2016. https://scroll.in/article/808148/nomore50-why-animal-lovers-are-rising-to-demand-harsher-punishments-for-cruelty-to-animals; PETA urges anti-speciesist approach in Prevention of Cruelty to Animals Act revision. *The Hindu*, April 27th 2021. https://www.thehindu.com/news/national/peta-urges-anti-speciesist-approach-in-prevention-of-cruelty-to-animals-act-revision/article34423056.ece.

25 The (Draft) Animal Welfare Act (DAWA) of 2011.https://www.downtoearth.org.in/blog/an-urge-to-act-34444http://www.indiaenvironmentportal.org.in/files/animal-welfare-preamble.pdf The Prevention of Cruelty to Animals (Amendment) Bill, 20162http://164.100.47.4/billstexts/lsbilltexts/asintroduced/33LS.pdf.

26 Animal abusers to finally face justice? Why it is high time India bins its outdated animal cruelty law. *Mirrornownews*, August 25th 2020. https://www.timesnownews.com/mirror-now/in-focus/article/animal-abusers-to-finally-face-justice-why-it-is-high-time-india-bins-its-outdated-animal-cruelty-law/642571; Priya Menon, Amendment to Prevention of Cruelty to Animals Act: Union ministry invites suggestions from public. *Times of India*, April 22nd 2021. https://timesofindia.indiatimes.com/india/amendment-to-prevention-of-cruelty-to-animals-act-union-ministry-invites-suggestions-from-public/articleshow/82192380.cms.

27 See: https://indianexpress.com/article/research/india-republic-constitution-cow-constituent-assembly-article-48-7161165/;https://indianexpress.com/article/explained/directive-principle-not-right-how-cow-protection-became-part-of-constitution-4683383/; Also see: https://harvardpolitics.com/protecting-cows/.

28 Aayush Akar, Beef Ban in India: A Legal Perspective. Latestlaws.com, November 18th, 2019. https://www.latestlaws.com/articles/beef-ban-in-india-legal-perspective-by-aayush-akar/.

29 Id. Note: The Hindu majority formerly treated the marginalized classes as "untouchables".

30 Article 51A(g) in The Constitution of India. https://indiankanoon.org/doc/1644544/?__cf_chl_jschl_tk__=pmd_40q4CKmS4_PYsI0eFTaH9FhH9JeTm266QQ04WfQiXb8-1631549003-0-gqNtZGzNAiWjcnBszQIR.

environment, forests, and wildlife.[31] Article 48 of the Indian Constitution states, *The state shall endeavour to organise agriculture and animal husbandry on modern and scientific lines and shall, in particular, take steps for preserving and improving the breeds and prohibiting the slaughter of cows and calves and other milch and draught cattle.*[32] Under 51A(h), the Directive Principle of State Policy (DPSP) seeks citizens to develop the scientific temper, humanism, and the spirit of inquiry and reform.[33] The DPSPs, such as Articles 48 and 51, act as guiding principles for state action and framing of the state's policies.[34] However, the Fundamental Duties laid under the DPSP are non-binding and unenforceable against the people and the Indian government.[35]

Following the lead of DPSP, under the ambit of Article 48 of the Indian Constitution, the majority of the state legislatures have enacted cow slaughter bans[36] which appease the influential religious sentiments of the Hindus.[37] Only in a minority of states are there no prohibitions on cow slaughter.[38] Among the states that have cow slaughter bans, there is a partial allowance for the slaughter of economically unviable, old, and /or "unfit cows" deemed so by the "fit for slaughter" certification process.[39]

In contrast to the various state beef ban laws, the Indian parliament has been unsuccessful in enacting a nationwide cow slaughter ban.[40] The absence of a national cow slaughter law is because India has held itself out as a religious and culturally tolerant land of unity in diversity. A secular, democratic republic form of governance is supposed to allow people to exercise their Fundamental Right to Life and liberty. Furthermore, the constitutionality of cow slaughter bans has been challenged ever since India gained independence from colonialism. Hence, the Indian Supreme Court has regularly addressed whether the cow protection laws violate the fundamental

---

31 Article 48A in The Constitution of India. https://indiankanoon.org/doc/871328/.

32 https://indiankanoon.org/doc/1452355/?__cf_chl_tk=S3skVnqeT5r5vIDl036oMXh5vIoKqDUMDg3rMye_8e8-1638291507-0-gaNycGzNCv0.

33 Article 51A(h) in The Constitution of India. https://indiankanoon.org/doc/560422/?__cf_chl_jschl_tk__=pmd_ZDJEtKupL3yKdeB_QFMxSXVTMohwUjIjg23Fae8LjVg-1631550911-0-gqNtZGzNAhCjcnBszQrl.

34 https://www.animallaw.info/article/constitutional-scheme-animal-rights-india#s2.

35 https://www.legalserviceindia.com/legal/article-916-directive-principles-of-state-policy-dpsps-.html; Also see: https://indianexpress.com/article/explained/directive-principle-not-right-how-cow-protection-became-part-of-constitution-4683383/.

36 https://indianexpress.com/article/research/india-republic-constitution-cow-constituent-assembly-article-48-7161165/; https://www.hrw.org/sites/default/files/report_pdf/india0219_appendix_1.pdf.

37 See: https://scroll.in/article/998735/cow-protection-was-a-sensitive-subject-in-india-even-when-the-constitution-was-being-framed; https://scroll.in/article/998735/cow-protection-was-a-sensitive-subject-in-india-even-when-the-constitution-was-being-framed; https://indianexpress.com/article/research/india-republic-constitution-cow-constituent-assembly-article-48-7161165/.

38 https://factly.in/a-review-of-the-prevention-of-cow-slaughter-laws-across-the-country/; https://www.hrw.org/sites/default/files/report_pdf/india0219_appendix_1.pdf; https://www.indiatvnews.com/news/india-over-99-indians-now-live-in-areas-under-cow-protection-laws-377209.

39 See: https://factly.in/a-review-of-the-prevention-of-cow-slaughter-laws-across-the-country/; https://www.hrw.org/sites/default/files/report_pdf/india0219_appendix_1.pdf.

40 See: https://indianexpress.com/article/research/india-republic-constitution-cow-constituent-assembly-article-48-7161165/; https://www.latestlaws.com/articles/beef-ban-in-india-legal-perspective-by-aayush-akar/.

rights of the challenging petitioners.[41] Also often, the Apex Court has upheld the bans on cow slaughter as not unconstitutional and still found the legal slaughter of economically unproductive bulls, buffaloes, and bullocks as acceptable.

In *Mohd. Hanif Quereshi v. the State of Bihar*, the petitioner, contended that the cow slaughter ban was an infringement of his Fundamental Right to Practice Religion—as the ban prohibited religious cow sacrifice during an Islamic festival.[42] The Supreme Court ruled against the petitioner by reasoning that none of the religious texts mandate cow slaughter and that the petitioner could instead use a camel or goat to perform religious sacrifices.[43] By reasoning so, the Supreme Court held that a total ban on cow slaughter did not infringe on the religious freedom of Muslims.[44] However, the Court noted that Article 48 could not be read to prohibit the slaughter of all cows or cattle, but instead that the Directive Principle is meant to protect milch and draught cattle that hold an economic value for agriculture.[45] The petitioners, in this case, were from the Muslim community engaged in butcher trade but were prohibited from slaughtering cattle due to the state legislature. The petitioners claimed that the state law violated their fundamental rights guaranteed under the Constitution to Freedom of Trade and Religion.[46] The Court dismissed this argument and held that the total ban of cow and calves slaughter was a fair and reasonable restriction on the petitioners' fundamental rights and was as per Article 48A of the Constitution.[47] Nevertheless, the Court held that a general rule that burdened butchers with keeping she-buffaloes, old bull, or bullocks with no economic value was an unreasonable restriction on cattle traders' Right to Freedom of earning a livelihood and against the public interest.[48] This holding was replaced by the 2005 *State of Gujarat v. Mirzapur Moti Quereshi* ruling, where the Constitutional Bench of the Supreme Court interpreted Article 48 to imply a total ban on the slaughter of cows and their progeny.[49] The Supreme Court upheld the complete slaughter ban, even on bulls and bullocks for the entirety of their lives, despite them becoming economically devalued.[50] The Court reasoned that cattle must be treated with compassion as emphasized under Article 51A(g) and that Article 48 explicitly requires protection of cows, calves, and other milch and draught cattle.[51] Although the Court justified its position to be based on compassion, note the distinction made within the cattle protection policies and judicial agreement with the bans.[52] As we see here, slaughter laws are drawn on the economic value of bovines and the religious labels assigned to cows—the stark

---

41 https://www.latestlaws.com/articles/beef-ban-in-india-legal-perspective-by-aayush-akar/.

42 Mohd. Hanif Quareshi & Others vs The State of Bihar, 1958 (1958 AIR 731, 1959 SCR 629) https://indiankanoon.org/doc/93885/.

43 Mohd. Hanif Quareshi & Others vs The State of Bihar, 1958 (1958 AIR 731, 1959 SCR 629) https://indiankanoon.org/doc/93885/.

44 Id. 1959 SCR 629.

45 Id.

46 Id.

47 Id.

48 Id.

49 State of Gujarat vs. Mirzapur Moti Qureshi Kasab AIR 2005 SC 212.

50 2005 (8) SCC 534.

51 Id.

52 https://www.article-14.com/post/the-dark-chronology-of-india-s-cow-slaughter-laws.

difference in what constitutes animal cruelty toward cows and what is considered otherwise against other bovines and non-bovines.

In the last decade alone, most state laws have been amended to impose stricter punishments for cow slaughter-related activities and severely criminalize the cow slaughter trade.[53] Although the minority religious groups do not follow the same religion and food practices as the Hindus, they are still required to abstain from the cow slaughter trade and targeted by the cattle slaughter bans.[54] In states with cow slaughter bans, persons found violating the law are often arbitrarily imprisoned for trading, transporting, or possessing protected cattle.[55] In sharp contrast, these state laws are silent on the inhumane practices in the dairy industry, whose by-products are consumed by most of the citizens and yet normalized as part and parcel of animal agriculture.[56]

Some state cow protection laws have dismantled the fairness of the criminal justice system where fundamental rights, equal protection of the law, non-discrimination, and liberty principles have been severely compromised.[57] What acts as draconian cattle ban laws are allowed to be enforced by private vigilante groups who often engage in Kangaroo court proceedings where the emboldened cow vigilantes are usurping the fundamental rights of the minority with impunity.[58] Under the guise of enforcing cow protection, vigilantes and masses susceptible to religious indoctrination have engaged in human rights violations against the minority religious groups.[59] Apart from outsourcing the enforcement of laws to volunteer groups, some state laws have shifted the burden of proof to the defendants. Accused now need to clearly show that the meat they possess is not sourced from slaughtering protected cattle,

---

[53] For instance, in the State of Gujarat, the slaughter of cows is punishable with life imprisonment. https://factly.in/a-review-of-the-prevention-of-cow-slaughter-laws-across-the-country/; Also see: Aakar Patel, The Dark Chronology of India's Cow-Slaughter Laws. In Article 14. https://www.article-14.com/post/the-dark-chronology-of-india-s-cow-slaughter-laws; By Rachit Garg, author Harmanpreet Kaur, An analysis of the Cow Slaughter Act. On June 18th 2021, I Pleaders. https://blog.ipleaders.in/analysis-cow-slaughter-act/.

[54] Aakar Patel, The Dark Chronology of India's Cow-Slaughter Laws. In Article 14. https://www.article-14.com/post/the-dark-chronology-of-india-s-cow-slaughter-laws.

[55] Id.

[56] https://www.hrw.org/report/2019/02/18/violent-cow-protection-india/vigilante-groups-attack-minorities#; https://thecsrjournal.in/csr-reality-dairy-industry-cows/; https://www.mapsofindia.com/my-india/india/the-animal-cruelty-in-our-milk-industries.

[57] See: Aakar Patel, The Dark Chronology of India's Cow-Slaughter Laws. In Article 14. https://www.article-14.com/post/the-dark-chronology-of-india-s-cow-slaughter-laws; https://www.thequint.com/voices/opinion/cow-protection-laws-slaughter-cattle-smuggling-beef-ban-modi-government-yogi-adityanath-lynching-incidents#read-more.

[58] Violent cow protection in India, vigilante groups attack minorities. *Human Rights Watch*, February 18th 2019. https://www.hrw.org/report/2019/02/18/violent-cow-protection-india/vigilante-groups-attack-minorities#; Cow protection legislation and vigilante violence in India. *ACLED*. https://acleddata.com/2021/05/03/cow-protection-legislation-and-vigilante-violence-in-india/; Also see: https://www.theguardian.com/world/2016/oct/27/on-patrol-hindu-vigilantes-smuggling-protect-india-cows-kill; https://www.thequint.com/voices/opinion/cow-protection-laws-slaughter-cattle-smuggling-beef-ban-modi-government-yogi-adityanath-lynching-incidents#read-more.

[59] Nimisha Jaiswal, Why India's beef lynching is not about a dead cow. *Global Post*, October 12th 2015. https://www.pri.org/stories/2015-10-12/why-indias-beef-lynching-not-about-dead-cow; https://www.theguardian.com/world/2016/oct/27/on-patrol-hindu-vigilantes-smuggling-protect-india-cows-kill.

making it easier for states to convict suspects.[60] Hence, these arbitrary laws reversed the presumption of the *innocent until proven guilty* principle and shifted the burden of proof from the state prosecution to the defendants, who are now burdened with the onus of proving beyond reasonable doubt that they did not commit the cognizable, non-bailable offense.[61]

### 5.1.2 Cattle Protection Laws often Have No Foothold in Animal Welfare

The vested political interest of the ruling party is furthered by supporting cattle ban laws as it garners the majoritarian religion's support due to vote bank politics.[62] As yet, attempts to impose a nationwide cattle slaughter ban has failed as it goes against the secular principles of the Indian Constitution—which aims to protect the religious, cultural, and economic rights of a diverse population in the country.[63]

For instance, *The 2017 Prevention of Cruelty to Animals (Regulation of Livestock Markets) Rules* indirectly attempted to supplement the state beef ban laws with a de-facto ban on cattle sale for slaughter purposes[64]; under the Livestock Rules of 2017, only agriculturists were allowed to trade in cattle, and Section 22(b)(iii) & Section 22(d)(ii) specifically banned the sale or purchase of cattle in live animal markets for slaughter purposes.[65] The Ministry of Environment, Forest, and Climate Change (MoEFCC) declared that its Order primarily protects animals from uncontrolled and unregulated animal trade.[66] The notified rules state that their purpose is to ensure animal welfare, where adequate facilities are provided for cattle housing, upkeep, and maintenance, and proper enclosures and veterinary care for cattle in existing

60 Cow slaughter prevention laws in India, how the law not just protects cow vigilantes, but sanctifies lynchings. *CJP*, July 2nd, 2018. https://cjp.org.in/cow-slaughter-prevention-laws-in-india/.

61 https://cjp.org.in/cow-slaughter-prevention-laws-in-india/; Cow protection legislation and vigilante violence in India. *ACLED* (Note: The Supreme Court addressed the escalating communal violence and mob lynching by directing the Central government and state governments to prevent mob violence and lynching.) https://acleddata.com/2021/05/03/cow-protection-legislation-and-vigilante-violence-in-india/.

62 See: India bans sale of cows for slaughter, a move designed to appease conservative Hindus. *LA Times*. https://www.latimes.com/world/la-fg-india-cow-slaughter-20170526-story.html; Also see: https://www.firstpost.com/politics/bjp-has-ensured-cow-protection-becomes-universal-brand-of-politics-congress-led-govt-invoking-nsa-in-mp-a-typical-example-6036511.html.

63 See: The Constitutional Principles of India: http://164.100.47.193/Refinput/Research_notes/English/04122019_153433_1021204140.pdf; http://www.mati.gov.in/docs/Constitution_of_India.pdf.

64 https://www.thequint.com/explainers/breaking-down-sc-stay-cattle-slaughter#read-more#read-more#read-more.

65 https://www.thequint.com/explainers/breaking-down-sc-stay-cattle-slaughter#read-more#read-more#read-more; https://timesofindia.indiatimes.com/india/centre-bans-sale-of-cattle-for-slaughter-at-livestock-markets/articleshow/58861631.cms;https://veterinary.assam.gov.in/sites/default/files/swf_utility_folder/departments/ahvetdir_webcomindia_org_oid_4/menu/document/Regulation%20of%20Livestock%20Market%20rules.pdf.

66 https://www.firstpost.com/india/ban-on-cattle-trade-govt-clarifies-that-rules-modified-to-end-animal-cruelty-3488365.html; India Supreme Court suspends cattle slaughter ban. *BBC news*, July 11th 2017. https://www.bbc.com/news/world-asia-india-40565457.

animal markets.[67] However, in effect, its provisions imposed blanket restrictions only on the sale and purchase of cattle for consumption and religious sacrifices.[68] This specific provision of the 2017 Livestock Rules was challenged before the Madras High Court. The petitioners contended that the cattle slaughter rules violate the Right to Equality, the Right to Freedom to make personal choices in food, privacy, and personal liberty.[69] Accepting the petitioners' grounds, the Court made an interim stay of the operation of the 2017 PCA (Regulation of Livestock Market) Rules.[70] Moreover, state governments are the primary regulators of cattle slaughter in India.[71]

A similar PIL challenging the Cattle slaughter ban was filed before the Supreme Court during that time.[72] The highest Court affirmed and extended the Madras High Court's Order by noting that the 2017 Rules contradict the exceptions laid down under PCA because the Central Act itself protects the Fundamental Right to trade in animals (PCA permits slaughter and religious sacrifices).[73] The Supreme Court agreed that the restrictive provisions interfered with a citizen's Fundamental Right to carry on trade, profession, business, and religious practice and discriminated against the Right to livelihood of the minority population.[74] Affected state governments challenged the Central Rules as unconstitutional because the Rules outright banned sale transactions for slaughter purposes and parallelly encouraged cattle trade for other animal agriculture purposes.[75] Under the state list of the Indian Constitution, the regulation of animal markets and animal slaughter comes under the state's jurisdiction. Only the various state governments have the exclusive authority to pass laws and regulations relating to such state subjects.[76] Hence, the Court agreed with the petitioner's arguments and affirmed the High Court's stay of the specific provisions of the 2017 Livestock Rules.[77]

---

67 https://www.firstpost.com/india/ban-on-cattle-trade-govt-clarifies-that-rules-modified-to-end-animal-cruelty-3488365.html.

68 https://www.thequint.com/news/india/cow-sale-animal-markets-banned-slaughter-muslim-dalit#read-more.

69 https://www.thehindu.com/news/national/tamil-nadu/madras-hc-refuses-to-extend-interim-stay-on-cattle-sale/article19232358.ece.

70 https://www.hindustantimes.com/india-news/madras-hc-stays-centre-s-notification-on-sale-of-cattle-for-slaughter/story-VwIHeacouPVj7wauniC9eN.html.

71 Schedule 7, Entry 15, Constitution of India. https://blogs.loc.gov/law/2015/11/falqs-beef-bans-in-india/.

72 Cattle trade for slaughter: Supreme Court suspends ban across India. *Hindustan Times*. https://www.hindustantimes.com/india-news/supreme-court-puts-on-hold-ban-on-cattle-trade-for-slaughter-govt-says-new-rules-by-august-end/story-WypElt9CMgFZP1wxPsLdgJ.html.

73 Krishnadas Rajagopal, Law itself allows cattle slaughter, S.C. tells govt. *The Hindu*, on August 11th 2017. https://www.thehindu.com/news/national/law-itself-allows-cattle-slaughter-sc-tells-govt/article19476483.ece As noted above: The Prevention of Cruelty to Animals Act, 1960 allows animal slaughter for food and religious sacrifices.

74 Krishnadas Rajagopal, Law itself allows cattle slaughter, S.C. tells govt. *The Hindu*, on August 11th 2017. https://www.thehindu.com/news/national/law-itself-allows-cattle-slaughter-sc-tells-govt/article19476483.ece.

75 List II, State List, Entry 15 & Entry 28, Constitution of India. http://constitutionofindia.etal.in/schedule_7_2/; https://www.thehindu.com/news/national/tamil-nadu/madras-hc-refuses-to-extend-interim-stay-on-cattle-sale/article19232358.ece.

76 Id.

77 https://www.hindustantimes.com/india-news/supreme-court-puts-on-hold-ban-on-cattle-trade-for-slaughter-govt-says-new-rules-by-august-end/story-WypElt9CMgFZP1wxPsLdgJ.html.

When the affected parties challenged this Central notification, the Center defended the Restrictive Rule as not intended to regulate cattle slaughter.[78] Instead, it framed the Livestock Rules of 2017 to respond to a 2014 Supreme Court Order.[79] In *Gauri Maulekhi v. Union of India* (a Writ Petition of 2010), the Supreme Court took note of the rampant trafficking of cattle and buffaloes to Nepal for large scale animal sacrifices that occurred during the annual Gandhimai festival.[80] The Apex Court in 2014 directed the Central Government to frame appropriate guidelines to prevent cattle smuggling from India to Nepal.[81] Subsequently, the Animal Welfare Board of India (AWBI) prepared the draft rules that the Ministry adopted in framing the Livestock Rules of 2017.[82] Nevertheless, the Livestock Rules of 2017 did not limit itself to addressing the illegal sale and smuggling of cattle or the betterment of cattle's living and transport conditions.[83] Instead, *The Restrictions on the Sale of Cattle* provision targeted certain sections from selling and purchasing cattle for slaughter purposes while not addressing animal cruelty in the dairy industry. The Central Government could have primarily passed the PCA Rules to uniformly and impartially protect cattle and improve cattle welfare. Instead, the Central Government exceeded its scope of authority. Due to this, other cattle protection provisions' enforcement (aiming to prevent animal cruelty and promote cattle welfare) is thwarted due to their association with a contentious legal provision. To date, the remaining Rules have not been effectively enforced, leaving much to be desired in improving the living conditions of cattle in animal agriculture. Such undue interference by the Central Government hurts cattle welfare as it politicizes and jeopardizes the enforcement of animal welfare standards in animal agriculture.

## 5.2 ANIMAL CRUELTY SLIDES INTO OBSCURITY IN THE GIGANTIC INDIAN DAIRY INDUSTRY

The country has the greatest number of livestock animals and cattle population globally.[84] India is also the largest dairy producer in the world, with over 300 million

---

78 New restrictions on cattle slaughter. *The Hindu*, Updated May 30th, 2017. https://www.thehindu.com/news/national/government-bans-sale-purchase-of-cattle-from-animal-markets-for-slaughter/article18585018.ece

79 See the statement of Environmental Minister, Harshvardhan: https://www.downtoearth.org.in/news/agriculture/centre-to-roll-back-controversial-ban-on-trade-of-cattle-for-slaughter-59259.

80 Gauri Maulekhi v. Union of India, Writ Petition (PIL) No. 77 of 2010. https://www.firstpost.com/india/ban-on-cattle-trade-govt-clarifies-that-rules-modified-to-end-animal-cruelty-3488365.html.

81 Humane Society International, Supreme Court of India Intervenes to Save Thousands of Animals from Nepal's Brutal Gadhimai Festival Sacrifice, HSI, October 20th 2014. https://www.hsi.org/news-media/india-supreme-court-gadhimai-ruling-102014/After much international public outcry, the animal sacrifice festival in Nepal has stopped. Animal sacrifice banned in Nepal festival. *The Hindu*, July 29th 2015. https://www.thehindu.com/news/international/animal-sacrifice-banned-in-nepal-festival/article7474680.ece.

82 See: https://indianexpress.com/article/india/cattle-trade-laws-is-central-government-competent-to-make-rules-under-state-list-5272460/.

83 See Section 22, Regulation-of-Livestock-Markets-Rules-2017. https://veterinary.assam.gov.in/sites/default/files/swf_utility_folder/departments/ahvetdir_webcomindia_org_oid_4/menu/document/Regulation%20of%20Livestock%20Market%20rules.pdf.

84 Smita Sirohi & Axel Michaelowa (2007). Sufferer and Cause: Indian Livestock and Climate Change, *Climatic Change*, 85, 285–298, DOI: 10.1007/s10584-007-9241-8 (p. 287).

bovines.[85] The dairy industry alone accounts for 4.2 percent of the national gross domestic product, and over 150 million people are involved in the Indian dairy farming sector.[86] India takes pride in being a dairy-intensive country and has over 75 million dairy farms that produce over 22 percent of the global milk production.[87] With Indians drinking over 85 million metric tons of milk in 2020, India was the largest milk consumer in 2022 alone. It produced over 187 million metric tons of milk in 2019 alone.[88] Clearly, the beef ban laws have not deterred India's increasing appetite for dairy products. This form of surplus production and consumption comes at the cost of sacrificing animal welfare standards where cattle are kept under distress conditions and in violation of the PCA and its associated rules and regulations.[89]

Indians treat cows not just as sacred animals but also as movable property in the agricultural sector. As opposed to punishing crimes against animals, the Indian Penal Code primarily protects the proprietorial rights of animal owners. Sections 428 and 429 make it a criminal offense to kill, maim, poison, or render an owned animal with economic value useless to the animal owner.[90] Section 378 makes it a criminal offense to commit animal theft against the animal's owner.[91] Section 503 prescribes intimidation against the animal caretaker as a criminal offense.[92] Thus, these criminal provisions in the Criminal Code affirm the rights of animal owners over animals with a specific economic value. As seen, the current set of slaughter ban laws categorically criminalizes the slaughtering of cow progeny by arguing that it is necessary to protect cattle.[93] However, milch animals are subjected to even harsher conditions filled with prolonged suffering, yet the intensive exploitation of cattle in the dairy industry is overlooked.[94]

Animal cruelty is implicit in the deregulated dairy industry, where the participants are insulated from public scrutiny. Dairy producers often do not comply with farm animal welfare rules, standards, and guidelines and still continue their operations without adequate review and oversight.[95] The PCA has laid down guidelines for the

---

85 https://www.statista.com/statistics/1181408/india-cattle-population/.

86 https://www.downtoearth.org.in/blog/climate-change/climate-crisis-way-forward-for-dairy-giants-in-india-78226.

87 Gateway to dairy production and products, Food and Agriculture Organization of the United Nations. http://www.fao.org/dairy-production-products/production/en/; https://qz.com/235085/india-75-million-dairy-farms-now-make-more-milk-than-all-of-the-european-union/.

88 https://www.statista.com/statistics/272003/global-annual-consumption-of-milk-by-region/.

89 The Indian Penal Code, 1860. https://www.indiacode.nic.in/bitstream/123456789/4219/1/THE-INDIAN-PENAL-CODE-1860.pdf

90 Id. https://legislative.gov.in/sites/default/files/A1860-45.pdf.

91 Id.

92 Id.

93 https://scroll.in/article/759157/a-short-account-of-indias-long-history-of-hypocrisy-on-cow-slaughter-laws.

94 See: Rajesh Kasturirangan, Krithika Srinivasan, & Smitha Rao, Dark and dairy: The sorry tale of the milch animals. *The Hindu*, November 08th 2014. https://www.thehindu.com/opinion/open-page/the-sorry-tale-of-the-milch-animals/article6578663.ece; Yamini Narayanan (2018). Cow Protection as 'Casteised Speciesism': Sacralisation, Commercialisation and Politicisation, *South Asia: Journal of South Asian Studies*, 41(2), 331–351, DOI: 10.1080/00856401.2018.1419794; https://www.mapsofindia.com/my-india/india/the-animal-cruelty-in-our-milk-industries.

95 https://www.thehindu.com/opinion/open-page/the-sorry-tale-of-the-milch-animals/article6578663.ece.

humane handling of cattle.[96] The Department of Animal Husbandry is the regulatory body responsible for monitoring cattle welfare, upkeep, and housing conditions.[97] However, unlike the slaughter industry being regularly on the regulatory radar,[98] the law enforcement authority is less motivated to oversee the dairy industry's compliance with the non-binding guidelines.[99]

The White Revolution massively transformed the animal husbandry practices in India.[100] The dairy industry adopted an industrialized production model, where animal welfare standards are often neglected to make way for intensive dairy production.[101] Industrial-scale dairy farming is not compatible with the humane handling of cattle, as untold violence is inflicted on the species to meet the growing demand for dairy products.[102] The World Organization for Animal Health (OIE) states that animal welfare requires due regard to the individual animal's overall well-being—from the time of an animal's birth to its death. The OIE goes on to describe animal welfare as *an animal experiences good welfare if the animal is healthy, comfortable, well-nourished, safe, is not suffering from unpleasant states such as pain, fear, and distress, and can express behaviors that are important for its physical and mental state.*[103] Freedom from discomfort; Freedom from hunger and thirst; Freedom from pain, injury, or disease; Freedom to express natural behavior; and Freedom from fear and distress are considered basic animal welfare standards, found essential in maintaining animals without undue duress.[104] Although these animal welfare principles are internationally recognized as essential in maintaining animal well-being, the Indian dairy industry falls short of protecting cattle welfare.[105]

The documentary *Deadly Dairy* conducted by an animal non-profit in India revealed the inhumane conditions in which the Indian cattle are kept for dairy production.[106] An undercover investigation conducted countrywide by Animal Equality showed how most dairy producers were in rampant violation of the PCA 1960 and

96 The Transport of Animals Rules, 1978. http://www.awbi.org/awbi-pdf/APL.pdf.

97 Department of Animal Husbandry and Dairying. https://dahd.nic.in/.

98 Chapter III - Administration of Cattle Laws, Department of Animal Husbandry and Dairying. https://dahd.nic.in/node/86828.

99 https://www.theindiaforum.in/article/india-s-deregulated-dairy-sector-signposts-future-our-food.

100 Yamini Narayanan (2018). Cow Protection as 'Casteised Speciesism':Sacralisation, Commercialisation and Politicisation, *South Asia: Journal of South Asian Studies*, 41(2), 331–351, DOI: 10.1080/00856401.2018.1419794 (p. 338).

101 Id.

102 Id; Animal Equality, Reveals the Hidden cost of dairy. https://animalequality.in/india-deadly-dairy/.

103 World Organization for Animal Health, OIE. Chapter 7.1. Introduction to the Recommendations for Animal Welfare, Article 7.1.1. World Organization for Animal Health, OIE https://www.oie.int/fileadmin/Home/eng/Health_standards/tahc/2018/en_chapitre_aw_introduction.htm.

104 Id. World Organization for Animal Health, OIE. https://www.oie.int/en/what-we-do/animal-health-and-welfare/animal-welfare/.

105 See: Bobins Abraham, Let us take a look at how we treat our 'Sacred' cow and how safe is the milk we drink. *India Times*, on November 27th 2017. https://www.indiatimes.com/news/india/on-national-milk-day-let-us-take-a-look-at-how-the-cattle-are-mistreated-and-how-safe-the-milk-we-drink-is-334436.html.

106 Animal Equality, https://animalequality.in/issues/dairy/; Badri Chatterjee, Two-year undercover study reveals cruel side of India's dairy industries. *Hindustan Times*, November 25th 2017. https://www.hindustantimes.com/mumbai-news/two-year-undercover-study-reveals-cruel-side-of-india-s-dairy-industries/story-7icLDyv1Rq2tVV2kbYKccN.html.

the Slaughterhouse Rules, 2002.[107] The housing, medical care, upkeep, living conditions, and transport of bovine animals are supposed to be regulated under the comprehensive framework provided under the PCA, including the Registration of Cattle Premises Rules 1978, Draught and Pack Animals Rules 1965, Transport of Animal Rules 1978, and Transport of Animals on Foot Rules 2001.[108]

In order to maximize profits and meet the demand for milk and meat, cows are frequently impregnated and often subjected to artificial impregnation.[109] Growth and reproductive hormone are prohibited from unregulated uses. The use of Oxytocin on milch cows is illegal and amounts to cruelty under the PCA.[110] The use of the substance to induce milking is a cognizable, criminal offense under the PCA and severely restricted as a schedule H1 drug under the Food & Drugs Cosmetics Act.[111] Nonetheless, dairy cows are often injected with Oxytocin, antibiotics, and other unnatural substances to enable the dairy industry to induce milking in overused cows.[112] Not surprisingly, bovines often succumb to such appalling conditions due to inexperienced staff, lack of medical care, improper upkeep, and inhumane housing.[113] Despite the prevalent animal cruelty in the dairy industry, the zeal for enforcing animal welfare laws is severely lacking.

## 5.3 THE SLAUGHTER INDUSTRY IS A BY-PRODUCT OF THE DAIRY INDUSTRY

Amidst all the beef ban laws, the dairy industry breeds the slaughter industry—behind the door. The lifecycle of bovine trade involves the breeding, rearing, and use of bovines, first, in the dairy industry.[114] Once the bovine animals are rendered

[107] Animal Equality, https://animalequality.in/issues/dairy/; https://www.mapsofindia.com/my-india/india/the-animal-cruelty-in-our-milk-industries.

[108] The Prevention of Cruelty to Animals (Registration of Cattle Premises) Rules, 1978. http://awbi.in/awbi-pdf/(REGISTRATION%20OF%20CATTLE%20PREMISES)%20RULES,1978.pdf; The Prevention of Cruelty to Draught and Pack Animals Rules, 1965. http://awbi.in/awbi-pdf/Draught_and_Pack_Animal_Rules_1965_English.pdf; Prevention of Cruelty to Animals (Transport of Animals on Foot) Rules, 2001. http://awbi.in/awbi-pdf/(TRANSPORT%20OF%20ANIMALS%20ON%20FOOT)%20RULES,%202001.pdf; The Transport of Animals Rules, 1978. http://www.awbi.org/awbi-pdf/APL.pdf.

[109] Yamini Narayanan (2018). Cow Protection as 'Casteised Speciesism': Sacralisation, Commercialisation and Politicisation, *South Asia: Journal of South Asian Studies*, 41(2), 331–351, DOI: 10.1080/00856401.2018.1419794 (pp. 333,344) https://www.thenewsminute.com/article/filth-confinement-death-how-indian-dairies-treat-cows-72199; https://www.thehindu.com/opinion/open-page/the-sorry-tale-of-the-milch-animals/article6578663.ece.

[110] Section 12, PCA 1960. http://www.awbi.org/awbi-pdf/APL.pdf.

[111] Section 12 of the Prevention of Cruelty to Animals Act, 1960 & Food and Drug Adulteration Act of 1960. http://www.awbi.org/awbi-pdf/APL.pdf.

[112] https://www.thehindu.com/opinion/open-page/the-sorry-tale-of-the-milch-animals/article6578663.ece.

[113] See: https://www.thehindu.com/news/cities/Vijayawada/usage-of-banned-oxytocin-drug-rampant-in-dairying/article4723934.ece; https://www.thehindu.com/features/magazine/whats-behind-that-glass-of-milk/article4675921.ece; https://www.theguardian.com/commentisfree/2017/mar/30/dairy-scary-public-farming-calves-pens-alternatives.

[114] https://www.thehindu.com/features/magazine/whats-behind-that-glass-of-milk/article4675921.ece.

"unfit" for milking purposes, they are certified as fit for slaughter.[115] Consequentially, dairy industry primarily harbors the bovine meat (beef) industry which generates huge sums of revenue for the country by exporting the tabooed beef overseas.

Although 28 of the 29 states in India have some form of restriction on cow slaughter and cow exports,[116] India is still the leading beef exporter and one of the largest market shareholders of beef produced globally.[117] Amidst the slaughter bans, beef exports have significantly increased, due to which India annually draws over four billion USD from exporting beef to over 65 countries.[118] The hypocrisy of the nation's domestic and international economic policies is quite apparent.[119] On one end, the administration has condemned the local consumption of beef. On the other end, the administration has ramped up beef production for overseas consumption. The result is India earning a staggering INR 25,168.33 crores in revenues from exporting beef in 2018–2019 alone.[120]

While cows are often spared due to religious sentiments and state laws, most beef meat comes from water buffaloes mercilessly slaughtered and incessantly traded.[121] The rest come from cows and other bovines slaughtered in illegal abattoirs that have alarmingly multiplied in unsupervised markets.[122] Unlike what appears to be a highly scrutinized cattle protection sector (at least on paper), the buffalo slaughter industry is a huge cash cow for India.[123] Due to this, buffaloes' well-being and medical care are disregarded, and the large-scale slaughtering of the animals goes unregulated. The informality of the carabeef sector leads to legal regulations (requiring sanitation and humane conditions) during the slaughtering of buffaloes to be widely ignored.[124]

---

115 https://www.hrw.org/sites/default/files/report_pdf/india0219_appendix_1.pdf; https://cjp.org.in/cow-slaughter-prevention-laws-in-india/.

116 https://www.thehindu.com/features/magazine/whats-behind-that-glass-of-milk/article4675921.ece; https://cjp.org.in/cow-slaughter-prevention-laws-in-india/; https://www.hrw.org/sites/default/files/report_pdf/india0219_appendix_1.pdf.

117 https://www.bbc.com/news/world-asia-india-40565457; https://www.theatlantic.com/business/archive/2015/02/selling-the-sacred-cow-indias-contentious-beef-industry/385359/.

118 https://www.theatlantic.com/business/archive/2015/02/selling-the-sacred-cow-indias-contentious-beef-industry/385359/.

119 https://www.theatlantic.com/business/archive/2015/02/selling-the-sacred-cow-indias-contentious-beef-industry/385359/; https://www.statecraft.co.in/article/an-overview-of-india-s-beef-export-industry.

120 https://www.statecraft.co.in/article/an-overview-of-india-s-beef-export-industry.

121 See: https://www.bbc.com/news/world-asia-india-40565457; https://www.statecraft.co.in/article/an-overview-of-india-s-beef-export-industry; https://www.theatlantic.com/business/archive/2015/02/selling-the-sacred-cow-indias-contentious-beef-industry/385359/.

122 https://www.thequint.com/explainers/cattle-market-rules-are-farmers-bane-but-its-more-complicated-than-that-detailed-q-and-a#read-more;https://www.theatlantic.com/business/archive/2015/02/selling-the-sacred-cow-indias-contentious-beef-industry/385359/; Uttara Kennedy, Arvind Sharma, & Clive Phillips (2018). The Sheltering of Unwanted Cattle, Experiences in India and Implications for Cattle Industries Elsewhere, *Animals: An Open Access Journal from MDPI*, 8(5), 64, DOI: 10.3390/ani8050064.

123 See: Livestock and Products Semi-annual – 2021, USDA March 2021 Report. https://apps.fas.usda.gov/newgainapi/api/Report/DownloadReportByFileName?fileName=Livestock%20and%20Products%20Semi-annual_New%20Delhi_India_03-01-2021.

124 See: Yamini Narayanan (2018). Cow Protection as 'Casteised Speciesism': Sacralisation, Commercialisation and Politicisation, *South Asia: Journal of South Asian Studies*, 41(2), 331–351, DOI: 10.1080/00856401.2018.1419794 (pp. 341–351)

Additionally, the polarized protection of cattle has created a corrupt market where illegal abattoirs function in an unscrupulous manner. The state laws on cow protection are anything but uniform. State laws regulating cow slaughter vary in scope and stringency, leading to conflicting law enforcement, consequential illegal trafficking, and illegal cattle slaughter.[125] In the article titled "Dark Chronology of India's Cow Slaughter Laws," writer Aakar Patel summarizes it as *It is no exaggeration to say that today nobody fully understands India's cow slaughter laws. There is confusion arising from a full ban in some states to no ban in some and partial bans in others. There are bans on beef-eating in some states, but not in other states. There are bans on the slaughter of some species of bovine but not others, on some cattle of a particular age, but not others, on some cattle of a particular gender but not the rest. There are bans on beef possession in some states and no clarity on this in other states, bans on transport in some places, regulations on time and temperature, and carrying conditions elsewhere. There is little logic or coherence to any of it.*[126] Since there are so many conflicting cow slaughter laws, it prevents a regulated, humane disposition of overused cattle deemed an economic burden to the milking industry.[127] Hence, "unproductive cows" and other bovine animals are released into Gaushalas or abandoned in public roads, where they are treated as nuisance animals and subjected to harsh conditions, acute starvation, and dehydration.[128]

Ensuingly, the cow slaughter bans leave large vacuums, leading to milch cows and milch buffaloes, first being used as reproductive and dairy machines and subsequently trafficked to slaughterhouses in states where cow protection laws partially exist or are simply non-existent.[129] Consequently, the slaughter ban laws and generous subsidies to the conjoined industries further breeds unregulated abattoirs.[130] Thus, the output-driven governmental policies undercut Animal Law's ability to enforce cattle welfare laws and remove farm animal welfare from the economic purview.

---

125 See: https://www.thehindu.com/features/magazine/whats-behind-that-glass-of-milk/article4675921.ece; Uttara Kennedy, Arvind Sharma, & Clive Phillips (2018). The Sheltering of Unwanted Cattle, Experiences in India and Implications for Cattle Industries Elsewhere, *Animals: An Open Access Journal from MDPI*, 8(5), 64. DOI: 10.3390/ani8050064; https://www.thehindu.com/features/magazine/whats-behind-that-glass-of-milk/article4675921.ece

126 https://www.article-14.com/post/the-dark-chronology-of-india-s-cow-slaughter-laws.

127 https://www.article-14.com/post/the-dark-chronology-of-india-s-cow-slaughter-laws.

128 Uttara Kennedy, Arvind Sharma, & Clive Phillips (2018). The Sheltering of Unwanted Cattle, Experiences in India and Implications for Cattle Industries Elsewhere, *Animals: An Open Access Journal from MDPI*, 8(5), 64. DOI: 10.3390/ani8050064 (PART 4. Welfare Issues for India's Surplus Cattle); How India's sacred cows are creating havoc on the streets. Made for Minds, Asia, https://www.dw.com/en/how-indias-sacred-cows-are-creating-havoc-on-the-streets/a-46771484; Nimisha Jaiswal, Should India get its cows off the streets? *Global Post*. November 15th, 2015, https://www.pri.org/stories/2015-11-15/should-india-get-its-cows-streets.

129 See: https://www.thehindu.com/features/magazine/whats-behind-that-glass-of-milk/article4675921.ece; Uttara Kennedy, Arvind Sharma, & Clive Phillips (2018). The Sheltering of Unwanted Cattle, Experiences in India and Implications for Cattle Industries Elsewhere, *Animals: An Open Access Journal from MDPI*, 8(5), 64. DOI: 10.3390/ani8050064/; https://www.theatlantic.com/business/archive/2015/02/selling-the-sacred-cow-indias-contentious-beef-industry/385359/.

130 See: https://www.thehindu.com/features/magazine/whats-behind-that-glass-of-milk/article4675921.ece; Also see: https://asiaconverge.com/2020/04/for-milk-subsidies-is-the-path-to-perdition/; https://swarajyamag.com/politics/beef-industry-in-india-political-mottos-notwithstanding-pink-revolution-continues-to-get-redder.

In this fashion, the Indian government implicitly incentivizes the dairy industry to dispose of unwanted cattle in legal and illegal slaughterhouses.[131] The unwavering governmental support gives the dairy industry an unethical leg up against its counterpart and animal welfare.

The resulting illegal bovine smuggling business is a billion-dollar transboundary mafia operation where bovine animals are transported under egregious conditions for slaughter.[132] For instance, thousands of cattle are illegally smuggled and transported every year for Bangladesh's mass massacre and meat consumption purposes.[133] Moreover, the country's stringent transport rules are blatantly violated in clandestine trafficking operations where bovine animals, including calves and other bovines, some drowned, pregnant, or otherwise incapacitated, are all shoved into overcrowded trucks.[134] The livestock animals are moved across state and national borders without regard to their need for food, water, or bodily urges, often crushing and killing the weaker animals during their ghastly and cramped journey to slaughterhouses.[135] Bovines are often prodded with sharp objects to forcefully squeeze them into trucks, violating animal welfare laws, rules, and legally permitted transport limits.[136] Further, red chili powder is often rubbed into their eyes, ears, genitals, and open wounds to make bovines yield to such abhorrent conditions.[137] By the time these bovine animals reach their next destination of torture, most would have succumbed to the unbelievable conditions or end up with broken tails, limbs, and stabbed body parts.[138]

### 5.3.1 The Disregarded Bovine Animals

The oppression of vulnerable animals occurs by reaffirming patriarchal views on farm animals, where sexual and reproductive violence is repetitively inflicted on livestock animals based on their breed, age, and gender.[139] Bullocks are castrated to suppress their natural urges and over 14 million bullocks are used to drag carts across

---

131 https://www.thehindu.com/features/magazine/whats-behind-that-glass-of-milk/article4675921.ece, https://www.downtoearth.org.in/indepth/the-meat-you-eat-13283; https://www.statecraft.co.in/article/an-overview-of-india-s-beef-export-industry.

132 https://www.fortuneindia.com/polemicist/the-near-billion-dollar-trade-behind-cattle-violence-in-india/102956;https://indianexpress.com/article/opinion/columns/india-bangladesh-cattle-smuggling-border-west-bengal-6515253/.

133 Uttara Kennedy, Arvind Sharma, & Clive Phillips (2018). The Sheltering of Unwanted Cattle, Experiences in India and Implications for Cattle Industries Elsewhere, *Animals: An Open Access Journal from MDPI*, 8(5), 64, DOI: 10.3390/ani8050064.

134 Yamini Narayanan (2018). Cow Protection as 'Casteised Speciesism': Sacralisation, Commercialisation and Politicisation, *South Asia: Journal of South Asian Studies*, 41(2), 331–351, DOI: 10.1080/00856401.2018.1419794.

135 Id.

136 Id.

137 Id.

138 Id.

139 Yamini Narayanan (2018). Cow Protection as 'Casteised Speciesism': Sacralisation, Commercialisation and Politicisation, *South Asia: Journal of South Asian Studies*, 41(2), 331–351, DOI: 10.1080/00856401.2018.1419794.

India.[140] The tamed buffaloes and bullocks are also extensively used as draught animals to plow agricultural lands, and these male bovines are further bullied for recreational animal sports in India.

Anthropocentrism, sectarianism, and speciesism reflect through the commercial objectification of bovine bodies in India, where caste-based hierarchy is imposed on livestock animals.[141] In the article titled *Cow Protection as Casteised Speciesism*, Yamini Narayan demonstrates how cows are regarded as sacred animals. Nevertheless, the cow protection laws are covertly used to further the Hindu majority's political objectives and commercial interests.[142] Although Article 48 seeks the protection of all bovines, the cow protection laws permit the categorization of bovines into three different classes, where non-milch, non-native cows are treated as an inferior class. Therefore, buffaloes, cross-bred cows, Jersey cows, and non-bovine animals are openly exploited in animal agriculture.[143] They are considered to belong to the "lowest classes" and as "impure animals"—not worthy of protection from slaughter prohibition laws.[144] This approach reflects the casteism-based discrimination that the Hindus denyingly follow toward vulnerable bovines and marginalized communities.[145] Buffaloes, cross-bred cows, and non-native bovines are economically valuable in the dairy and meat production industry; hence the casteized derecognition of their experiences serves capitalistic interests.[146] Furthermore, what seems to be the sparing of cows in the meat industry is overcompensated by the extensive milking of she-buffaloes and killing four times as many buffaloes as cows and bullocks combined.[147]

Narain rightly suggests that the cow slaughter bans do not concern themselves with improving animal welfare standards or the living conditions of the "sacred animals."[148] Instead, the prohibitions on cattle slaughter appease the power-wielding religion's traditional values and political objectives. Thus, despite the passing of numerous cow protection laws and majoritarian policies, animal welfare standards have not been maintained even for handling cow progeny. Rather, these purported

---

140 https://www.article-14.com/post/the-dark-chronology-of-india-s-cow-slaughter-laws; Bullock cart in its new avatar, DowntoEarth, 2004 Report. https://www.downtoearth.org.in/coverage/bullock-cart-in-its-new-avatar-12232. The report highlights *that Bullock carts account for 56 percent of the country's transport of goods and personnel. Also, over 80 percent of farm produce is transported in animal-drawn carts.*

141 See: Yamini Narayanan (2018). Cow Protection as 'Casteised Speciesism': Sacralisation, Commercialisation and Politicisation, *South Asia: Journal of South Asian Studies*, 41(2), 331–351, DOI: 10.1080/00856401.2018.1419794 (p. 341); Also see: https://www.article-14.com/post/the-dark-chronology-of-india-s-cow-slaughter-laws.

142 Id.

143 Id.

144 Id.

145 Id.

146 Also see: Rakhi Jagga, Explained: The American vs. Indian cow debate. *The Indian Express*, on September 26th, 2019. https://indianexpress.com/article/explained/explained-the-american-vs-indian-cow-debate-6029873/.

147 https://www.article-14.com/post/the-dark-chronology-of-india-s-cow-slaughter-laws.

148 Yamini Narayanan (2018). Cow Protection as 'Casteised Speciesism': Sacralisation, Commercialisation and Politicisation, *South Asia: Journal of South Asian Studies*, 41(2), 331–351, DOI: 10.1080/00856401.2018.1419794.

cow protection laws harbor sectarianism, casteism, speciesism, and communal disharmony.[149]

Animal agriculture and the dairy industry thrive by separating mother bovines from their children and keeping overworked, abused, and traumatized cattle in distress conditions throughout their lifecycles.[150] Yamini Narayan's ethnographic study of Gaushalas in India documents how mother cows are further exploited after the dairy industry abandoning them for becoming "unproductive," "uneconomic," and "useless."[151] The term Gaushalas gives the impression that they take care of the old, sick, and overused cows in "animal sanctuaries." In contrast, they continue subjecting cows to zoo psychosis and forced motherhood.[152] Gaushalas also reinforce the casteized cattle hierarchy by not taking in other bovine and non-bovine animals that are equally in need of care and shelter.[153] In Narayan's article on Gaushalas rendering selective cow protection, Narayan summarizes how Gaushalas replicate casteized speciesism as *This speciesist appeal to motherhood underpins gaushalas' refusal to shelter other mothers enslaved in dairying in India, such as buffaloes, goats, sheep, and camels.*[154]

Thus, the Speceists' objectification of animals as livestock or sacred animals does not mitigate the pain and suffering the livestock animals go through. Instead, it sanctifies violence inherent in animal agriculture, where the utilitarian economic model overlooks the acute emotional and physical trauma inflicted on bovines throughout their lifecycle.[155]

## 5.4 BULLS AND BUFFALOES ARE THE UNWILLING PARTICIPANTS IN ANIMAL SPORTS OF INDIA: THE NULLIFICATION OF A LANDMARK JUDGMENT

As seen, the legal journey of cattle protection laws has been an uphill battle where a volatile political atmosphere gives way to religious and cultural practices of the majority religion.

In 2009, the Tamil Nadu (T.N.) government passed the Tamil Nadu Regulation of Jallikattu Act, which broadly permitted bull racing and bull-taming practices.[156]

---

149 https://www.article-14.com/post/the-dark-chronology-of-india-s-cow-slaughter-laws.

150 Id.

151 Yamini Narayanan (2019). "Cow Is a Mother, Mothers Can Do Anything for Their Children!" Gaushalas as Landscapes of Anthropatriarchy and Hindu Patriarchy, *Hypatia*, 34(2), 195–221. DOI: 10.1111/hypa.12460.

152 Id.

153 Id. Also see: Rakhi Jagga, Explained: The American vs Indian cow debate. *The Indian Express*, on September 26th, 2019. https://indianexpress.com/article/explained/explained-the-american-vs-indian-cow-debate-6029873/.

154 Yamini Narayanan (2019). "Cow Is a Mother, Mothers Can Do Anything for Their Children!" Gaushalas as Landscapes of Anthropatriarchy and Hindu Patriarchy, *Hypatia*, 34(2), 195–221. DOI: 10.1111/hypa.12460.

155 See: Yamini Narayanan (2018). Cow Protection as 'Casteised Speciesism': Sacralisation, Commercialisation and Politicisation, *South Asia: Journal of South Asian Studies*, 41(2), 331–351, DOI: 10.1080/00856401.2018.1419794.

156 http://www.bareactslive.com/TN/tn984.htm.

In 2011, the MoEF included bulls in the animals protected from the commercial exhibition and forced performances list.[157] The Ministry clarified that Jallikattu, bullock-cart racing, and all other forms of exhibition, training, or performances of bulls are illegal everywhere in India.[158]

In *AWBI V. A. Nagaraj*, the Supreme Court upheld animal rights by interpreting the Constitution, nation's laws, and societal values.[159] The Court upheld the Central Government's Order and struck down the Tamil Nadu Regulation of Jallikattu Act of 2009. Further, the Court banned similar animal sports due to the unreasonableness of the religious and cultural connotations against animals.[160] By considering animal behavior and animal anatomy, the Supreme Court recognized the inherent cruelty in engaging bulls in events like the Jallikattu.[161] The Court reasoned that the bulls are not innately or anatomically suited to engage in such performing activities. Hence, forcefully using them to participate inflicts bulls with unnecessary pain and suffering.[162] By finding so, it held that forcing bulls to work as performing animals was "inherently cruel" under the tenets of the PCA and as a direct violation of Section 3 and Section 11(1) of the PCA.[163]

The *AWBI V. A. Nagaraj* decision was the first progressive animal law judgment on many fronts. The Apex Court, for the first time, included the protection of animal life within the ambit of the Fundamental Right to life under Article 21 of the Constitution.[164] The Supreme Court interpreted compassion for all living creatures under Article 51A(g) to mean having concern for animals' pain, suffering, and well-being.[165] What followed the landmark ruling was political and social unrest in Tamil Nadu and repeated challenges by the state government to the Supreme Court's 2014 Judgment.[166] The supporters of animal sports undermined the Rule of Law with massive protests and refused to adhere to the binding Apex Court's ruling.[167] A review petition was filed by the T.N. government urging the Supreme Court to reconsider its 2014 Judgment.[168] In November 2016, the Court dismissed the state's

157 http://www.bwcindia.org/web/Info&Action/Legislation/ClarificationonBullsunderbanonperformin ganimalsNotification.pdf.

158 http://www.bwcindia.org/web/Info&Action/Legislation/ClarificationonBullsunderbanonper forminganimalsNotification.pdf.

159 Animal Welfare Board of India vs. A. Nagaraja & Ors, Judgment dated 7th May 2014 in Writ Petition (C) No.145 of 2011. https://indiankanoon.org/doc/88304811/?type=print.

160 Animal Welfare Board of India v. A. Nagaraja (2014) 7 SCC 547.

161 Id.

162 Id.

163 Id.

164 Id.

165 Id.

166 https://www.indiatoday.in/india/story/tamil-nadu-jallikattu-protests-madurai-coimbatore-954536-2017-01-12; https://indianexpress.com/article/explained/jallikattu-protests-in-tamil-nadu-and-the-many-legal-twists-and-turns-4482563/.

167 Id.

168 https://www.hindustantimes.com/india-news/protests-erupt-in-tn-over-sc-refusal-to-expedite-jallikattu-order/story-LcPXdC0fTzC7fKQYBDqUAP.html.

plea by rejecting their repetitive argument that Jallikattu was a religious and cultural right.[169] The Supreme Court resisted uprooting its holding by dismissing the T.N. government's arguments as untenable by reasoning that the cultural practice cannot be indulged merely because it was a century-old tradition.[170] The top Court reiterated that inflicting cruelty on bulls is not necessary to practice one's rights and reiterated that Animal Welfare Law supersedes age-old cultural and traditional practices.[171]

Giving in to this political pressure, the MoEF, on January 7, 2016, passed a contrary notification that enabled bull races, bullfights, and events like Jallikattu with certain restrictions.[172] Soon after that, animal welfare petitioners filed a Writ Petition before the Supreme Court, seeking the quashing of MoEF notification and compliance with the law laid down in *AWBI V. Nagaraj*.[173] Subsequently, the 2017 T.N. amendment removed Jallikattu from the PCA's radar by explicitly protecting it from legal challenges.[174] In the wake of the T.N.'s state amendment to the PCA, the Central Government withdrew its January 7th notification that had acted as an interim solution to the political unrest situation.[175] Despite the matter being sub-judice before the Constitutional Bench of the Supreme Court, the principles of separation of power were violated by the Executive body of India and the state Legislative body of Tamil Nadu.[176] With the prior consent of the President, the Governor of the state promulgated this ordinance—amending the PCA to permit the conduct of Jallikattu events in Tamil Nadu.[177]

The ordinance was immediately replaced by a state legislative bill, passed with bipartisan support in the state assembly to exempt Jallikattu from the Central animal protection Act's purview.[178] By amending the central legislation on animal welfare and gaining the assent of the President, the state legislation hurriedly overturned the Supreme Court ban and Central Law itself.[179] First, it amended Section 3 of the PCA to permit Jallikattu to be conducted, subject to the state's rules and regulations. Second, the amended Act adds the conduct of Jallikattu to the list of excep-

---

169 https://indiankanoon.org/doc/43612276/; https://indiankanoon.org/doc/146430583/; https://economictimes.indiatimes.com/news/politics-and-nation/sc-refers-pleas-against-jallikattu-to-constitution-bench/articleshow/62752587.cms?from=mdr; https://www.thehindu.com/news/national/Jallikattu-ban-SC-dismisses-T.N.-plea/article16643787.ece.

170 https://indiankanoon.org/doc/146430583/; https://www.news18.com/news/india/the-jallikattu-ban-what-are-the-options-now-1337569.html.

171 https://indiankanoon.org/doc/146430583/; https://www.news18.com/news/india/the-jallikattu-ban-what-are-the-options-now-1337569.html.

172 Notification No. G.S.R.13(E) dated January 07th, 2016, published by the Union of India, in The Gazette of India.

173 See: https://thewire.in/culture/jallikattu-supreme-court-2.

174 https://cms.tn.gov.in/sites/default/files/acts/20_Ex_IV_2.pdf.

175 https://indiankanoon.org/doc/45344307/; https://www.firstpost.com/india/jallikattu-case-centre-to-withdraw-january-2016-notification-days-ahead-of-sc-judgement-on-it-3218192.html.

176 Id.

177 https://www.thehindu.com/news/national/tamil-nadu/Tamil-Nadu-ordinance-to-permit-jallikattu-an-explainer/article17082141.ece2; https://www.indialegallive.com/top-news-of-the-day/analysis-updates/what-a-reprieve/.

178 https://www.thehindu.com/news/national/tamil-nadu/Jallikattu-Act-passed-unanimously/article17084322.ece.

179 Id.

tions that are not legally deemed as animal cruelty under Section 11 of the PCA.[180] Third, it amended Section 22 that restricted using certain animals like the bulls as performing animals, and now Section 22 is inapplicable to Jallikattu.[181] Fourth, it added Jallikattu to the list of "exemptions" from the rules that barred using some animals as performing animals. The other exemptions include using animals for scientific, education, and law enforcement purposes by the police and military.[182] Lastly, it added Section 28A to provide blanket legal protection to the conduct of Jallikattu by saying that nothing in the Act would apply to Jallikattu.[183] Thereby, the (2017 Tamil Nadu Amendment) Act of the PCA and its (Conduct of Jallikattu) Rules of 2017 once again legalized the bull-taming sport by disregarding the 2014 ban by the Supreme Court.

These amendments to the Central Law have been challenged before the Supreme Court.[184] The principal petitioner, the AWBI, had initially challenged the T.N. amendment as ultra vires to the 1960 Act and the Constitution by arguing that only the Central Government is vested with the power to frame rules under Section 38 of the PCA. Hence, the T.N. amendment permitting Jallikattu under Section 3(2) of the T.N. Act was invalid and inapplicable.[185] The AWBI, in its application, pointed out that the T.N. amendment was a surreptitious way of disrespecting the Supreme Court's 2014 Judgment, whereby a binding Apex Court ruling was nullified by enacting state legislation that substantially overrules the intended effect of the judgment.[186] Subsequently, an AWBI chairman, who also happened to be the Special Secretary (Forest) in the MoEF, told an Indian press agency that the statutory body would not be challenging the T.N. amendment.[187] The AWBI is an autonomous body affiliated with the MoEF, where it advises the Central Government on adequately implementing the 1960 Act.[188] The animal welfare activists also assailed the amendments as a fraud on the Constitution, passed by the state to patently bypass the Apex Court's 2014 holding the bull-taming sport as illegal.[189]

The state created a contrary state law by a "State amendment" to the Central Act as the PCA falls under Entry 17 of the Concurrent List of the Constitution.[190] Since the Act is in the Concurrent List, both the State governments and Central

---

180 https://www.thehindu.com/news/national/tamil-nadu/Tamil-Nadu-ordinance-to-permit-jallikattu-an-explainer/article17082141.ece2.

181 Id.

182 Id.

183 Id.

184 Id; https://thewire.in/culture/jallikattu-supreme-court-2.

185 https://www.thehindu.com/news/national/tamil-nadu/Tamil-Nadu-ordinance-to-permit-jallikattu-an-explainer/article17082141.ece2; https://thewire.in/culture/jallikattu-supreme-court-2.

186 https://www.thenewsminute.com/article/awbi-still-opposing-jallikattu-supreme-court-divided-within-56583.

187 https://www.thenewsminute.com/article/awbi-still-opposing-jallikattu-supreme-court-divided-within-56583.

188 https://www.thenewsminute.com/article/awbi-still-opposing-jallikattu-supreme-court-divided-within-56583.

189 Krishnadas Rajagopal & Jacob Koshy, T.N.'s new jallikattu law faces challenge in Supreme Court. *The Hindu*, January 26th, 2017. https://www.thehindu.com/news/national/tamil-nadu/T.N.%E2%80%99s-new-jallikattu-law-faces-challenge-in-Supreme-Court/article17092209.ece.

190 List III, Concurrent List of The Indian Constitution. https://www.mea.gov.in/Images/pdf1/S7.pdf.

Government have concurrent powers to enact laws on the common subject—the PCA.[191] Generally, under Article 254(1) of the Indian Constitution, when there is a conflict between the Central Law and the State Law, the provisions of the Central Law must prevail.[192] Nevertheless, the State Law can still prevail over the Central Law if the State Law procures the President's assent as laid down under Article 254(2).[193] Here the President intervened under Article 213 of the Constitution by encouraging the T.N. state governor's power to promulgate an Ordinance permitting Jallikattu.[194] Subsequently, the President also assented to the contrary state law under Article 254(2), which was passed by the state assembly to bypass the 2014 ban on Jallikattu.[195] Hence, by nullifying the 2014 ban on animal cruelty, the country leaders demonstrated that the rule of law and ethical and constitutional principles could be sacrificed and circumvented to uphold ancient animal cruelty practices.

In the name of culturally preserving domesticated bulls, mental and physical harassment of the bulls is highly prevalent.[196] Undercover investigations of the cultural sport have consistently revealed that bulls are often inflicted with barbaric cruelty during Jallikattu events.[197] At the end of such mistreatment, bulls have open wounds, lacerated bodies, broken tails, and more. During these bull races, bulls are pushed beyond the point of exhaustion, leading to the deaths of bulls, participants, and the audience.[198] Yet, throughout this period, the T.N. government has defended Jallikattu as a cultural practice that promotes age-old traditional practices and projects it as a means of ensuring the preservation of native cattle breeds.[199]

Following the suit of T.N., Karnataka and Maharashtra's passed similar amendments to the PCA, allowing bullock-cart races within the state.[200] Thus, the states have pushed Jallikattu and bullock-cart races as their religious and cultural rights and demanded protection under Articles 25 and 29(1) of the Indian Constitution.[201] Whether engaging in such animal sports is to be protected as a constitutionally protected cultural rights of Indian citizens is pending before the Constitutional Bench of

---

191 The State government may amend central laws or have a separate set of laws on the common subject so long as they comply with procedural requirements, with a few limitations on their power.

192 https://indiankanoon.org/doc/344383/.

193 https://indiankanoon.org/doc/1930681/.

194 https://indiankanoon.org/doc/860592/.

195 https://www.thehindu.com/news/national/tamil-nadu/Jallikattu-Act-passed-unanimously/article17084322.ece.

196 See: https://www.petaindia.com/features/new-jallikattu-investigation/; https://www.indiatoday.in/india/story/peta-india-expose-shows-jallikattu-bulls-tortured-calls-for-ban-yet-again-1542815-2019-06-04; https://www.thenewsminute.com/article/armed-evidence-cruelty-2018-jallikattu-peta-seeks-ban-sport-77444.

197 Id.

198 Id.

199 https://www.indialegallive.com/top-news-of-the-day/analysis-updates/what-a-reprieve/; https://thewire.in/culture/jallikattu-supreme-court-2.

200 https://economictimes.indiatimes.com/news/politics-and-nation/maharashtra-passes-legislation-to-resume-bullock-cart-race/articleshow/58050560.cms?from=mdr.

201 Arun Janardhanan, Chennai, Jallikattu in Tamil Nadu: pride & politics. *Indian Express*, Updated: January 18th, 2021; Ahead of Tamil Nadu Assembly polls, Jallikattu has become the subject of politics between Congress and BJP. A look at what it means for Tamil culture and the controversies leading to an ongoing legal battle. https://indianexpress.com/article/explained/jallikattu-tamil-nadu-assembly-election-bjp-congress-7148150/.

the Supreme Court.[202] The Apex Court is now considering whether the new laws align with the principles of the central legislation animal welfare standards while catering to communal interests.[203] Given the inherent cruelty in using animals for racing, the question of whether the animal sport permitting laws frustrate the Central Act's objective of preventing cruelty to animals can be answered in the affirmative. Considering the circumvention of the landmark judgment, the Constitutional Bench should hold these amendments as not in consonance with the basic tenets of the 1960 Act. [204]

The supremacy of the central legislation over the conflicting state amendments is yet to be struck down by the Supreme Court. Meanwhile, Jallikattu participants blatantly disregard even the recent guidelines that conditioned the conduct of Jallikattu with the humane treatment of bulls.[205] Due to the political, religious, and unsettled legal issues, Jallikattu events are being conducted unfettered.[206] In 2019 alone, a staggering number of 1,353 bulls were compelled by over 500 bull tamers who had promised to implement improvements in animal welfare standards.[207]

At the time this Chapter is going for publication, the batch of petitions have been reserved for judgment by the Constitutional Bench.[208] The lawyer for the petitioners argued that the new laws are patently against the principles of the animal protection legislation.[209] The Tamil Nadu's legal counsel argued that since PCA permits animal cruelty when its necessary for humans, Jallikattu is also legally justifiable because certain regional communities consider the bull taming sport as a cultural practice.[210] Since PCA only criminalizes and penalizes "unnecessary infliction of pain and suffering" the domestication of cattle for human beings' economic use, religious beliefs, food habits and cultural practices is therefore acceptable.[211]

---

202 Krishnadas Rajagopal, Supreme court refers jallikattu challenge to constitution bench. *The Hindu*, Updated: February 02nd, 2018. https://www.thehindu.com/news/national/sc-refers-jallikattu-challenge-to-constitution-bench/article22630214.ece; https://www.latestlaws.com/latest-news/jallikattu-a-cultural-right-or-not-now-the-constitution-bench-to-decide-the-issue/.

203 https://www.thehindu.com/news/national/Jallikattu-ban-SC-dismisses-T.N.-plea/article16643787.ece.

204 Id.

205 See: http://awbi.in/awbi-pdf/guidelines_jallikattu_jan2018.pdf; Also see: https://www.hindustantimes.com/india-news/peta-urges-tamil-nadu-government-to-withdraw-jallikattu-order/story-SCPKLIHTo8sIy8I10R7A2L.html.

206 Sanya Dhingra, Jallikattu begins today — the 'barbaric sport' that celebrates hard-working Tamil farmer. *The Print*, January 15th, 2020. https://theprint.in/theprint-essential/jallikattu-begins-today-the-barbaric-sport-that-celebrates-hard-working-tamil-farmer/349508/; Also see: https://www.hindustantimes.com/india-news/peta-urges-tamil-nadu-government-to-withdraw-jallikattu-order/story-SCPKLIHTo8sIy8I10R7A2L.html.

207 Shubhendu Shekhar, Jallikattu: A tradition of torture. U.K. Centre for Animal Law. On October 20th, 2019. https://www.alaw.org.uk/blog/2019/10/20/jallikattu-a-tradition-of-torture-by-shubhendu-shekhar/.

208 https://www.thestatesman.com/india/2023-awaits-judgments-likely-to-impact-govts-decision-making-process-1503142724.html

209 https://www.livelaw.in/top-stories/supreme-court-constitution-bench-laws-permitting-jallikattu-214971

210 https://www.livelaw.in/top-stories/animals-have-no-rights-under-constitution-jallikattu-not-a-mere-sport-has-a-societal-purpose-tn-govt-tells-supreme-court-215561

211 See Id.

Thus, the State's utter contempt for the highest court's opinion is just so that its people can continue the forceful domestication of bulls in the name of cultural sports and regional pride. To surmise, despite the landmark judgment, the broad framework of the PCA, animal welfare guidelines, and the most recent beef ban laws, there has been no measurable improvement in the treatment of cows, cattle, or other animals used as livestock and for animal sports in India.[212]

### 5.4.1 Animal Agriculture Is the Key Contributor to Environmental Pollution and Climate Change

The adverse impact of industrial animal agriculture on the environment and worldwide communities is well documented.[213] The dairy industry is one of the leading causes for climate change and environmental pollution.[214] India is the third-largest GHG emitter in the world. A recent study indicates that the Indian livestock industry can significantly increase global temperatures due to rapid and unsustainable growth.[215] Cattle and buffaloes engaged in animal agriculture already contribute to over 74 percent of the country's methane gas emissions.[216]

The cow protection laws are silent on the dairy industry's adverse environmental impact and climate change. Animal farming puts an enormous strain on the country's heavily polluted and scarce water resources and contributes to over 74 percent of India's methane emissions.[217] The shortsighted animal agriculture practices significantly contribute to an impending water crisis in India, where groundwater resources and aquifers are fast depleting.[218] Despite this, the National Policy for the dairy industry hyper focuses on increasing milk production.[219] The result is the unprecedented consumption of animal products and a resulting disregard for animal

---

212 https://helpanimalsindia.org/news/library/restoring-the-sacredness-of-indias-cows; https://www.petaindia.com/features/latest-jallikattu-investigation-proves-laws-fail-bulls/.

213 https://www.newindianexpress.com/opinions/2020/nov/07/factory-farming-of-animals-and-climate-change-2220463.html.

214 The Food Print of the Dairy. https://foodprint.org/reports/the-foodprint-of-dairy/.

215 Nanditha Krishna, Factory farming of animals and climate change. *Indian Express*, November 07[th], 2020. https://www.carbonbrief.org/the-carbon-brief-profile-india.

216 https://www.carbonbrief.org/the-carbon-brief-profile-india; Shilpi Kumari, Moonmoon Hiloidhari, Nisha Kumari, Satya Narayan Naik, & Raj Pal Dahiya (2018). Climate Change Impact of Livestock $CH_4$ Emission in India: Global Temperature Change Potential (GTP) and Surface Temperature Response, *Ecotoxicology and Environmental Safety*, 147, 516–522, ISSN 0147-6513, DOI: 10.1016/j.ecoenv.2017.09.003; Somnath Singh. Methane from Indian livestock adds to global warming. SciDev.Net's, on 17th January 2017. https://www.scidev.net/asia-pacific/news/methane-from-indian-livestock-adds-to-global-warming/.

217 https://www.downtoearth.org.in/blog/climate-change/climate-crisis-way-forward-for-dairy-giants-in-india-78226;https://www.scidev.net/asia-pacific/news/methane-from-indian-livestock-adds-to-global-warming/; https://www.newindianexpress.com/opinions/2020/jan/15/why-we-should-ban-meat-exports-2089635.html.

218 https://www.downtoearth.org.in/blog/climate-change/climate-crisis-way-forward-for-dairy-giants-in-india-78226; https://www.globalcitizen.org/en/content/india-farmers-water-loss-food-security/.

219 https://economictimes.indiatimes.com/news/economy/agriculture/economic-survey-milk-production-rises-by-five-percent-to-198-4-million-tonnes-in-2019-20/articleshow/80585416.cms?from=mdr; https://www.fao.org/3/i0588e/I0588E05.htm.

welfare and the associated environmental impact.[220] Large-scale confinement of cattle not only harms animals' well-being, but the residual dairy waste is a severe health hazard to the public as well.[221]

Despite being the largest employer of livestock animals, the country has not yet adopted a comprehensive climate change mitigation policy or environmental risk management plan in the animal agriculture sector.[222] Animal exploitation, particularly cattle raising, contributes to the loss of natural habitats, deforestation, and climate change and indirectly leads to biodiversity extinction and the outbreak of zoonotic diseases.[223] The intensive animal labor practices aggravate the climate change crisis and further worsen the living conditions of livestock animals, which are most susceptible to the adverse effects of climate change.[224] In the absence of an appropriate disease management plan, the high livestock population density makes India susceptible to major biological threats, including agricultural infestations and pathogenic outbreaks from zoonotic diseases.[225]

## 5.5 CONCLUSION

As seen above, despite numerous cow protection laws, the cruelty inflicted on cattle is one of the most significant agrarian crimes in the country. The historical wear and tear of cow protection laws have become a political landmine issue. The Hindu majoritarian's advocacy for cow protection has caused unacknowledged distress to bovine animals while polarizing animal welfare, animal sentience, and animal protection movements in India. To date, the Indian government has paid much attention to the socio-economic gains of indulging in animal agriculture and harboring Hindus' religious sentiments. The pointless legal provisions keep up with the anthropogenic policies of the country by obscurely exploiting animals in the dairy industry—where there lies a total disregard for the well-being of milch animals. At the same time,

---

220 https://www.dairyglobal.net/industry-and-markets/market-trends/indias-demand-for-dairy-products-increasing/; Yamini Narayanan (2018). Cow Protection as 'Casteised Speciesism': Sacralisation, Commercialisation and Politicisation, *South Asia: Journal of South Asian Studies*, 41(2), 331–351, DOI: 10.1080/00856401.2018.1419794.

221 https://www.humanesociety.org/sites/default/files/docs/hsus-report-animal-welfare-cow-dairy-industry.pdf; B. V. Raghunath, A. Punnagaiarasi, G. Rajarajan, A. Irshad, A. Elango, & G. Mahesh Kumar (2016) Impact of Dairy Effluent on Environment—A Review. In: Prashanthi M., Sundaram R. (eds) *Integrated Waste Management in India. Environmental Science and Engineering*. Springer, Cham. DOI: 10.1007/978-3-319-27228-3_22.

222 Smita Sirohi & Axel Michaelowa (2007). Sufferer and Cause: Indian Livestock and Climate Change, *Climatic Change*, 85, 285–298. DOI: 10.1007/s10584-007-9241-8; https://www.scidev.net/asia-pacific/news/methane-from-indian-livestock-adds-to-global-warming/; http://content.time.com/time/world/article/0,8599,1890646,00.html.

223 See: https://www.theguardian.com/commentisfree/2017/dec/04/animal-agriculture-choking-earth-making-sick-climate-food-environmental-impact-james-cameron-suzy-amis-cameron; https://www.downtoearth.org.in/blog/agriculture/india-s-cow-crisis-a-rebuttal-to-sunita-narain-62763; https://www.fao.org/3/a0701e/a0701e.pdf; https://www.downtoearth.org.in/factsheet/how-livestock-farming-affects-the-environment-64218.

224 Id.

225 https://carnegieendowment.org/2020/12/09/biological-risks-in-india-perspectives-and-analysis-pub-83399; https://www.downtoearth.org.in/news/health/75-emerging-infectious-diseases-zoonotic-un-report-72173; https://www.statecraft.co.in/article/an-overview-of-india-s-beef-export-industry

draconian beef bans are enforced against the marginalized communities of India, causing severe social disharmony in Indian society. Due to this, protecting cattle is a frustrating event and has created a cynical approach to animal welfare and mired the animal rights movement in India. Thus, impartial legal enforcement of animal welfare standards is critical to maintaining communal peace and the moral fabric of the diverse land.

The atrocious conditions continue to prevail in animal agriculture as it churns out animal by-products by treating them as fungible and nonliving commodities. Animal welfare laws would be severely underenforced as long as we continue depending on animal meat and animal milk for our nutritional needs. Because the dairy industry is so closely tied to the slaughterhouse industry, the cow protection advocates need to take a step back to consider how dairy farming significantly pushes millions of cattle to slaughterhouses every year. In all probability, switching to a dairy and meat-free diet would lessen the excessive abuse of cattle and reduce the harmful impact of animal agriculture on the environment, climate, and public health.[226] It is time India starts paying close attention to securing animal welfare for cattle while mitigating the ecological footprint of animal agriculture. Hence, instead of resisting beneficial changes, the Indian government must transition to non-dairy alternatives and vegan meat.

An exception for utilitarianism reveals the systemic failure of animal welfare laws in protecting animals. The strident demand of Indians seeking legal protection for their archaic religious beliefs, cultural activities, and unsound animal agriculture practices often trumps the protection of the subjugated animals. The partial protection of cattle violates human rights, permits environmental pollution, and overlooks animal cruelty, all at the same time. Perhaps, it is time to rethink the cow protection policies by bearing in mind the half-baked effect of animal protection laws that causes more harm than good.

To conclude, all animals' right to moral personhood must be upheld as they are sentient, vulnerable, and valuable beings who deserve equal and equivalent rights to life. Given the vast number of ecological services animals provide and their innate capacity to feel pain, suffering, and emotions, animals especially deserve unflinching legal personhood rights. Furthermore, animal abusers' vociferous opposition to recognizing the moral rights of certain animals is evidence that legal personhood to all animals needs to be promptly conferred. Recognizing all animals' inherent and instrumental value is a pressing need to preserve the moral fabric of our society and protect public and environmental health for generations to come.

---

[226] https://www.downtoearth.org.in/news/agriculture/changing-your-diet-can-reduce-green-house-emissions-by-25--60353; Andrew Joyce, et al. (2012). Reducing the Environmental Impact of Dietary Choice: Perspectives from a Behavioural and Social Change Approach, *Journal of Environmental and Public Health*, 2012, 978672, DOI: 10.1155/2012/978672; Also see: A report from the FSSAI reveals the adulteration of the Indian dairy products posing a serious threat and causer of lifestyles diseases in humans. https://fssai.gov.in/upload/uploadfiles/files/Report_Milk_Survey_NMQS_18_10_2019.pdf.

# 6 Dairy Cows and Goats
## *Animal Welfare, Sustainability, and the Global Regulatory Environment*

*Amanda G. Verkest*
Northeastern University

## CONTENTS



"The way we eat drives ecosystems around the world."

*-Bill Keener, Sequatchie Cove Farm*
*Chattanooga, Tennessee, USA*

DOI: 10.1201/9781003137733-12

## 6.1 INTRODUCTION

Milk and dairy products are some of the most valuable agricultural commodities in the world.[1] These treasured goods provide high-quality nutrition (protein, vitamins, minerals) to people from all walks of life regardless of their socioeconomic status or geographic location. Dairy-based products are rooted in culinary delights around the world, from cottage cheese to paneer to crème brûlée, to name a few. Many cultures around the world depend on dairy for their recipes and well-being, including comfort foods and even staples for children and infants. Another aspect of the dairy industry is that dairy operations are viewed as a tangible way for individuals and families to pull themselves out of poverty in various regions across the globe.[2] Conversely, the production and transportation of these goods are proven to have significant impacts on the environment and the welfare of the ruminant animals necessary to develop them.[3] Depending on the method used, the production of animal products can ameliorate or harm global food security.[4]

### 6.1.1 The Global Dairy Sector

The Food and Agriculture Organization of the United Nations (henceforth FAO) reports that in 2019, 883 million tons of milk were produced globally, a 64 percent increase compared to 1989.[5] Milk and dairy products account for approximately 14 percent of global agricultural trade and are produced in nearly all of the worlds' countries.[6] Within the dairy sector, skim milk powder (SMP) and whole milk powder are the most globally traded agricultural commodities, while highly perishable fresh dairy products are the least traded.[7] The top ten milk-producing nations in 2019 were India, the U.S., Pakistan, Germany, the Russian Federation, Brazil, the People's Republic of China (henceforth "China"), France, Turkey, and New Zealand, with 81 percent produced by cattle, 15 percent by buffaloes, 2 percent by goats, 1 percent by sheep, and 0.4 percent by camels.[8] The majority of dairy products consumed globally are in the form of unprocessed or minimally processed fresh dairy products. Cheese consumption primarily occurs in Europe and North America, with rising consumption in other regions.[9]

Many socioeconomic benefits are offered by the dairy industry. The FAO discusses how dairy products are an affordable source of sustenance and are key to nutrition and health because they supply significant amounts of protein and

---

[1] Food and Agriculture Organization of the United Nations (2016).
[2] Ibid., p. 2.
[3] The current chapter relies heavily on "The Political Ecology of the Dairy Industry", in Steier, G., & Patel, K. K. (Eds.), *International Farm Animal, Wildlife and Food Safety Law*. This author thanks C. Gupta for her work in the field upon which we expand in this chapter.
[4] Foley et al. (2011).
[5] Food and Agriculture Organization of the United Nations (2021).
[6] Food and Agriculture Organization of the United Nations (2016).
[7] Ibid., p. 1.
[8] Food and Agriculture Organization of the United Nations (2021).
[9] Organization for Economic Co-operation and Development - Food and Agriculture Organization of the United Nations (2021).

micronutrients, including calcium, magnesium, selenium, riboflavin, and vitamins B5 and B12 (essential to reduce hunger and malnutrition, especially among children and pregnant women).[10] The global dairy sector creates over 240 million jobs, allowing dairy farmers to develop wealth and a sense of socioeconomic resilience. Notably, this sector does more than others to empower female dairy workers through the 37 million dairy farms spearheaded by women.[11] The global dairy sector plays an additional role in public health given dairy products are highly subject to zoonotic and foodborne diseases. Indeed, approximately 45 zoonotic bovine pathogens are known to present potential hazards in dairy products produced from cows,[12] and the consumption of unpasteurized raw milk and dairy products from both cows and goats can result in foodborne illnesses. Moreover, evidence suggests dairy animals with an excessive intake of antibiotics can contribute to antimicrobial-resistance in human beings.[13]

From an environmental perspective, feeding dairy animals (cows, buffaloes, goats, etc.) requires around 7 percent of the total land on earth, or 1 billion hectares (ha).[14] The global dairy herd consumes 2.5 billion tons of dry matter feed annually, approximately 40 percent of the global livestock feed intake. However, 77 percent of this material (grass and straw) is inedible to humans, so the global dairy herd is transforming these otherwise indigestible plant materials into high-quality protein and providing essential micronutrients.[15] Regardless, the materials required to produce enough feed for these animals have significant impacts on the environment, particularly because most feed is produced through monocropping. Thus, the production of dairy products requires extraordinary amounts of energy, nutrients, and water and is associated with land degradation, air and water pollution, loss of biodiversity, and deforestation.[16] Additionally, dairy animals produce approximately 3.1 gigatons of $CO_2$ equivalent per year, or 40 percent of all global livestock emissions.[17] The majority of the dairy sector's greenhouse gas (GHG) emission production is through the enteric fermentation process. Depending on the production system, 51–67 percent of all dairy emissions are enteric methane.[18] After methane is released into the air, it traps 84 times more heat in the atmosphere over the first two decades when compared to carbon dioxide. Although carbon dioxide is a longer-lived pollutant, up to 200 years atmospheric residence time, the amplified heating potential of methane is reason for concern.[19] Further, improperly managed manure on dairy farms has been known to generate water pollution that can result in eutrophication – excessive plant and algae growth that harms water quality and impacts biodiversity.[20] In the U.S., the Clean Water Act helps monitor these pollution sources, yet there are still routine

[10] Food and Agriculture Organization of the United Nations (2016).
[11] Ibid., p. 2.
[12] Ibid., p. 3.
[13] Ibid., p. 3.
[14] Ibid., p. 3.
[15] Ibid., p. 3.
[16] Ibid., p. 3.
[17] Ibid., p. 3.
[18] Ibid., p. 3.
[19] Ibid., p. 3.
[20] Steier & Patel (Eds.) (2017).

cases of eutrophication and algae blooms.[21] With all of this in mind, it is essential for the dairy industry to reduce its negative impacts on the climate through increased ruminant productivity, stronger land and manure management, improved feed efficiency, and overall strengthened herd performances via enhanced animal health and husbandry.[22]

#### 6.1.1.1 Dairy Market Projections 2021–2030

Despite being a highly perishable and vulnerable agricultural commodity, the global dairy sector was impressively resilient to supply chain disruptions and various other impacts of the COVID-19 pandemic. The chief impact was on butter prices due to loss in demand for milk fat from the hospitality sector.[23] Global butter price is forecasted to persist at post-pandemic levels and will remain significantly higher than the SMP price as it has since 2015 due to stronger demand for milk fat compared to other milk solids and the EU's SMP intervention.[24] Fresh dairy product consumption is expected to increase by 2030 due to greater demand from India, Pakistan, and Africa – countries where demand is driven by increases in income and population.[25] In Africa, cheese and whole milk powder currently account for the majority of processed dairy consumption, but SMP is expected to have the highest growth over the next decade.[26] World milk production is expected to increase at an annual rate of 1.7 percent throughout the decade, a figure higher than most other agricultural commodities.[27] The majority of growth will be seen in India and Pakistan, although their impact on the global dairy market will be minimal since very little of their production is traded internationally.[28] China is expected to remain the most important importer of milk products, despite a slight increase in domestic milk production.[29]

There is potential for international trade agreements to create opportunities for further trade growth in the dairy sector as they contain specific arrangements for dairy products.[30] As a result of Brexit and the United States-Mexico-Canada Agreement (USMCA), changes to trade flows between the EU and the United Kingdom (UK) as well as the U.S., Canada, and Mexico should be expected during the projection period.[31] Various regulation programs being implemented to address pollution and improve sustainability, particularly in the EU, could have a significant impact on the dairy sector, specifically in the Netherlands, Denmark, and Germany. Pressures from these programs also have the potential to drive innovation and improve competitiveness. [32] Furthermore, consumer attraction to vegan diets and general concerns about

---

[21] Ibid., p. 14.
[22] Food and Agriculture Organization of the United Nations (2016).
[23] Organization for Economic Co-operation and Development - Food and Agriculture Organization of the United Nations (2021).
[24] Ibid., p. 179.
[25] Ibid., p. 179.
[26] Ibid., p. 179.
[27] Ibid., p. 179.
[28] Ibid., p. 179.
[29] Ibid., p. 180.
[30] Ibid., p. 179.
[31] Ibid., p. 179.
[32] Ibid., p. 179.

the environment are projected to continue to increase the consumption of plant-based milk and dairy products throughout the next decade, particularly in East Asia, Europe, and North America.[33] The long-term effects of plant-based dairy replacements on the dairy industry are uncertain at this time.[34] It is known, however, that environmental legislation could have a strong impact on the dairy sector's future development, particularly as it relates to water access, GHG emissions, and manure management.[35] Climate change and unforeseen weather events (droughts, floods, disease threats, etc.) have the potential to constrain world milk production through price volatility, milk yields, and cow inventory adjustments, among other variables, over the next decade.[36]

## 6.2 DAIRY COWS

Milk from cattle accounts for the largest share, 81 percent, of total world milk production.[37] When comparing cows to other dairy animals, cattle have numerous advantages such as ease of milking, udder size, the animal's ability to store milk, and overall milk yield.[38] However, there are variations in these parameters when comparing dairy cows in developed and developing countries. For example, there are more milking cows in developing countries than in developed countries, but the cows in developing countries typically have lower milk yields and shorter lactations than their counterparts in developed countries.[39] Developing countries tend to produce milk through small-scale producers with indigenous cattle breeds. Indigenous breeds have low production and typically need to be milked with the calf at foot but are well adapted to the local climate (temperature, available food and water, etc.), making them less susceptible to diseases and parasites.[40] Dairy cows in developed countries, however, have been bred using large-scale agribusiness standards which prioritize the output of farm animals, resulting in cattle with giant-sized udders able to produce tremendous quantities of milk for human consumption rather than for their own offspring.[41] These standards have resulted in industrial dairy farming and concentrated animal feeding operations which maintain horrific conditions. Further, the lagoon and spray field system and other post-green revolution methodologies commonly used by industrial dairy cattle farms cause many significant environmental and land degradation issues, including catastrophic water pollution from petroleum-based fertilizer and manure runoff; air pollution through airborne particulate matter, endotoxins, and gases like hydrogen sulfide and ammonia; soil erosion due to the incredibly heavy and large hooved nature of cattle; and GHG emissions from enteric

33 Ibid., p. 179.
34 Ibid., p. 188.
35 Ibid., p. 188.
36 Ibid., p. 188.
37 Food and Agriculture Organization of the United Nations (2021).
38 Food and Agriculture Organization of the United Nations (n.d.a).
39 Ibid.
40 Ibid.
41 Steier & Patel (Eds.) (2017).

fermentation and manure.[42] These various impacts of the dairy sector are not only harmful to the environment but also human and animal welfare.

### 6.2.1 Animal Welfare

A key aim of animal welfare science is to determine which aspects of natural living are fundamental for animals and how producers can incorporate these needs into best management practices.[43] However, the unfortunate reality for many of the cows living on dairy farms around the world today is a life spent confined to a cage with limited access to fresh air and sunlight, in addition to continual pregnancy and birthing without ever being able to have a relationship with their young. Only 15–20 percent of all dairy cows have access to pasture for part of the year, leaving 80–85 percent to be kept indoors year-round.[44] These startling statistics are quite obviously contradictory to the fundamental aspects of natural living for bovine animals. Knaus highlights that "when pasture management is site-appropriate, pasturing positively influences animal health and welfare, reduces exogeneous energy input (fossil energy), and increases consumer acceptance, and biodiversity"; Knaus goes on to say the current "use of year-round indoor feeding of high-yielding dairy cows with total mixed rations containing substantial amounts of human-edible crops from arable land puts global food security at risk and fails to utilize the evolutionary advantages of ruminants".[45] Thus, the consideration of animal welfare should be synonymous with environmental sustainability and food security. When raising dairy cattle on pasture, which is proven to be ideal for both animal welfare and land management, the herd must have access to shade and/or shelter depending on the climate in which the animals are living. Dairy herds also require consistent access to clean, fresh water. Polsky and von Keyserlingk evaluated the effects of heat stress on dairy cattle welfare, an expensive obstacle in the U.S. that results in annual economic losses of $900 million.[46] These losses are due to decreased milk production, compromised reproduction, and increased culling in dairy cows as a consequence of external forces causing an increase in body temperature.[47] Their study concluded that there are gaps in the literature that highlight the need for research into the pain, frustration, aggression, and malaise associated with heat stress, specifically the short-term impacts that cause increased hunger and thirst, and long-term impacts that cause foot lesions and lameness.[48] Moreover, there is need to examine the substantial deviation of the modern dairy cow, particularly the *Bos taurus* animal, from their evolutionary ancestors and consider how this has negatively affected their ability to manage heat stress.[49] Further research in the response of dairy cows and calves to separation

42 Ibid.
43 Polsky & von Keyserlingk (2017).
44 Knaus (2016).
45 Ibid., p. 15
46 Polsky & von Keyserlingk. (2017).
47 Ibid., p. 8645.
48 Ibid., p. 8654.
49 Ibid., p. 8654.

and how it impacts animal welfare is also needed.[50] Stěhulová et al. analyzed the effects of early separation of dairy cows and calves and determined it was optimal to remove the calf on day one rather than day four or seven based on behavioral and physiological responses.[51] However, the study determined that delayed separation may influence the social behavior of calves which can enhance their coping abilities later in life.[52]

### 6.2.2 Food Safety and Human Health Considerations

There are significant variations around the world on the approach to food safety regulations. In the EU, for example, food regulation policies are based on a precautionary principle, enabling consumers to have high levels of confidence in the products they choose to consume; conversely, in the U.S., the Food and Drug Administration (FDA) "interprets food safety and health from the perspective of acute effects of contamination and long-term effects of nutritional deficiencies".[53] Thus, there is no regard for the long-term effects of exposure to foods containing genetically engineered technologies, antibiotics, hormones, or other chemicals used in production, processing, or packaging.[54] The U.S. currently lacks any regulations relevant to the use of milk from cloned animals and their offspring, so these products may be sold to consumers without any required labeling or preapprovals.[55] Further, the FDA has not historically required labeling of genetically modified products or products made with ingredients from genetically modified organisms.[56] However, U.S. Congress passed The National Bioengineered Food Disclosure Law in July of 2016 which went into full effect on January 1, 2022.

Specific to the dairy cow industry, there exists concern regarding the use of recombinant bovine somatotropin (rBST) and its effects on human health. A synthetically produced hormone, rBST, combines with naturally occurring bovine somatotropin (bST) to increase milk production in cattle up to 10 percent.[57] The FDA required Monsanto, an agrochemical powerhouse that invented the controversial hormone, to determine whether rBST was absorbed into the human body. After only one study that presumptively concluded the hormone was not absorbed by humans, the FDA considered milk from cows treated with rBST as safe for human consumption.[58] The agency approved Monsanto's use of rBST in 1993 after deciding there was no significant difference between milk from cows treated with the drug and those that were not.[59] However, the approval of this compound had no consideration for its impact on the welfare of the animals which can lead to harmful health impacts in humans including cancer. Specifically, there was no cogitation over the fact that cows treated

[50] Stěhulová et al. (2008).
[51] Ibid., p. 164.
[52] Ibid., p. 145.
[53] Gupta (2017).
[54] Ibid., p. 232.
[55] Ibid., p. 232.
[56] Ibid., p. 232.
[57] Ibid., p. 233.
[58] Ibid., p. 233.
[59] Ibid., p. 233.

with rBST were more likely to contract mastitis which is treated with antibiotics that can persist in milk.[60] Regardless, the FDA found no fault in the use of rBST and its application continues throughout the U.S. today, even though the EU[61] and Canada[62] banned the drug before the turn of the 21st century. Within the U.S., "several court suits regarding state regulations that both prohibited (Ohio) and required (Vermont) the labeling of milk derived from rBST have resulted in court orders concluding that requiring a producer to label- or to not label- a product goes against first-amendment rights".[63]

Considering the aforementioned, another human health and food safety concern is the tendency to create antimicrobial-resistant bacteria as a result of over-prescribed antibiotics for mastitis. Mastitis is the most frequent bacterial disease in dairy cattle and regular use of antibiotics to treat the disease has been linked to antimicrobial-resistance (AMR).[64] "In France, as well as other European, American, and African countries, *Streptococcus uberis, Escherichia coli,* and coagulase-positive staphylococci are the three most frequent causative pathogens of clinical mastitis".[65] The most helpful preventions for AMR on dairy farms determined by Boireau et al. were proper biosecurity, improved hygiene (post-milking teat dipping, hygienic milking procedures, etc.), and minimized exposure to antibiotics.[66] There is also a need for national surveillance networks including the harmonization of laboratories, farms, and regulatory bodies capable of monitoring AMR of pathogens in dairy animals.[67] This is particularly important seeing as Kümmel et al. determined that *Staphylococcus aureus* can persist through the dairy chain into cheese products. This pathogen is known for its tendency to be AMR, specifically through resistance to methicillin, a common antimicrobial drug.[68]

Since the early 1900s, there have been governmental policies requiring milk pasteurization to reduce and ultimately prevent milk-borne diseases. For example, in the U.S., the U.S. Public Health Service prescribed the Pasteurized Milk Ordinance in 1924 to assist states and municipalities with creating effective programs for the safe production of dairy products.[69] The Pasteurized Milk Ordinance contains provisions governing the production, processing, and sale of "Grade A" milk and milk products and has been used to set the standards for domestic trade and retail sales within the country.[70] However, there are now 30 U.S. states that allow raw milk sales within their borders, and approximately 3 percent of the U.S. population consumes raw milk despite these recommendations.[71] Further, "under EU hygiene rules, Member States can prohibit or restrict the placing on the market of raw milk intended for human

[60] Ibid., p. 233.
[61] European Commission (1999).
[62] Holstead (2007).
[63] Gupta (2017).
[64] Boireau et al. (2018).
[65] Ibid., p. 9451.
[66] Ibid., p. 9460.
[67] Ibid., p. 9460.
[68] Kümmel et al. (2016).
[69] Ibid., p. 234.
[70] Ibid., p. 234.
[71] Ibid., p. 234.

consumption".[72] This conundrum has sparked continued debate between global public health authorities and consumers since many individuals prefer the taste and enzymatic benefits of raw milk despite the microbial risk factors.

### 6.2.3 Case Study: Kisan Dairy: Tarawadi, Haryana, India

Kisan Dairy in Tarawadi, Haryana, a prominent agricultural area in India, was started by the Kumar family in 1987 with 20 cows – 10 indigenous and 10 Holstein Friesian crosses.[73] Thanks to continual growth, as of 2014, the Kisan Dairy farm had 90 dairy cows and a total of 222 cattle.[74] Almost all of their cows are cross bred to be better suited for the local climate while maintaining high levels of milk production and are fed with a variety of crops grown on the family's 50 acres (approximately 20 ha).[75] Kisan Dairy has a wallowing pool for the cows to cool down during the hot summer months, and there are numerous water troughs throughout the farm.[76] In addition to green fodder, the family uses a polycropping methodology to feed their cows by growing maize, sorghum, barsim grass, and oats in rotation.[77] Any excess food materials are converted into silage (a preserved form of fodder produced through fermentation) and fed to the cows when there is little fresh food available, typically during rainy and lean seasons.[78] The family also supplements the cows' food with high energy and protein supplements made on site with locally produced ingredients like cotton seeds, soy, rice bran, mustard, barley, and maize.[79] The rice bran in this formula is noteworthy as it is a low-cost by-product from the rice farms prevalent in the area.[80] Paddy straw is another local agricultural by-product used at the dairy.[81] The key use for this product, however, is for bedding for the cows at calving, and it is later incorporated back into the fields as fertilizer.[82] Manure from the cattle is also used as fertilizer for the multiple crops grown on the farm and any excess is sold to neighboring farmers to be used as organic fertilizer.[83]

There are four family members employed at Kisan Dairy, as well as 12 other individuals recruited locally.[84] These employees, the majority of whom have been employed at the dairy for over 14 years, are responsible for milk production and the marketing of milk in the local village.[85] Kisan Dairy has a similar marketing model to the other dairy farms in India; their milk is sold directly to consumers at the farm gate twice a day and at various sales points in the village.[86] However, in the winter

---

72 European Food Safety Authority (2015).
73 World Society for the Protection of Animals (2014).
74 Ibid., p. 2.
75 Ibid., p. 2.
76 Ibid., p. 2.
77 Ibid., p. 2.
78 Ibid., p. 2.
79 Ibid., p. 2.
80 Ibid., p. 2.
81 Ibid., p. 2.
82 Ibid., p. 2.
83 Ibid., p. 2.
84 Ibid., p. 2.
85 Ibid., p. 2.
86 Ibid., p. 2.

when there are surpluses, the dairy sells their excess milk to a local dairy cooperative.[87] By having a short supply chain, the producer can maintain maximum income and the freshness of milk is guaranteed.[88] In the summer, however, the milk must be distributed quickly and efficiently to avoid spoilage due to lacking refrigeration access.[89] The Kisan family and dairy employees also consume the milk produced at the Kisan Dairy.

The Kisan dairy produces 1,100–1,200 liters of milk per day during the wet summer months and 1,800–2,000 liters per day during the dry winter months.[90] This level of production is almost six times the national average in India.[91] A significant contributor to this high production level is the longevity of the cows at Kisan Dairy. On average, Kisan Dairy cows tend to have seven to eight lactations, which the farmers attribute to the "good quality feeding, genetics appropriate to the local environment and the high level of care, which all lead to good health and welfare".[92] Moreover, having healthy, long-lived cows saves the Kisan Dairy on replacement costs and expensive veterinary treatments while simultaneously guaranteeing a stable milk supply.[93] Milking at the Kumar's dairy is carried out using portable milking machines twice a day and takes 2 hours for each milking cycle when using five machines.[94] The employee or family member performing the milking will clean the cows' teats before attaching the machines to the udders, and they will apply an approved anti-bacterial solution to the udders after the milking has commenced.[95]

Kisan Dairy also has a proper mechanism in place for the welfare of calves born on their farm. The newborn calves, both male and females, are able to suckle from their mothers for a few days before being transitioned to being bucket fed.[96] This early feeding is not only emotionally beneficial for the cows, but it allows the calves to build natural immunity through their mother's colostrum.[97] Female calves are either kept on the farm to replace older cows for milking or are sold to neighboring farms.[98] There is typically a challenge that arises when it comes to male dairy cows, especially in India since many states have justly banned the slaughter of dairy cattle, and beef is not a common meat consumed in the country due to religious and cultural preferences.[99] However, the Kumar family also runs a gaushala – a retirement home for cattle. The male cows who are not able to be sold to neighboring farms for breeding are moved to this sanctuary to peacefully live out their days.[100] Older female cows, or those who are no longer producing milk, are also moved to the gaushala to

87 Ibid., p. 2.
88 Ibid., p. 2.
89 Ibid., p. 2.
90 Ibid., p. 2.
91 Ibid., p. 2.
92 Ibid., p. 2.
93 Ibid., p. 2.
94 Ibid., p. 3.
95 Ibid., p. 3.
96 Ibid., p. 3.
97 Ibid., p. 3.
98 Ibid., p. 3.
99 Ibid., p. 3.
100 Ibid., p. 3.

maintain the productivity of Kisan Dairy.[101] The Kumar family acknowledges that this method has its limitations, and they are currently working toward being able to use sexed semen, a much more expensive alternative to typical bull semen, to reduce the number of male cows being born on their farm.[102]

Kisan Dairy employs a local veterinarian to perform their inspections, vaccinations, de-worming, and artificial insemination.[103] The veterinarian typically visits the farm each evening to carry out these tasks; however, in his absence, a member of the Kumar family who has been trained on the basics of animal care, artificial insemination, and pregnancy diagnosis by the National Dairy Research Institute (NDRI) in Karnal will ensure the animals are properly cared for.[104] A senior scientist at the NDRI said the following about the Kisan Dairy:

> "The breeding, feeding, housing and healthcare practices followed at Kisan Dairy are as per current scientific recommendations. The farmer is well-trained and experienced in cow management. The cows are able to realize most of their natural behaviors due to proper housing, climatic protection and good feeding and therefore farm productivity is better with minimal health problems. There is a proper mechanism for dealing with old and other unwanted animals. There are a few areas where there is scope for further improvement, specifically with respect to proper calf housing facilities and mechanized sanitary milking practices, which need financial investment, but the farmer plans to add these in the near future. Still, as the cows appear comfortable, well-fed and satisfied and are performing well without major health or behavioral problems, this farm could be rated as very good from a cattle welfare perspective. The farm may therefore be considered a model, which could be replicated under most Indian farming conditions to promote sustainable commercial dairy farming."

This case study illustrates how a small farm in India became a prime example of sustainable commercial dairy farming.[105] As discussed by the NDRI scientist, Kisan Dairy is now viewed as a model for other dairy farms in India, and it could be argued that this model is generally transferrable to dairy farms around the globe.

## 6.3 DAIRY GOATS

Goat milk production represents 2 percent, or 17.66 of the 883 million tons, of milk produced globally in 2019.[106] However, due to its unique biochemical and nutritional properties, it is increasingly being consumed by individuals with cow milk allergies and other gastrointestinal disorders.[107] When managed well, goats can prove beneficial to the environment through weed control, fire prevention, maintenance of biodiversity, and mitigation of some climate change effects.[108] Moreover, goats

---

101 Ibid., p. 3.
102 Ibid., p. 3.
103 Ibid., p. 3.
104 Ibid., p. 3.
105 This author thanks World Society for the Protection of Animals for their work to compile the included case study.
106 Food and Agriculture Organization of the United Nations (2021).
107 Animal and Plant Health Inspection Services (2012).
108 Miller & Lu (2019).

and sheep produce 20–25 percent fewer GHG emissions when compared to beef and dairy cattle.[109] It should be noted, however, that some studies suggest small ruminants actually have higher GHG production per unit of milk when compared to larger ruminants.[110] Ultimately, dairy goats are not well studied and their production systems are far from uniform, so it is challenging to make generalizations about their carbon footprint and overall environmental impact.[111] Nearly 60 percent of the world's goats are found in Asia, with China, India, Pakistan, and Bangladesh having the highest populations.[112] The majority of global income from goat farming is a derivate of meat sales, but there has been an increase in goat milk production and consumption due to its unique flavor and its greater digestibility and lipid metabolism.[113] Unlike previous trends, goats are no longer associated solely with low-income countries or dry areas; they are found in nearly all countries around the world.[114] The majority of global goat milk production and consumption is in India, Bangladesh, Pakistan, and Turkey, but it is also very prominent in the countries of the Mediterranean, the Middle East, Eastern Europe, and parts of South America.[115] France is the most well-known country for high-quality goat cheese production, but many other countries (especially in Europe and North America) are also producing these premiere quality products.[116] Most goats are maintained by small-scale producers, so it is difficult to determine their impact on personal livelihoods; however, it is known that goat farmers in areas without well-developed banking systems use the animals as an important asset.[117] Generally speaking, goat milk and its products are more likely to be consumed locally when compared to cow's milk.[118]

Though the most established market for goat milk production is in Europe, dairy goat numbers in the U.S. have been rapidly rising since the 1980s, doubling in size from 1997 to 2012.[119] In these two regions, Europe and North America, genetic selection of dairy goats has created an increase in production and longer lactation.[120] As a result of these desired traits, specialized dairy goat breeds have been sold to developing countries (both as live animal transport and as frozen semen/embryo) where they have been crossbred with indigenous breeds of dairy goats.[121] However, the results of this process have been mixed, specifically as it relates to feeding and managing crossbred goats to reach their genetic potential.[122] As the effects of climate change take hold and drier areas of the globe expand, there is a growing interest in indigenous goat breeds thanks to their drought and parasite resistance.[123] Further, many

109 Ibid., p. 1222.
110 Ibid., p. 1222.
111 Ibid., p. 1222.
112 Ibid., p. 1220.
113 Ibid., p. 1220.
114 Ibid., p. 1220.
115 Ibid., p. 1221.
116 Ibid., p. 1221.
117 Ibid., p. 1221.
118 Ibid., p. 1221.
119 Ibid., p. 1222.
120 Ibid., p. 1222.
121 Ibid., p. 1222.
122 Ibid., p. 1222.
123 Ibid., p. 1222.

desert breeds of goats produce milk with higher fat and protein content despite the lower volume, empowering the animals to be highly valued by cheese makers and processers alike.[124] It is important to note here that some dairy goat breeds are in danger of disappearing as a result of indiscriminate crossbreeding.[125]

The future of goat milk farming seems promising especially for low and mid dle-income countries; however, there is a need for investments to integrate inputs, markets, research, and production infrastructure.[126] Prices to producers are forecasted to be 19 percent higher by 2027 compared to the base period of 2015–2017.[127] Thus, government policy and action are critical to ensure that producers can benefit from rising demand for dairy goat products.[128] Since 1982, the International Goat Association has been the leading global supporter of the goat sector and has trail blazed the expansion of goat knowledge and practice through social media, conferences, and the academic journal *Small Ruminant Research*.[129]

### 6.3.1 Animal Welfare, Human Health, and Food Safety Considerations

Dairy goat farming systems are typically viewed as being a high welfare environment by consumers.[130] However, similar to dairy cows, there is an ever-increasing tendency for these farms to become industrialized and for the animals to be confined to cages or have limited access to pasture. Moreover, there are similar risks for zoonotic diseases as well as foodborne illnesses in dairy products from goats.

A key concern for dairy goat welfare is *Peste des Petits Ruminants* (PPR) (the sheep and goat plague), a highly contagious disease affecting small ruminants.[131] The FAO discusses that once PPR is introduced to a herd, the virus can infect up to 90 percent of the animals and 30–70 percent of the cases are fatal.[132] Although the PPR virus does not infect humans, it is clear that this disease is severely problematic for dairy goat farmers and is a significant disquietude for animals welfare.[133] The FAO, the World Organization for Animal Health, and individual countries have organized a PPR Eradication Campaign which has helped bring awareness of the deadly disease, as well as improvements to the goat sector in Africa, Asia, and the Middle East.[134] The aim of the campaign is to eradicate PPR by 2030, and there is current research being conducted on novel vaccines.[135] In the same vein, there is research being conducted on novel vaccines for other microbiologically induced goat diseases like Rift Valley Fever, Caprine Arthritis Encephalitis Virus, Goat Pox,

---

124 Ibid., p. 1222.
125 Ibid., p. 1222.
126 Ibid., p. 1221.
127 Ibid., p. 1222.
128 Ibid., p. 1221.
129 Ibid., p. 1219.
130 Tiezzi et al. (2019).
131 Food and Agricultural Organization of the United Nations (n.d.b).
132 Ibid.
133 Ibid.
134 Miller & Lu (2019).
135 Ibid, p. 1224.

and Contagious Caprine Pleuropneumonia.[136] Unlike PPR and other goat diseases, *Brucella melitesis* is a bacterial disease of goats and sheep that can infect humans through the consumption of raw milk or contact with aborted fetal tissue.[137] This disease has been nearly eradicated in developed countries thanks to the Rev. 1 vaccine; however, it is still a prominent illness in developing countries because of high cost to farmers and low priority of goats to policymakers.[138] According to Miller and Lu, dairy goats are more likely to be confined to a cage when compared to goats raised for meat, increasing their likelihood to contract brucellosis, thereby raising further concern for dairy goat welfare and human health impacts.[139]

Similar to dairy cattle, dairy goats, especially those that are high producing, are prone to mastitis and therefore also have the potential to be overprescribed antibiotics and contribute to AMR.[140] Mastitis is particularly prevalent when milking is not conducted via sanitary practices, i.e. absence of pre- and post-milking udder sanitation. "A Guide to Udder Health for Dairy Goats" is a free online manual that provides the most up-to-date information to help producers, veterinarians, extension, and dairy support personnel with the production of quality goat milk.[141] However, the spread of information in low-income countries continues to be a challenge.[142]

## 6.4 SUSTAINABILITY OF THE DAIRY SECTOR

Climate change is one of the defining obstacles of the 21st century, and the global dairy sector is both having an impact on and is subject to its devastating effects. With the global population expected to reach 9.8 billion in 2050, the global demand for nutritious food is predicted to double. [143] More than 80 percent of the global population consumes dairy products for their affordable, highly nutritious properties which illuminates the potential for extraordinary growth in the sector.[144] Therefore, it is paramount that as the dairy sector grows to match these demands, it does so sustainably in terms of the environment, human, and animal health, and in terms of development, poverty alleviation, and social progress.[145]

### 6.4.1 Sustainable Development Goals (SDGs)

In 2015, all United Nations (UN) Member States adopted the 2030 Agenda for Sustainable Development.[146] Within this Agenda, there are 17 goals, the SDGs, which are necessary to change the world as we know it. In 2021, the International Dairy Federation played a role in the UN Food Systems Summit to discuss how

---

136 Ibid., p. 1224.
137 Ibid., p. 1224.
138 Ibid., p. 1224.
139 Ibid., p. 1224–1225.
140 Ibid., p. 1225.
141 Ibid., p. 1225.
142 Ibid., p. 1225.
143 Food and Agricultural Organization of the United Nations (2019).
144 Ibid., p. 6.
145 Ibid., p. 6.
146 United Nations (n.d.).

the dairy industry is implementing various programs around the globe and working toward the achievement of the SDGs.[147] The International Dairy Federation's Dairy Sustainability Outlook highlighted how the global dairy sector is working to achieve SDGs 2 (zero hunger), 3 (good health and well-being), 4 (quality education), 8 (decent work and economic growth), 11 (sustainable cities and communities), 12 (responsible production and consumption), 13 (climate action), 15 (life on land), and 17 (partnership for the goals) through projects ranging from Denmark's GAIN Access to Better Dairy project to develop safe, nutritious, and affordable yogurt in Ethiopia to point system climate protection programs for Swiss farms and the annual progress reporting of The Dairy Sustainability Framework.[148] Further, García-Yuste provides a succinct yet comprehensive discussion on the various components for dairy farms to consider, including ruminant biology and metabolism, animal husbandry, GHG emission reduction, manure and land management, and milking and production systems to achieve a sustainable and environmentally friendly dairy farm.[149] Despite there being an extraordinarily long journey to improve the sustainability of the dairy sector, the programs that have been initiated, along with the expansion of literature available on the topic, provide a glimmer of hope for the industry's future and the future of its impact on our beloved planet.

## 6.5 GLOBAL REGULATORY ENVIRONMENT

Dairy products are some of the most regulated commodities across the globe. Gupta discusses the six reasons why this is true in most developed countries.[150] The first reason is that milk is one of the most basic foods, so governments around the world want to ensure there is an adequate (and mostly self-sufficient) supply. Second, milk products are highly perishable, bulky agricultural commodities. Because of this, the market power has historically been positioned in favor of milk handlers. However, and fortunately for dairy farmers, governments have put policies and institutions in place to protect them from being forced to choose between accepting the handler's price and dumping their products. Technological advancements, specifically in refrigerated transportation, have also enabled dairy products to stay fresh for longer periods of time. The third reason for heavy regulation "is due to the seasonal imbalance between milk demand and supply".[151] For example, North American demand for milk is typically highest in the fall and lowest in the summer, the inverse of production; yet, consumer pressure expects milk producers to keep supply constant throughout the year.[152] Fourth, "minor changes in the supply or demand of milk can cause severe changes in milk prices due to the high inelasticity of supply and demand", and fifth, "milk markets are oligopsonistic (many sellers relative to very few buyers of raw milk)".[153] The sixth and final reason for substantial regulation

---

147 International Dairy Federation (2021).
148 Ibid.
149 García-Yuste (2020).
150 Gupta (2017).
151 Ibid., p. 216.
152 Ibid., p. 216.
153 Ibid., p. 216.

of the dairy industry in developed countries is due to the various types, or grades, of milk products which typically depend on certain sanitary health standards like somatic cell counts, bacterial counts, and conditions of farm facilities. Thus, government regulations subsist to grade milk and dairy products accordingly.

Milk is one of the most supported agricultural commodities, and market price support policies have traditionally been the most common support mechanism used to insulate milk producers from market changes in Organization for Economic Co-operation and Development countries (excluding New Zealand).[154] Payments based on input use is the second most prevalent support provided in all Organization for Economic Co-operation and Development countries, but there are considerable variations between countries in level of support provided to milk producers, especially outside of these two support mechanisms.[155] Trade measures like tariffs, import quotas, and export subsidies are common methods used to protect dairy producers from traded products and to enable domestic pricing arrangements. However, market price support policies have faced a decline since the 1990s. This decline has been chiefly a result of the World Trade Organization (WTO) Uruguay Round Agreement on Agriculture (URAA) commitments in 1994 to decrease the support provided through trade measures to prevent dairy shortages and surpluses. Additional intentions of the WTO URAA commitments for the dairy industry were to maintain fair pricing and improve global cooperation in the dairy sector to expand and liberate world trade.[156] WTO agreements from the URAA 1994 (terminated in 1997) and 2004 Doha Round resulted in further reduction of agricultural supports, specifically by calling for most trade-distorting supports to be substantially reduced with product-specific capping of spending.[157] In light of the United Nations SDGs and the Paris Climate Agreement, among other general concerns, there has been an increase in the number and strength of policies addressing environmental issues in agriculture and this trend is expected to continue.[158]

Bi- and multi-lateral trade agreements are also an impactful component to consider for the international dairy sector. Most notable are the United States-Mexico-Canada Agreement (which replaced the North American Free Trade Agreement in 2020), the Closer Economic Relations Agreement between Australia and New Zealand, the U.S. and Australia Free Trade Agreement, the Comprehensive and Progressive Agreement for Trans-Pacific Partnership, the EU-Canada Comprehensive Economic and Trade Agreement, and the EU-Japan Economic Partnership Agreement. Many of these trade agreements are in their infancy, so their ultimate effect is yet to be determined. Nevertheless, these types of agreements have historically proven to be significant players in the global dairy industry.[159]

Additionally, there are international organizations responsible for the management of food safety regulations relevant to the dairy industry. These organizations

154 Ibid., pp. 216–217.
155 Ibid., pp. 216–217.
156 Ibid., p. 217.
157 Ibid., p. 217.
158 Organization for Economic Co-operation and Development - Food and Agricultural Organization of the United Nations (2021).
159 Steier & Patel (Eds.) (2017).

include the World Health Organization, the FAO, the WTO, the Codex Alimentarius Commission, and the International Epidemic Animal Disease Office.[160] Among the many goals of these organizations is a collective ambition to standardize food safety applications globally. The WTO's Sanitary and Phytosanitary Measures and the Agreement on Technical Barriers to Trade are two important administration bodies which establish hygiene rules and standards.[161] International dairy trade standards are determined by various Codex Committees including the Codex Committee on Milk and Milk Products, the Codex Committee on Food Additives, the Codex Committee on Food Hygiene, the Codex Committee on Pesticide Remains, and the Codex Committee on Residues of Veterinary Drugs in Foods.[162] The Codex Committees including the Codex Committee on Milk and Milk Products objective is to establish fundamental safety codes, international standards, and other guidelines for the dairy industry.[163] The key food safety standards are to have a low bacteria count, low or zero presence of pathogens, minimal chemical pollutants and microbial toxins, and low or zero presence of veterinary drug remnants.[164] Hazard Analysis at Critical Control Points, also known as HACCP, is a common system used to monitor food safety concerns in developed and developing countries alike. However, these systems are time consuming and difficult to implement for dairy and dairy products due to their high perishability and the need for monitoring on the farm and in production facilities.[165] This is especially true for developing countries and countries with weak dairy industries.[166]

### 6.5.1 Recent Developments in the EU

Milk and dairy products contribute a sizeable portion of the value of EU agricultural output every year.[167] Milk is produced in all EU countries with Germany, France, Poland, the Netherlands, Italy, and Spain accounting for almost 70 percent of the estimated 155 million tons produced annually.[168] The EU is also known as the biggest exporter of cheese and SMP in the world.[169] The legal bases for EU milk and dairy products are covered by EU Regulation 1308/2013 in the common market organization for agricultural products and through EU implementation regulation 511/2012 and EU Regulation 880/2012.[170] To protect the EU dairy sector during times of market volatility, there are a number of intervention mechanisms in place. The two key mechanisms used for market intervention are public intervention and aid for private storage.[171] There are additional legal measures for ad-hoc,

160 Ibid., p. 218.
161 Ibid., p. 218.
162 Ibid., p. 218.
163 Ibid., p. 218.
164 Ibid., pp. 218–219.
165 Ibid., p. 219.
166 Ibid., p. 219.
167 European Commission (n.d.).
168 Ibid.
169 Ibid.
170 Ibid.
171 Ibid.

emergency situations provided for in Articles 219 to 222 of the common market organization EU Regulation 1308/2013, which proved helpful during the COVID-19 pandemic.[172]

There are also marketing standards in place for the EU dairy sector, particularly as they relate to spreadable fats, specific definitions, designations and sales descriptions for drinking milk and milk products. In 2012 the EU introduced "a series of instruments to improve the supply chain in the dairy sector and to increase its resilience following the end of the quota system in 2015" known collectively as the Milk Package.[173] The Milk Package provisions include written contracts between milk producers and processors, collective negotiation through producer organizations, rules for interbranch organizations, and regulation of supply of protected designation of origin/protected geographical indication cheeses.[174] For trade with non-EU countries, regulation exists for both imports and exports. For example, dairy exports under certain quotas are subject to the issuing of an export license, and a specific import regime applies to dairy products through which preferential imports are subject to import licenses and tariffs, i.e. import duties, [175] although various trade agreements allow preferential imports at reduced or zero duty, typically through import quotas.[176] Market data and information on the milk sector is provided by the Milk Market Observatory and includes global and EU trends on production, supply and demand balance, production costs, and market perspectives.[177] There are various committees composed of government representatives and chaired by a European Commission representative that meet regularly to discuss and ensure smooth operations of the agricultural markets in the EU.[178]

As 2019 came to a close, the EU presented the European Green Deal, "a roadmap for making the EU's economy sustainable by turning climate and environmental challenges into opportunities across all policy areas and making the transition just and inclusive for all".[179] Within the European Green Deal, the Farm to Fork strategy is the vehicle for ensuring the transition to a more sustainable, equitable agricultural sector and food system.[180] This strategy includes the development of a mandatory sustainable food labeling framework that covers the nutritional, environmental, and social aspects of food products to ensure that producers are held responsible and consumers are able to make sustainability informed decisions.[181] Further, in 2021, the European Commission came to a proposed agreement on the future of the Common Agricultural Policy (CAP) for the period of 2023–2027 based on the impacts of the COVID-19 pandemic and the Farm to Fork Strategy.[182]

---

172 Ibid.
173 Ibid.
174 Ibid.
175 Ibid.
176 Ibid.
177 Ibid.
178 Ibid.
179 European Commission (2019).
180 European Commission (2020).
181 Ibid., p. 2.
182 European Commission (2021).

The primary goal of the CAP changes is to make the policies greener and fairer.[183] Member States will have to take account of the analysis, objectives, and targets of key EU laws on climate, energy, water, air, biodiversity, and pesticides and will be required to update their plans where necessary to align with changes to the relevant legislation.[184] Within the framework, there is a "new obligation to protect wetlands and peatlands, two major carbon sinks", as well as the inclusion of relevant elements of the Water Framework Directive and the Directive on the Sustainable Use of Pesticides.[185] A new feature of the CAP when compared to the previous period, is a "no backsliding" clause which will require Member States to "aim higher" with regard to the environment and climate in the way they use CAP funds.[186] Additionally, Member States will be required to spend a minimum of 35 percent of their rural development budgets on action over the environment, climate, and animal welfare.[187] Member States will also set aside 10 percent of their direct payments to the collective redistributive income support tool for small farmers among other redistributive needs.[188] The new CAP will include a never before seen measure that ensures a link between receiving CAP income support and rural development funding to farmers' respect of basic social and labor rights of farm workers.[189] There are also specific initiatives for young farmers and women farmers in the new CAP.[190] Further, improvements in the knowledge base and advancement of innovation and scientific research for the sector are included in the CAP by allowing transfer of CAP funds to the Erasmus+ program and allocating €9 billion for Food, Bioeconomy, Natural Resources, Agriculture and Environment as part of the Horizon Europe program 2021–2027.[191] Thus, the EU is trailblazing the global efforts in policy and regulation to support animal welfare and environmental, social, and economic sustainability.

### 6.5.2 U.S. Dairy Industry

The structure of the dairy industry in the U.S. is and has been dynamic and nuanced. Born in the early 1900s, the industrial vision of dairying resulted in an unrealistic goal of the "perfect dairy farm".[192] This vision prioritized worker efficiency, the technical education of the farmer, the ability to produce high-quality milk year round, a favorability of large herds, substantial amounts of machinery, well-trained employees, and a high level of production per cow.[193] In reality, the industrialization of dairy gave rise to larger debt loads for farmers, increased prevalence of disease in dairy animals, and overall added risks and costs.[194] The idealized system has only

[183] Ibid., p.2.
[184] Ibid., p. 4.
[185] Ibid., p. 4.
[186] Ibid., p. 4.
[187] Ibid., p. 6.
[188] Ibid., p. 7.
[189] Ibid., p. 9.
[190] Ibid., p. 10.
[191] Ibid., pp. 10–11.
[192] Ibid., p. 220.
[193] Ibid., p. 220.
[194] Ibid., pp. 220–221.

come close to fruition in California and various other Western dry-lot farms, but these farms now account for a significant amount of U.S. milk production as they are able to outcompete other dairy regions that use predominately family labor.[195] The top ten milk-producing U.S. states in 2019 and 2020 were California, Wisconsin, Idaho, New York, Texas, Michigan, Pennsylvania, Minnesota, New Mexico, and Washington.[196] The majority of dairy farms in the U.S. belong to producer-owned cooperatives which combine members' milk and transfer it to processors and manufacturers.[197] The number of dairy farms that are family owned or family cooperatives made up 70 percent of all dairy farms in 2002, but this number has perpetually been challenged as the number of large-scale dairy farms, including concentrated animal feeding operations, continuously increases.[198] The labor-intensive dairy jobs in the U.S. rely heavily on foreign-born workers, many of whom are undocumented.[199]

#### 6.5.2.1 Federal Support for the U.S. Dairy Sector

The U.S. dairy industry receives a significant amount of assistance from the federal government through various programs like the Federal Milk Marketing Orders (FMMOs), the Dairy Product Price Support Program (a price support program), the Milk Income Loss Contract Program (direct payment to producers), and the Dairy Export Incentive Program.[200] The two primary objectives of U.S. dairy policy are (1) to provide price supports to establish a minimum farm income, and (2) to incorporate counter-cyclical price stabilization systems so as to ensure an orderly supply and marketing of farm milk.[201] These programs have been a considerable expense for the federal government and have contributed to U.S. budget deficits for the past several decades.[202] However, these programs have proven to have limited impact on the financial viability of dairy farms, causing the programs to be heavily criticized.[203] Three pertinent government dairy programs, which have shifted over recent years, are import quotas on foreign dairy products (border measures), FMMOs, and the dairy price support programs; these programs were previously less directly involved and are now focused on assisting with milk marketing through export enhancement (encouraged disposal of excess dairy acquired through the dairy support program), low-income feeding programs, dairy research, and advertising.[204]

Border measures like tariff-rate quotas (TRQs) prevent the U.S. market from being overrun with inexpensive foreign dairy products.[205] TRQs enforce a relatively low tariff on imports up to a determined quota and then a relatively high tariff on any quantity over the established quota.[206] These TRQs have restricted dairy imports

195 Ibid., p. 221.
196 U.S. Department of Agriculture (2021).
197 Ibid., p. 221.
198 Steier & Patel (Eds.) (2017).
199 Ibid., p. 221.
200 Ibid., p. 221.
201 Ibid., p. 221.
202 Ibid., p. 222.
203 Ibid., p. 222.
204 Ibid., p. 222.
205 Ibid., p. 222.
206 Ibid., p. 222.

to less than 6 percent of total U.S. consumption and have contributed to the higher price of U.S. dairy products when compared to the international market.[207] FMMOs were authorized by Congress under the Agricultural Act of 1937 and are administered by the United States Department of Agriculture (USDA).[208] "The legislation states that the purposes of FMMOs are (1) to establish orderly marketing, rather than relying on chaotic marketing conditions; (2) to establish pricing mechanisms that are fair to farmers, distributors, and consumers; and (3) to provide consumers with an adequate supply of high-quality milk" (Gupta, 2017).[209] FMMOs allowed the federal government to organize and regulate markets and led to the development of dairy territories.[210] FMMOs segregate the country into geographic regions, and the manufacturers or processors in each region must pay farmers at least the minimum price for each of the four grades of milk.[211] Despite these intentions, the three major effects of FMMOs are price discrimination, revenue pooling, and regionalization.[212] Regardless, dairy farmers typically support FMMOs since they set minimum prices and help balance the marketing power with dairy processors.[213] The Federal Agriculture Improvement and Reform Act of 1996 mandated reforms to the FMMO program by changing the way minimum prices paid to farmers is determined and by consolidating the number of FMMOs.[214] Currently, about two-thirds of the U.S.'s fluid milk is regulated under FMMOs.[215]

The Dairy Price Support Program was legislated by the Agricultural Act of 1949 which stated that farm milk prices must be supported at between 75 and 90 percent of parity.[216] This program has operated as a market intervention program where the federal government offers to purchase nonperishable dairy products from manufacturers at a designated intervention price level.[217] The Secretary of Agriculture is authorized by the Act to determine the specific price support level within this range, and parity was based on the index of prices paid by farmers for commodities, services, interest, taxes, and wages relative to the base period of 1910–1914.[218] The Act also called for farm milk prices to be indirectly supported by the Commodity Credit Corporation through federal government purchases of cheese, butter, and nonfat dry milk from the processors of the products.[219] The aim of this measure was to ensure that farm prices of manufactured milk remained above the legislated support price, but in actuality led to large dairy surpluses held by the government.[220] The Agriculture and Food Act of 1981, the 1983 Dairy and Tobacco Adjustment Act, and the Food Security Act of

[207] Ibid., p. 222–223.
[208] Ibid., p. 223.
[209] Ibid., p. 223.
[210] Ibid., p. 223.
[211] Ibid., p. 223.
[212] Ibid., p. 224.
[213] Ibid., p. 224.
[214] Ibid., p. 224.
[215] Ibid., p. 224.
[216] Ibid., p. 224.
[217] Ibid., p. 224.
[218] Ibid., pp. 224–225.
[219] Ibid., p. 225.
[220] Ibid., p. 225.

1985 all lowered the support price and allowed for additional price reductions in an attempt to address the dilemma of dairy surpluses.[221] The dairy support price has since remained steady.[222] Recent Agricultural Acts, also known as U.S. Farm Bills, have been advancing toward a less regulated agricultural sector which would eventually indicate the need for reformation of FMMOs and the measured elimination of dairy price supports.[223] However, the aspirations to reduce these programs have been delayed due to the low commodity prices in recent years, as most recent policy efforts have been aimed to provide impromptu emergency assistance to farmers.[224]

#### 6.5.2.2 Impact of Agricultural Acts of 2014 and 2018 on U.S. Dairy

The 2014 Farm Bill brought several noteworthy changes to the U.S. dairy sector. The overarching theme was to update the safety nets that have been in place since the mid-twentieth century.[225] The specific policies relevant to the dairy industry in the 2014 Agricultural Act include the creation of two new programs to benefit dairy producers and low-income populations, plus the removal or reauthorization of several other dairy-related provisions.[226] The Dairy Product Price Support Program was discontinued while the more permanent Dairy Price Support Program from the 1949 Agricultural Act was retained; however, the Dairy Price Support Program was suspended for the duration of the 2014 Farm Bill.[227] The Act also discontinued the Milk Income Loss Contract program once the new Margin Protection Program (MPP) begins.[228] The MPP intends to pay operators a margin protection payment when de facto dairy production margins fall below the threshold levels for margin protection payment.[229] The new MPP is connected with another new program, the Dairy Production Donation Program, which requests that the government purchase dairy products to provide nutritional assistance for low-income individuals.[230] The Act also terminated the Dairy Export Incentive Program and repealed the Federal Milk Marketing Order Review Commission, which was entrusted with conducting a comprehensive review and evaluation of the federal and non-federal milk marketing order systems currently in effect. There were a number of dairy programs that remained in effect after the 2014 Farm Bill. These include the Dairy Forward Pricing Program which allows non-cooperative buyers of milk who are regulated under FMMOs to offer farmers forward pricing on Class II, III, or IV milk rather than the minimum FMMO blend price for pooled milk.[231] Moreover, the Act kept the Dairy Indemnity Program which provides payments to dairy producers if a public regulatory agency instructs them to remove raw milk from the commercial market due to contamination issues.[232]

[221] Ibid., p. 225.
[222] Ibid., p. 225.
[223] Ibid., p. 225.
[224] Ibid., p. 225.
[225] Ibid., p. 225.
[226] Ibid., p. 225.
[227] Ibid., p. 225.
[228] Ibid., p. 225.
[229] Ibid., p. 226.
[230] Ibid., p. 226.
[231] Ibid., p. 226.
[232] Ibid., p. 226.

Lastly, the 2014 Farm Bill held onto some provisions to increase the development of export markets under the National Dairy Promotion and Research Program.[233] These programs collectively secure farmers ability to sell their products and maintain their income.[234]

The 2018 Farm Bill had fewer significant changes for the dairy industry as compared to the 2014 Act discussed above.[235] The key difference for the dairy sector surrounds the renaming and changes of the MPP for Dairy Producers, now called the Dairy Margin Coverage program.[236] The chief change in the new program is that the Dairy Margin Coverage has stronger coverage and lower premiums compared to the MPP.[237] The Secretary of Agriculture is required to repay each dairy that participated in the previous program part of the difference between total MMP payments and premiums and there are other minor stipulations that come along with the changes of this program.[238] For FMMOs, the Class I price mover is now defined as the average of Class III and Class IV price plus $0.74/cwt under the new Act.[239] Further, the Dairy Product Donation Program was repealed and replaced with a new Milk Donation Program.[240] The aim of this change was to make it easier for dairy producers, processors, and cooperatives to donate fluid milk to public or private nonprofits that provide food assistance to low-income individuals.[241] The 2018 Farm Bill will expire in 2023.

## 6.6 CONCLUSION

The global dairy industry plays a vital role in fueling the world's economy and providing sustenance to billions of people across the globe. However, these monumental achievements are not attained without repercussions. Pollution of air and water, loss of biodiversity, land degradation, lack of animal welfare, and human health risks are all also associated with this sector's role in the world. Therefore, the laws and policies in which the sector operates are vital for its smooth global operations, as well as its realization in developing countries and transition toward sustainability in those that are already developed. A shift in the global mindset of dairy production and consumption are fundamental necessities to ensure the welfare of animals, people, and planet. Further, a delicate balance must be struck between ensuring there are regulations that encourage ethical agricultural practices, ensure food safety measures, and enable responsible consumer choices while also maintaining simplicity and affordability to ensure inclusivity of small farms. The EU's current regulatory framework can arguably be viewed as the best current model for this equilibrium. Ultimately, the future of food security, animal welfare, and climate stability rests in this quandary.

---

233 Ibid., p. 226.
234 Ibid., p. 226.
235 See U.S. Congress. *Agricultural Improvement Act of 2018.*
236 Zulauf & Wolf (2018).
237 Ibid.
238 Ibid.
239 Ibid.
240 Ibid.
241 Ibid.

## REFERENCES

Animal and Plant Health Inspection Services. (2012). U.S. dairy goat operations: Info sheet. United States Department of Agriculture. https://www.aphis.usda.gov/animal_health/nahms/goats/downloads/goat09/Goat09_is_DairyGoatOps_1.pdf.

Boireau, C., Cazeau, G., Jarrige, N., Calavas, D., Madec, J.-Y., Leblond, A., Haenni, M., & Gay, É. (2018). Antimicrobial resistance in bacteria isolated from mastitis in dairy cattle in France, 2006–2016. *Journal of Dairy Science*, 101(10), 9451–9462. https://doi.org/10.3168/jds.2018-14835.

European Commission. (1999). Commission proposes ban on BST. https://ec.europa.eu/commission/presscorner/detail/en/ip_99_798.

European Commission. (2019). The European Green Deal sets out how to make Europe the first climate-neutral continent by 2050, boosting the economy, improving people's health and quality of life, caring for nature, and leaving no one behind. https://ec.europa.eu/commission/presscorner/detail/en/ip_19_6691.

European Commission. (2020). Factsheet: From farm to fork: Our food, our health, our planet, our future. https://ec.europa.eu/commission/presscorner/detail/en/fs_20_908.

European Commission. (2021). A greener and fairer CAP. Agriculture and Rural Development. https://ec.europa.eu/info/sites/default/files/food-farming-fisheries/key_policies/documents/factsheet-newcap-environment-fairness_en.pdf.

European Commission. (n.d.). Milk and dairy products. https://ec.europa.eu/info/food-farming-fisheries/animals-and-animal-products/animal-products/milk-and-dairy-products_en.

European Food Safety Authority. (2015). Raw drinking milk: What are the risks? https://www.efsa.europa.eu/en/press/news/150113.

Foley, J. A., Ramankutty, N., Brauman, K. A., Cassidy, E. S., Gerber, J. S., Johnston, M., Mueller, N. D., O'Connell, C., Ray, D. K., West, P. C., & Balzer, C. (2011). Solutions for a cultivated planet. *Nature (London)*, 478(7369), 337–342. https://doi.org/10.1038/nature10452.

Food and Agricultural Organization of the United Nations. (2016). The global dairy sector: Facts. https://www.fao.org/3/cb2992en/cb2992en.pdf.

Food and Agriculture Organization of the United Nations. (2019). Climate change and the global dairy cattle sector: The role of the dairy sector in a low-carbon future. *Global Dairy Platform*. https://www.fao.org/publications/card/en/c/CA2929EN/.

Food and Agricultural Organization of the United Nations. (2021). Milk facts. https://www.fao.org/3/I9966EN/i9966en.pdf.

Food and Agricultural Organization of the United Nations. (n.d.a). Gateway to dairy production and products. https://www.fao.org/dairy-production-products/production/dairy-animals/cattle/en/.

Food and Agricultural Organization of the United Nations. (n.d.b). Peste des petits ruminants. https://www.fao.org/ppr/en/.

García-Yuste, S. (2020). *Sustainable and Environmentally Friendly Dairy Farms*. Springer.

Gupta, C. (2017). The political ecology of the dairy industry, in Steier, G., & Patel, K. K. (Eds.), *International Farm Animal, Wildlife and Food Safety Law*. Springer.

Holstead, J. (2007). Recombinant bovine growth hormone. *Connecticut General Assembly*. https://www.cga.ct.gov/2007/rpt/2007-R-0159.htm.

International Dairy Federation. (2021). Dairy sustainability outlook. *United Nations Food Systems Summit 2021*. https://fil-idf.org/publications/free-of-charge/issue-4-dairy-sustainability-outlook/.

Knaus, W. (2016). Perspectives on pasture versus indoor feeding of dairy cows. *Journal of the Science of Food and Agriculture*, 96(1), 9–17. https://doi.org/10.1002/jsfa.7273.

Kümmel, J., Stessl, B., Gonano, M., Walcher, G., Bereuter, O., Fricker, M., Grunert, T., Wagner, M., & Ehling-Schulz, M. (2016). *Staphylococcus aureus* entrance into the dairy chain: Tracking *S. aureus* from dairy cow to cheese. *Frontiers in Microbiology*, 7, 1603–1603. https://doi.org/10.3389/fmicb.2016.01603.
Miller, B. A., & Lu, C. D. (2019). Current status of global dairy goat production: an overview. *Asian-Australasian Journal of Animal Sciences*, *32*(8), 1219–1232.
Organization for Economic Co-operation and Development - Food and Agricultural Organization of the United Nations. (2021). Dairy and dairy products. OECD-FAO Agricultural Outlook 2021–2030. pp. 178–189. https://www.fao.org/3/CB5332EN/Dairy.pdf.
Polsky, L., & von Keyserlingk, M. A. (2017). Invited review: Effects of heat stress on dairy cattle welfare. *Journal of Dairy Science*, 100(11), 8645–8657. https://doi.org/10.3168/jds.2017-12651.
Stěhulová, I., Lidfors, L., & Špinka, M. (2008). Response of dairy cows and calves to early separation: Effect of calf age and visual and auditory contact after separation. *Applied Animal Behaviour Science*, 110(1), 144–165. https://doi.org/10.1016/j.applanim.2007.03.028.
Steier, G., & Patel, K. K. (Eds.). (2017). *International Farm Animal, Wildlife and Food Safety Law*. Springer.
United Nations. (n.d.). The 17 goals. Department of Economic and Social Affairs: Sustainable Development. https://sdgs.un.org/goals.
United States Department of Agriculture. (2021). Top 10 milk producing U.S. states in 2019 and 2020 (in million pounds) [Graph]. In *Statista*. Retrieved October 20, 2021, from https://www-statista-com.ezproxy.neu.edu/statistics/194968/top-10-us-states-by-milk-production/.
United States Congress. *Agricultural improvement act of 2018*. https://www.congress.gov/bill/113th-congress/house-bill/2642.
World Society for the Protection of Animals. (2014). A case study of high welfare milk production in India. https://www.worldanimalprotection.org/sites/default/files/media/int_files/high-welfare-milk-production-india.pdf.
Zulauf, C., & Wolf, C. (2018). Dairy provisions in the 2018 farm bill. *Farmdoc Daily*, 8, 232.

# 7 Poultry Welfare Regulation

## *Lacking Protections for Chickens, Turkeys, Ducks, and Geese*

*Amanda G. Verkest*
Northeastern University

## CONTENTS



## 7.1 INTRODUCTION

Poultry products, including meat, eggs, and also organs, such as livers, are touted to be virtually a clean-conscience choice.[1] Even though there is deep science in favor of a plant-based diet, poultry continues to be widely popular. In fact, the world's primary source of animal protein, poultry meat,[2] is not only supposed to be rich in protein but also a good source of phosphorus, other minerals, and B-complex vitamins.[3] Eggs from poultry also contain many essential vitamins and minerals with folic acid being noteworthy for its role in reducing the incidence of neural tube defects during

[1] The current chapter relies heavily on "Still in the jungle: poultry slaughter and the USDA" *New York University Environmental Law Journal.* This author thanks B. Friedrich for his work in the field upon which we expand in this chapter.
[2] Food and Agriculture Organization of the United Nations (n.d.a).
[3] Food and Agriculture Organization of the United Nations (n.d.b).

DOI: 10.1201/9781003137733-13

pregnancy.[4] Since the early 1960s, global consumption of poultry eggs has doubled while consumption of poultry meat has increased fivefold due to growing population, urbanization, and rising income levels in developing countries.[5] Chicken, in particular, spearheads global meat consumption thanks to its affordability, low fat content, and lack of cultural and religious barriers.[6] Much of the support for poultry products comes from subsidies, but the poultry industry is also a blockbuster monolith with a tight control over the lucrative image in favor of consuming poultry products.

### 7.1.1 International Popularity

In 2019, 131.6 million tons of poultry meat were produced and 14.2 million tons of this product were traded internationally.[7] The U.S. is the world's largest poultry meat producer, with 17 percent of global output, and China is the world's largest egg producer, with 37 percent of global production.[8] Moreover, poultry meat is the main driver of the projected 13 percent increase in global meat production expected by 2026.[9] Low production cost and product price cause poultry to be the meat of choice for producers and consumers alike.[10] Overall, the global meat trade is currently being driven by China, with its total meat purchases exceeding 11 million tons in 2021, and COVID-19-related lockdowns, travel restrictions, and supply chain tightness continue to disrupt the sector.[11] Highly pathogenic avian influenza (HPAI) outbreaks have restrained poultry meat exports from Africa and Europe at the time of writing.[12] The Codex Alimentarius or "Food Code" developed by the World Health Organization and Food and Agriculture Organization of the United Nations provides noteworthy guidelines for the control of *Campylobacter* and *Salmonella*, two human-illness-inducing bacterium commonly found in poultry meat, as well as many other resources on food safety and risk management for the global poultry sector.[13] These resources are invaluable seeing as the Center for Disease Control attributes more food-borne illness-related deaths to poultry than any other commodity.[14]

### 7.1.2 Carbon Footprint of the Industrialization of the Poultry Industry

Depending on the production system and feed used, the poultry industry has the *potential* to be carbon neutral.[15] Similar to ruminant animals, poultry convert materials indigestible to humans into highly nutritious foods and beneficial organic fertil-

[4] Ibid.
[5] Food and Agriculture Organization of the United Nations (n.d.a).
[6] Food and Agriculture Organization of the United Nations (n.d.a).
[7] Food and Agriculture Organization of the United Nations (2021, June). p. 7.
[8] Food and Agriculture Organization of the United Nations (n.d.a).
[9] Organization for Economic Co-operation and Development - Food and Agricultural Organization of the United Nations (2021).
[10] Ibid., p. 110.
[11] Food and Agriculture Organization of the United Nations (2021, June). p. 7.
[12] Ibid., p. 46.
[13] Food and Agriculture Organization of the United Nations (2019, November).
[14] Painter, et al. (2013).
[15] Applachian State University (n.d.).

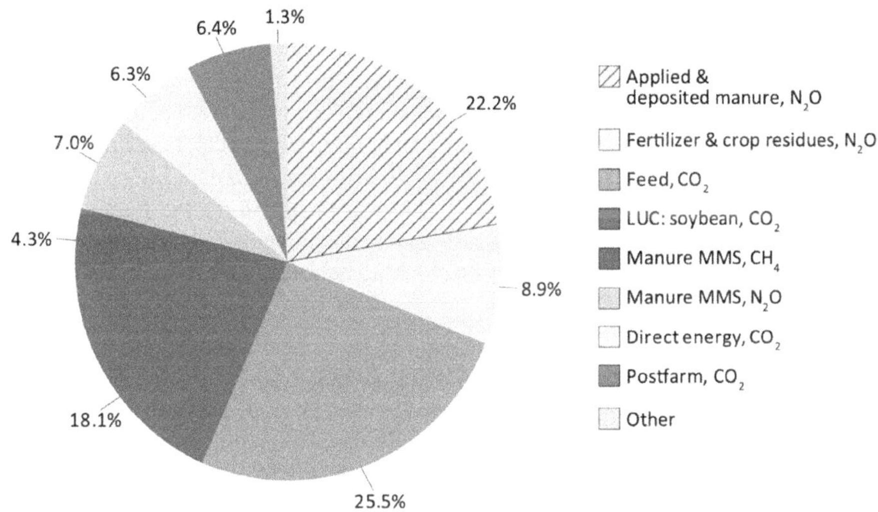


**FIGURE 7.1** Breakdown of total global GHG emissions by category for chicken meat and egg supply chains. (MacLeod, M., Gerber, P., Mottet, A., Tempio, G., Falcucci, A., Opio, C., Vellinga, T., Henderson, B. & Steinfeld, H. 2013. *Greenhouse Gas Emissions from Pig and Chicken Supply Chains – A Global Life Cycle Assessment.* Food and Agriculture Organization of the United Nations (FAO), Rome, https://www.fao.org/3/i3460e/i3460e.pdf.)

izer. However, poultry animals do not produce significant amounts of greenhouse gases, e.g., methane, and can be fed insects that are typically regarded as pests.[16] Unfortunately, carbon neutrality is not the reality for the majority of poultry production and animal welfare remains a topic of grave concern. The industrialization and intensification of poultry production have resulted in horrific living conditions for poultry animals – many of which spend their lives in a movement-restrictive cage where they can never see the light of day or breath fresh air. Further, the work environment for poultry production and processing staff members has been described as physically and emotionally harmful.[17] Researchers report simply using slow-growing broiler chickens rather than genetically modified fast-growing breeds results in healthier chickens that express more behavioral indicators of positive welfare.[18] Concentrated animal feeding operations that use agribusiness standards have dominated the poultry industry in developed countries illuminating the need to consider animal welfare regulations (Figure 7.1).

## 7.2 ANIMAL WELFARE REGULATION

The "five freedoms" have historically formed the basis of animal welfare standards around the globe.[19] These include freedom from hunger and thirst, freedom from

[16] Ibid.
[17] Friedrich (2015). p. 13.
[18] Rayner, et al. (2020).
[19] Nicol & Davies (n.d.).

discomfort, freedom from pain, injury, and disease, freedom to express normal behavior, and freedom from fear and distress.[20] In recent years, these concepts have been expanded and transformed into scientifically measurable concepts to ensure transferability to the species of animals raised on farms worldwide. The FAO reports, "Scientific research into the coping capacity and preferences of chickens has allowed the development of measures that can be used to audit the welfare of laying hens and broiler chickens on commercial farms."[21] These measures form the foundation of poultry welfare regulations which typically include banning certain housing systems like conventional cages, setting limits on stocking density,[22] as well as measures for proper transportation and slaughter.

### 7.2.1 International Perspectives on Poultry Welfare

Poultry animals have been scientifically proven to outperform dogs in ethological tests of intellectual and behavioral complexity, to anticipate the future and demonstrate self-control, to understand that an object continues to exist despite being taken away and hidden, and to take on the perspective of other birds.[23] These traits have solidified the cognition and sentience of poultry animals indicating their need for ethical treatment. Despite this clear dichotomy, many countries around the world have not promulgated regulations to protect these animals from inhumane treatment, transport, and/or slaughter.

The EU's regulatory environment for poultry welfare is currently the strongest and most comprehensive in the world.[24] Serbia and Turkey, official candidates for membership in the EU, both have similar welfare standards for poultry when compared to the current Member States.[25] However, although Turkey has broad legislation on the welfare of farm animals and their transport, it does not appear to have legislation on welfare during slaughter.[26] Egypt, Morocco, and the Russian Federation have relatively little legislation for farm animal welfare including poultry animals despite being members of the OIE (World Organization for Animal Health).[27] Ukraine, Brazil, and Thailand provide the EU with high-value poultry products which must be up to their premiere standards indicating the presence of animal welfare considerations in these countries.[28] The U.S., however, has woefully neglected to protect poultry welfare at slaughter despite there being numerous forms of legislation that contradict this neglect. This fact is particularly concerning since the U.S. is the global leader in poultry meat production.

20 Ibid., p. 110.
21 Food and Agriculture Organization of the United Nations (n.d.c).
22 Ibid.
23 Friedrich (2015). p. 3.
24 Stevenson, et al. (2014).
25 Ibid., pp. 63–64.
26 Ibid., p. 64.
27 Ibid., p. 64.
28 European Commission (n.d.a).

### 7.2.2 U.S.: Deficit in Poultry Slaughter Regulation

The U.S. Congress passed the first federal law designed to protect farm animals from abuse through the Humane Methods of Slaughter Act (HMSA) of 1958.[29] Friedrich denotes that the HMSA of 1958 has the following three key provisions: (1) "it dictates that no method of slaughtering or handling in connection with slaughtering shall be deemed to comply with the public policy of the U.S. unless it is humane." Therefore, "in the case of cattle, calves, horses, mules, sheep, swine, and other livestock, all animals [must be] rendered insensible to pain by a single blow or gunshot or an electrical, chemical or other means that is rapid and effective before being shackled, hoisted, thrown, cast, or cut…," (2) "it directs the USDA to designate humane methods of slaughter with respect to each species of livestock," and (3) "in a provision that was repealed in 1978 and replaced with a more broadly applicable enforcement mechanism, it required that any plant supplying meat to the federal government comply with the law."[30] After the HMSA was passed, the USDA promulgated regulations to designate acceptable methods of slaughter under the Act.[31] However, enforcement of the regulations was not discussed and no comments were solicited or incorporated.[32]

In 1978, Congress passed the HMSA of 1978 which directly incorporated humane slaughter into the Federal Meat Inspection Act of 1907 (FMIA).[33] The HMSA of 1978 did not repeal or replace the HMSA of 1958; they are two separate laws that are both still operative today.[34] The only legislative purpose of the HMSA of 1978 is "to require that meat inspected and approved under [the FMIA] be produced only from livestock slaughtered in accordance with humane methods."[35] However, the HMSA of 1978 removed the phrase "other livestock" and included a specific list of protected animals instead.[36] Moreover, the 1978 Act affected the HMSA of 1958 by "rescinding the prohibition on the federal government from purchasing meat slaughtered in violation of the Act and replacing the provision with a markedly stronger means of enforcement."[37] These stronger means of enforcement included the ability of the USDA to suspend inspection at a plant that is slaughtering animals inhumanely which essentially shuts the plant down.[38] Further, by incorporating humane slaughter into the FMIA, criminal enforcement became available under Section 676 of the FMIA which allows for either imprisonment of up to 1 year or a fine of up to $1,000.[39] However, if the violation involves "intent to defraud" or adulterated products, the criminal sanctions are increased allowing for imprisonment of up to 3 years, a fine up to $10,000, or both.[40] A notable variation of the 1978 Act when compared to the

[29] Friedrich (2015). p. 4.
[30] Ibid., pp. 4–5.
[31] Ibid., p. 5.
[32] Ibid., p. 5.
[33] Ibid., p. 5.
[34] Ibid., p. 5.
[35] Ibid., p. 5.
[36] Ibid., p. 5.
[37] Ibid., p. 5.
[38] Ibid., p. 5.
[39] Ibid., p. 5.
[40] Ibid., p. 5.

HMSA of 1958 is that the USDA promulgated regulations according to the agency's authority.[41] The final rule, including the comments of 16 individuals and organizations, established humane protections under the law at both the handling and slaughter stages and provided the first proper enforcement program.[42]

Then, in September of 2005, the USDA issued a notice titled "Treatment of Live Poultry before Slaughter" which was the first and only notice in the Federal Register concerning the treatment of poultry at slaughter.[43] In the Notice, the USDA stated that "there is no specific federal humane handling and slaughter statute for poultry. … The [Humane Slaughter Act] of 1978 (7 U.S.C. §§ 1901 et seq.) requires that humane methods be used for handling and slaughtering livestock but does not include comparable provisions concerning the handling and slaughter of poultry."[44] Following this Notice, in November of 2005, the U.S. Congress incorporated an amendment to the humane slaughter piece of the FMIA.[45] The amendment replaces each mention of "cattle, sheep, swine, goats, horses, mules, and other equines" with the phrase "amenable species."[46] The definitions section of the FMIA was also adjusted so that "amenable species" means any animals already covered under the FMIA as well as "any species of livestock that the Secretary considers appropriate."[47] This amendment, therefore, grants explicit authority to the USDA to regulate any additional species under the FMIA, however, the USDA has not added one new species to this list.[48]

As a result, in 2006, the Humane Society of the United States (HSUS) sued the USDA over its 2005 Federal Register Notice. The HSUS argued that the USDA violated the HMSA of 1958, abused its discretion, and acted arbitrarily and capriciously and not under the law set out in the Administrative Procedure Act by informing the public that protections do not extend to chickens, turkeys, and other poultry species.[49] The court denied the government's motion to dismiss, so the USDA was obligated to entertain the question of whether or not poultry was considered "other livestock" regarding the HMSA.[50] Friedrich notes, "The government argued that both the statutory text and legislative history were ambiguous as to the meaning of 'other livestock,' and that the Agency should be granted a deference under a *Chevron* analysis."[51] Despite HSUS's argument that the plain language of the Act should be interpreted to include poultry based on the dictionary definition of livestock, the district court determined that the term livestock was "ambiguous on its face" and "that legislative history unambiguously proved Congress' intent to exclude poultry from the scope of the HMSA."[52] However, the Ninth Circuit vacated the district court's

[41] Ibid., p. 5.
[42] Ibid., p. 5.
[43] Ibid., p. 5.
[44] Ibid., p. 5.
[45] Ibid., p. 6.
[46] Ibid. p. 6.
[47] Ibid., p. 6.
[48] Ibid., p. 6.
[49] Ibid., p. 6.
[50] Ibid., p. 6.
[51] Ibid., p. 7.
[52] Ibid., p. 7.

holding on standing grounds regarding their analysis of the legislative history.[53] The Ninth Circuit's perspective emphasized the fact that there was Congressional debate that held views favoring both interpretations – one that would include poultry within the HMSA and one that rejected these animals from inclusion.[54] Through this statement by the Ninth Circuit, there is an indication that the USDA could include poultry under the law if they wish.[55] Thus, there is only a need for a strong argument in favor of doing so to convince the USDA to reverse its position. Law professionals argue that there is palpable evidence for this to come to fruition and the USDA must eventually protect poultry welfare through both the HMSA and FMIA.

### 7.2.3 EU: Global Leader in Poultry Welfare Regulation

Although the EU is not the global leader of poultry meat production, it is still a large contributor of poultry meat and a net exporter of poultry products with an annual production of around 13.4 million tons.[56] The EU is, in fact, the global leader of poultry welfare regulation seeing as it was the first to include "Welfare Indicators" in animal welfare legislation as a means of scientific assessment in June 2007.[57] Council Directive 2007/43/CE lays down the minimum rules for the protection of chickens kept for meat production.[58] The aim of this directive is "to reduce the overcrowding of chicken holdings by setting a maximum stocking density and ensure better animal welfare by specifying requirements such as lighting, litter, feeding, and ventilation."[59] Moreover, in 2016, the European Commission (EC) produced a report to the European Parliament and the Council on the impact of genetic selection on the welfare of chickens kept for meat production.[60] In 2017, the EC concluded a study on the "Application of the broiler directive (DIR 2007/43/EC) and development of welfare indicators" which further expanded the work on animal welfare in the poultry industry.[61] These works were improved further in 2018 when the EC adopted a "Report to the European Parliament and the Council on the application of Directive 2007/43/EC and its influence on the welfare of chickens kept for meat production, as well as the development of welfare indicators."[62]

Regarding welfare at slaughter, the EU adopted Council Regulation (EC) No 1099/2009 on the protection of animals at the time of killing in 2009 which started to apply in January 2013.[63] This replaced the previous legislation to protect animals at the time of killing, Directive 93/119/EC, which was described as "outdated."[64] The 2009 regulation was then amended in 2018 as a result of the opinion of the European

[53] Ibid., p. 7.
[54] Ibid., p. 7.
[55] Ibid., p. 7.
[56] European Commission (n.d.a).
[57] European Commission (n.d.b).
[58] Ibid.
[59] Ibid.
[60] Ibid.
[61] Ibid.
[62] Ibid.
[63] European Commission (n.d.c).
[64] Ibid.

Food Safety Authority on low atmospheric pressure system for the stunning of broiler chickens.[65] Thus, Annexes I and II to Council Regulation (EC) No 1099/2009 have been amended by Commission Implementing Regulation (EU) 2018/723 to include the approval of low atmospheric pressure stunning.[66]

## 7.3 CONCLUSION

Even though a plant-based diet is more sustainable, poultry is a frontrunner in nutritious, economically beneficial, and potentially carbon neutral agricultural commodities produced from avian animals that are proven to have high levels of cognition and sentience. Yet, the world's highest producer of poultry meat and many other countries around the world have a deficit in poultry welfare regulation which prevents these intelligent and sensitive animals from having the respect they deserve. The USDA is perceivably obligated under the HMSA and FMIA to include poultry in its humane slaughter regulation moving forward. Further, the U.S. and any country attempting to expand their animal welfare regulation should look to the EU's current system for guidance, which appears to be more developed albeit not perfect. There remains much room to improve the welfare of poultry, from the hatcheries to transportation, to slaughtering.

## REFERENCES

Applachian State University. (n.d.). Sustainable poultry production. *Sustainable Development.* https://sd.appstate.edu/community/sustainable-poultry-production.

European Commission. (n.d.a). Poultry. https://ec.europa.eu/info/food-farming-fisheries/animals-and-animal-products/animal-products/poultry_en.

European Commission. (n.d.b). Poultry meat. *Food Safety.* https://ec.europa.eu/food/animals/animals-products-trade-imports/poultry-meat_en.

European Commission. (n.d.c). Slaughter and stunning. *Food Safety.* https://ec.europa.eu/food/animals/animal-welfare/animal-welfare-practice/slaughter-stunning_en.

Food and Agriculture Organization of the United Nations. (2019, November). Codex alimentarius. *Gateway to poultry production and products.* https://www.fao.org/poultry-production-products/products-processing/risks-to-human-health/codex-alimentarius/en/.

Food and Agriculture Organization of the United Nations. (2021, June). Food outlook biannual report on global food markets. https://www.fao.org/3/cb4479en/cb4479en.pdf.

Food and Agriculture Organization of the United Nations. (n.d.a). Products and processing. *Gateway to poultry production and products.* https://www.fao.org/poultry-production-products/products-processing/en/.

Food and Agriculture Organization of the United Nations. (n.d.b). Poultry in human nutrition. *Gateway to poultry production and products.* https://www.fao.org/poultry-production-products/products-processing/poultry-in-human-nutrition/en/.

Food and Agriculture Organization of the United Nations. (n.d.c). Animal welfare. *Gateway to poultry production and products.* https://www.fao.org/poultry-production-products/production/animal-welfare/en/.

---

[65] Ibid.

[66] Ibid.

Friedrich, B. (2015). Still in the jungle: poultry slaughter and the USDA. *New York University Environmental Law Journal*, *23*(2), 245.

MacLeod, M., Gerber, P., Mottet, A., Tempio, G., Falcucci, A., Opio, C., Vellinga, T., Henderson, B. & Steinfeld, H. (2013). Breakdown of total global GHG emissions by category for chicken meat and egg supply chains in *Greenhouse Gas Emissions from Pig and Chicken Supply Chains – A Global Life Cycle Assessment*. Food and Agriculture Organization of the United Nations (FAO), Rome, p. 47. https://www.fao.org/3/i3460e/i3460e.pdf.

Nicol, C. J. & Davies, A. (n.d.). Poultry welfare in developing countries. Poultry Development Review. School of Veterinary Science, University of Bristol, Bristol, United Kingdom. https://www.fao.org/3/i3531e/i3531e09.pdf.

Organization for Economic Co-operation and Development - Food and Agricultural Organization of the United Nations. (2021). Meat. OECD-FAO agricultural outlook 2021–2030. pp. 110–121. https://www.fao.org/3/BT089e/BT089e.pdf.

Painter, J. A., Hoekstra, R. M., Ayers, T., Tauxe, R. V., Braden, C. R., Angulo, F. J., & Griffin, P. M. (2013). Attribution of foodborne illnesses, hospitalizations, and deaths to food commodities by using outbreak data, United States, 1998–2008. *Emerging Infectious Diseases*, *19*(3), 407–415. https://doi.org/10.3201/eid1903.111866.

Rayner, A. C., Newberry, R. C., Vas, J. et al. (2020, September). Slow-growing broilers are healthier and express more behavioural indicators of positive welfare. *Scientific Reports*, *10*, 15151. https://doi.org/10.1038/s41598-020-72198-x.

Stevenson, P., Battaglia, D., Bullon, C., & Carita, A. (2014). Review of animal welfare legislation in the beef, pork, and poultry industries. *Food and Agriculture Organization of the United Nations*. https://www.fao.org/3/i4002e/i4002e.pdf.

# *Part 3*

## *Wildlife, Climate Change, Habitat, and Invasive Species*

# *Wildlife, Climate Change, Habitat, and Invasive Species*

*Sahana Ramdas*

Since the beginning of life on our Planet Earth, wild animals have played a pivotal role in creating a thriving environment. Wildlife is an integral part of the natural environment and worldwide ecosystems. However, humans are accelerating the Sixth mass extinction of wildlife, also known as the Anthropocene extinction. Over two-thirds of wildlife have been wiped out from their natural habitats in the last five decades alone.[1] Between 1970 and 2016, an average 68% drop in wildlife population sizes of amphibians, birds, aquatic animals, and reptiles occurred worldwide.[2] The most significant fall in wildlife was witnessed in tropical sub-regions of America, with over 94% of wildlife species plummeting in the last five decades.[3] Over One million wildlife species are now threatened with extinction; this includes 500,000 wild animals and wild plants.[4] In this Anthropocene era, wildlife species are dying off 1,000 times more frequently than during the 60 million years before the arrival of humans.[5] Wild animals are killed directly for wildlife trade and indirectly due to industrial food production systems. Humans' appetite for dairy, animal meat and wildlife by-products is the leading cause of the alarming decline of wildlife populations and the recurring outbreak of zoonotic diseases. Due to our disharmonious relationship with wildlife and their natural habitats, countries are also grappling with complex issues like climate change.

---

[1] https://www.smithsonianmag.com/smart-news/humans-wiped-out-two-thirds-worlds-wildlife-50-years-180975824/.

[2] https://f.hubspotusercontent20.net/hubfs/4783129/LPR/PDFs/ENGLISH-FULL.pdf; https://www.weforum.org/agenda/2020/11/extinction-facts-wildlife-attenborough/.

[3] https://f.hubspotusercontent20.net/hubfs/4783129/LPR/PDFs/ENGLISH-FULL.pdf.

[4] https://www.un.org/sustainabledevelopment/blog/2019/05/nature-decline-unprecedented-report/; Also see: https://www.bbc.com/news/science-environment-54091048.

[5] Brown University's 2014 study. https://news.brown.edu/articles/2014/09/extinctions.

DOI: 10.1201/9781003137733-15

So far, the sobering reality of massive biodiversity losses is not adequately addressed. Often, the need to protect wild animals comes to the forefront only after the critical time to save them has passed by. This is because many of the wildlife species that require legal attention are either unknown, or the threat levels to individual wild animals are not fully ascertained. At the pace in which wildlife is being extirpated, we are bringing in the Homogenocene—an era where the richness of biodiversity is severely diluted, ushering in monoculture-themed landscapes, in lifeless and nondiverse ecosystems.[6]

Wildlife habitats provide critical shelter, food, and nesting areas for biodiversity and help the larger environment by securing the resilient functioning of natural resources like clean air, water, and soil. Habitat loss affects most wildlife species, starting from terrestrial to marine wildlife. Marine species are adversely affected by overfishing, dumpage of water effluents, and pollution as water ecosystems are the only habitats where aquatic animals can survive and thrive. Since we use our waterways as dumping and sewage grounds, marine biodiversity's health, well-being, and survival are severely jeopardized.

Habitat losses not only endanger individual wild animal populations but have an adverse ripple effect on the global ecosystem. As apart from providing critical wildlife habitats for wild animals, natural habitats significantly contribute to carbon sequestration. Hence, the need for restoring wildlife habitats is critical to retaining the same life sources we are heavily dependent on. Destruction, fragmentation, and loss of wildlife habitats mainly occur from razing away wilderness for agriculture, deforestation, livestock grazing, logging, mining, urbanization, and water conversions. The immediate impact of land-use conversions is soil erosion, degradation, and desertification. Soil is an integral part of terrestrial landscapes, and soil habitats play an astounding number of beneficial roles in worldwide ecosystems. Often regulations governing land-use and spatial land-use planning impact the extent of soil protection a country adopts. Accordingly, small but progressive steps are taken internationally, like the United Nations Rio + 20 Conference final draft, which seeks member nations to combat land degradation, soil erosion, and loss of healthy ecosystems.[7] The chapter on Grounding Habitat and sustainability (Chapter 8)[8] first highlights the causes behind soil erosion, desertification, soil acidification, and other forms of soil degradation. Second, it analyses the global legal framework to assess whether phytoremediation can be integrated to combat environmental pollution and climate change; it examines the role of soil remediation in promoting soil stability and sustainable crop production. Third, it highlights the national and international legal landscape which affects soil management and crop cultivation practices.

According to the IPBES report, just between 1980 and 2000, over 100 million hectares of tropical forests were lost.[9] The main driver is cattle ranching operations

---

[6] See: Charles C. Mann on Homogenocene. https://longnow.org/seminars/02012/apr/23/living-homogenocene-first-500-years/.

[7] The Future We Want, Desertification, land degradation and drought, 205–207. http://rio20.net/wp-content/uploads/2012/06/N1238164.pdf.

[8] Grounding Habitat and Sustainability via Phytoremediation Strategies, Christian de Laire Mulgrew.

[9] https://www.un.org/sustainabledevelopment/blog/2019/05/nature-decline-unprecedented-report/.

in Latin America and plantations in Southeast Asia.[10] The overgrazing of pastoral lands has caused soil erosion, loss of wildlife habitats, and environmental degradation. However, the sustainable use of pastoral lands protects the ecological health of local ecosystems while contributing to livelihoods and the economy. The chapter on Pastoral lands (Chapter 12) explores the multilateral environmental agreements and national laws pertaining to pastoral land management and calls for legislative reform to sustain pastoralism.[11]

Globalization has led to humans introducing foreign wild animals outside their native ranges. These introduced animals act as ecological invaders, causing a detrimental impact on native plants, animals, and local environments. For instance, non-native wild animals deliberately introduced into foreign lands for farming, hunting, and wildlife management have caused severe ecological harm. When humans play God with nature, unintended consequences follow, like invasive species wrecking native biodiversity and natural ecosystems. Invasive, alien, and feral animals are notoriously known for outcompeting native animals for resources and successfully multiplying their populations, thereby seriously endangering native populations. Nations are undergoing ecological imbalances due to the ecological harms brought by invasive species in native ecosystems. The unfettered growth of invasive species costs each country billions of dollars every year.[12]

Incentivized to reduce disruptive animals in unwelcome lands, people are increasingly hunting and eating invasive species—this approach is called invasivorism. The chapter on Invasivorism (Chapter 9) emphasizes that since meat-eaters are taking to the consumption of destructive animals—to reduce the invasive population's impact on the environment—the indiscriminate killing of beneficial wild animals and domestic animals needs to become limited.[13] As an afterthought, commercializing the eradication of invasive species could cause unintended consequences, such as the demand for invasive animal meat outgrowing the supply—leading to invasive species becoming coveted animals.[14] Hence, the ecological tampering with undesirable wildlife introductions needs to stop.

The World Organization for Animal Health (OIE) has laid down the Five Freedoms every animal is naturally entitled to on this planet. However, promoting these animal welfare principles is severely undermined as they do not carry the weight of binding laws and regulations at the national, regional, or international level. Moreover the cataclysmic losses of wildlife and natural habitats disrupt ecosystem functioning and exacerbate the climate change crisis. Industrial agriculture has caused the loss of over 60% of global biodiversity and is directly attributed to about a quarter of $CO_2$ emissions worldwide.[15] Hence, halting and reversing wildlife extinction would help mitigate the impact of climate change.

---

[10] https://www.un.org/sustainabledevelopment/blog/2019/05/nature-decline-unprecedented-report/.

[11] Land as Carbon Sinks or Pollution Sources: International Pastoral Land Law, Christian De Laire Mulgrew.

[12] See: https://www.ecowatch.com/invasive-species-damage-cost-2651334657.html.

[13] Invasivorism as a Sustainable Strategy to Animal and Resource Exploitation, Morgan Boutellier.

[14] See: https://www.nytimes.com/2020/10/02/t-magazine/eating-invasive-species.html.

[15] https://www.weforum.org/agenda/2020/11/extinction-facts-wildlife-attenborough/.

Insects are the basis of biodiversity food webs, yet we are nearing an insect apocalypse because over 500,000 insects are seriously endangered.[16] The insect apocalypse is unsettling as the vast majority of animal pollinators are insects. Over 40% of insects face sharp declines due to habitat losses caused by Industrial agriculture.[17] The chapter on Pollinator Welfare (Chapter 10) briefly highlights the inhumane management of insects, like the honeybees in the industrial animal agriculture setup, and how the absence of Five Freedoms in beekeeping imperils bee welfare and pollinator biodiversity simultaneously.[18]

The unscrupulous hunting of wild animals for the international wildlife trade has been addressed through the Convention of International Trade in Endangered Species (CITES). However, the CITES does not address the domestic trade of wildlife or wildlife crimes that have a transboundary impact on public health and worldwide ecosystem health. Calls to reform the CITES are yet to be heeded. Since the spread of infectious diseases is a consequence of our increasingly adverse contact with wild animals, conservation advocates are calling for a One Health approach where the Food and Agricultural Organization (FAO), the World Organization for Animal Health (OIE), and the World Health Organization (WHO) work together to protect public health, wildlife health, and our planet's health.[19] The Convention on Biological Diversity aims to protect plant and animal biodiversity globally. Its associated Cartagena Protocol focuses on preventing genetically engineered organisms and other transgenic organisms from adversely impacting native ecosystems and endangering biodiversity and biosecurity. Although International Conventions like the Convention on Biological Diversity (CBD) help recognize and acknowledge issues that plague wildlife, member nations are yet to take substantive steps in enforcing wildlife conservation programs or abiding by environmental principles and wildlife protection policies.

The chapter on Brazil's Role in Food production (Chapter 11) signifies the need for the country to adopt a sustainable food production model so that it could honor its shared responsibility toward conserving and managing biodiversity hotspots like the Amazon Rainforest, which acts as the lungs of our planet. The chapter also examines the position of food security by dwelling on biotechnology-based cultivation, food wastage, and food insecurity issues. It also explores the beneficial qualities of plant-based food and approaches toward philanthropic food donation.[20]

The section mainly highlights that the disregard for biodiversity welfare and the anthropogenic nature of humans is causing the massive loss of wildlife and their natural habitats.

Finally, this book emphasizes that protecting animal welfare and biodiversity is critical to the healthy functioning of wildlife food webs, flora, fauna, and food production.

---

16 https://www.bbc.com/news/science-environment-54091048.

17 https://www.weforum.org/agenda/2020/11/extinction-facts-wildlife-attenborough/.

18 Managed Bees v. Pollinator Welfare, Sahana Ramdas.

19 See: https://www.scientificamerican.com/article/a-crucial-step-toward-preventing-wildlife-related-pandemics/.

20 Brazil's Role in Food Production – Food Security and Sustainability, Gerardo Figueiredo Junior and Paulo Bellonia.

# 8 Grounding Habitat and Sustainability via Phytoremediation Strategies[1]

*Christian Mulgrew de Laire*

## CONTENTS



## 8.1 INTRODUCTION[2]

Soils throughout the globe are under mounting environmental pressure driven by human activities. Agriculture, forestry, industrialization, tourism, and urban development accelerate soil degradation with no signs of amelioration. The main threats to soil health and ecosystems are erosion, chemical contamination, compaction, biodiversity loss, sealing, landslides, and flooding. While soils serve as a resource within and across countries, its degradation substantially impacts other resources

[1] Based on chapter 20 by Bernard Vanheusden in *International Farm Animal, Wildlife, and Food Safety Law* (Steier and Patel, Eds.) (Springer 2017).

[2] This contribution is partially based on the following research report within the Rejuvenate 2 project: Vanheusden et al. (2011). This project was funded by the Snowman network.

DOI: 10.1201/9781003137733-16

by affecting water quality and quantity, climate change, biological diversity, human health, food, feed security, and food security.

Soil is crucially important as the interface between water, air, and earth. It is the basis to grow food, provide biomass and raw materials, serve as a carbon sink, provide habitat, and house uncountable gene pools. Soil serves as a platform for human activities and as an archive of heritage. It stores, filters, and transforms many substances, including water and carbon. At 1,500 gigatonnes of carbon, soil is, in fact, the largest carbon store in the world.[3] These functions require protection for their socio-economic and environmental importance.

Thus, we must prevent soil degradation and restore degraded soils to levels functional for their intended use. This chapter examines the legal aspects of an innovation called phytoremediation, which simultaneously restores soils and uses them for agricultural cultivation. Phytoremediation uses plants and crops to remove pollutants from the environment or render the pollutants harmless.[4] These plants are then potentially valuable for food or non-food uses.

The combination of agricultural cultivation and soil remediation could become a significant part of sustainable risk-based land management and could address large-scale erosion and diffuse and moderate contaminations. Unfortunately, the legal framework to use phytoremediation or broader sustainable risk management is missing. This contribution offers an overview and analysis of potential legal bottlenecks that require consideration before starting a phytoremediation project.

## 8.2 WHAT IS PHYTOREMEDIATION?

Phytoremediation refers to the bioremediation of contaminated soils through the use of plants that are capable of removing or degrading organic or inorganic pollution in soil, water, and air. Maize, rapeseed, and sunflower are the most commonly used plants with these capabilities, along with various types of short rotation coppice (birch, maple, poplar, etc.)

The two main types of phytoremediation are stabilization and extraction. Both techniques use plants but differ in their processes. Stabilization focuses on immobilizing contaminants to prevent the uptake of contaminants into harvestable parts of plants or them leaching into the environment. However, the contaminants remain in the soil. Extraction aims at the uptake of the contaminants into harvestable parts of the crops. Subsequently, the contaminated biomass is harvested and removed from the site. This difference between the two techniques impacts the applicability of regulations for the phases of a phytoremediation project (Table 8.1).

The table below summarizes the advantages and disadvantages of conventional and phytoremediation techniques.[5]

3 European Commission (2006), p. 2

4 Refer to e.g. the Rejuvenate 2 project (Crop Based Systems for Sustainable Risk Based Land Management for Economically Marginal Degraded Areas). For more information refer to http://projects.swedgeo.se/r2/.

5 Witters et al. (2009).

**TABLE 8.1**
**Comparison of Conventional Remediation and Phytoremediation**

| Type of Remediation | Conventional | Phytoremediation |
|---|---|---|
| Characteristics | • Excavation<br>• *Ex-situ* treatment (usually)<br>• Treatment or disposal | • Use of plants<br>• *In situ* treatment<br>• Use for food/feed, energy, material |
| Advantages | • Good for serious contamination<br>• Quick remediation | • Good for diffuse contamination<br>• Suitable for large surfaces<br>• Keeps soil structure<br>• Cost similar to farming cost |
| Disadvantages | • Only for concentrated contamination<br>• Only suitable for small surfaces<br>• Loss of soil structure<br>• High cost | • Only for moderate contamination<br>• Slow remediation |

Phytoremediation could potentially enhance the degradation and removal of organic contaminants,[6] while simultaneously producing valuable crops (food and non-food). Phytoremediation thus simultaneously concerns soil remediation, soil usage, and improving the productivity of land and soil. Consequently, phytoremediation is environmentally sound and relates to a benefit for animals.

Combining agricultural cultivation and soil remediation could play an essential role in self-sustaining, risk-based land management remediation, resulting in other environmental benefits. Risk-based land management requires considering the impact of toxic substances before, during, and after their removal from the environment. Risk management aims to identify and break the contaminant-pathway-receptor chain. The pathways are diverse, but a typical example is the uptake of contaminants in the soil by vegetables that humans consume. After remediation, contaminants might end up at a disposal site. A remediation strategy is effective so long as it controls the health and environmental risks associated with a pollutant linkage by removing contaminant sources, breaking exposure pathways between source and receptor, or changing the receptor.[7] A management strategy, thus, may not necessarily remove the toxic substances.

Total contaminant removal rate and resulting remediation depend on soil characteristics, level of contamination, available contamination, crop extraction, and crop biomass production.[8] The removal of contaminants reduces the total amount of contamination and the number of available contaminants. In most cases, this relationship is logarithmic, meaning the number of available contaminants reduces faster than the level of contamination. However, some authors describe a replenishment of contaminants as is the case with sandy soils.[9] Biomass production of a plant might

[6] Vangronsveld et al. (2009) and Weyens et al. (2009, 2010).
[7] Mench et al. (2010).
[8] Koopmans et al. (2008).
[9] Van Nevel et al. (2007).

also change over time. For example, the biomass potential of short rotation coppice of willow might increase after several years within a rotation cycle. In contrast, the biomass production of maize or rapeseed might decrease over time (due to nutrient depletion). More so, the rooting depth might factor into the contaminant concentration in plants. Finally, there could be leaching losses to the groundwater.[10]

Phytoremediation becomes a more appealing method of land restoration when the produced plants/crops/biomass are valuable as food, feed, or biomass. This ability can mitigate some of the main drawbacks of phytoremediation—that it takes place on contaminated land and may require an extended remediation period. As a result, slower working phytoremediation schemes based on stabilization or gradual attenuation of the contaminants rather than short-term forced extraction may be envisaged.[11]

Conventional remediation techniques take less time but make it impossible to validate the soil. In contrast, the repeated cropping of plants in phytoremediation can produce high amounts of plants, crops, and biomass. So long as contaminant levels in these plants remain below allowable levels for food or feed, the plants can be used accordingly. Otherwise, they must be disposed of or properly treated to prevent risk to the environment.[12] Even when unusable for food or feed, the contaminated plants may be useable as an energy resource, which can be profitable and reduce carbon dioxide emissions.[13] Biomass can replace fossil fuels as sources of heat, electricity, transport fuel, and feedstock for material production.[14] Firbank (2008)[15] mentions the environmental benefit of alternative crops leading to their increased usage.

Energy conversion is a sustainable alternative for two reasons. First, energy production has more public approval than other destinations (e.g., paper mills). Farmers are rewarded for direct positive contributions to biodiversity, water quality, and soil health in many countries. Many countries also support bioenergy programs with the intent to produce and use cleaner sources of energy. Furthermore, on a global scale, increasing interest in carbon sequestering effects of agriculture indicates a growing number of programs will support certain farming practices as a path to improving overall air quality.[16] Second, energy conversion installation can trace heavy metals within their systems. As far as we know, there has been research conducted in the energy sector, but there has been no research on tracing heavy metals in other biomass.

Also, phytoremediation can quite possibly produce a combination of food, feed, and biomass. With maize, for example, contamination is taken up by the green parts but not the grains. As a result, the maize grains can be used for food, while the green parts can be harvested for fuel.

10 Van der Grift and Griffioen (2008).
11 Robinson et al. (1998).
12 Ghosh and Singh (2005).
13 Chaney et al. (1997) and Meers et al. (2005).
14 Dornburg and Faaij (2005).
15 Pg 579 No footnote.
16 De Vries (2000).

## 8.2.1 Goals and Structure of Phytoremediation

The combination of agricultural cultivation and soil remediation could be an essential part of sustainable risk-based land management and can feasibly address all kinds of large-scale diffuse and moderate contaminations. Unfortunately, there is no sound legal framework for phytoremediation and broader sustainable risk-based land management, to which this chapter offers an overview and common legal difficulties. This part examines decision-making chronologically before and during the phytoremediation project. The following two subsections divide the soil remediation process into crop selection and soil management, then growing and harvesting crops.

## 8.2.2 Selection of Plants for Soil Management

### 8.2.2.1 Plant Characteristics

The first question is whether any plant species are prohibited for use in the phytoremediation project. Several international, supranational, and national agreements and regulations impact plant legality. One should ensure that these instruments allow the planting of the selected species at the chosen site.

At the international level are the Convention on Biological Diversity (CBD) of 1992 and the International Plant Protection Convention (IPPC) of 1951. Within the European Union, there is the Habitats Directive[17] and the GMO Directive.[18] However, for each project, an assessment based on the site location and local circumstances remains necessary.

### 8.2.2.2 Invasive and Exotic Plants

Invasive and exotic plans are dangerous to biodiversity. Florida (USA) experienced first-hand the dangers of *Arundo donax*, also known as Giant Reed. This large clumping grass produces a lot of biomass usable for bioenergy feedstock[19] and is also used in phytoremediation to uptake arsenic,[20] mercury, and cadmium.[21] However, giant reed alters hydrology and displaces native species.[22] Its method of reproduction poses a risk to fish and amphibian wildlife, including rare species. Nonetheless, *Arundo donax* grows worldwide, planted for decorative purposes as well as its phytoremediation and biomass production capabilities. There is ongoing research on genetically modifying the grass for phytoremediation purposes.

---

[17] Council Directive 92/43/EEC of 21 May 1992 on the conservation of natural habitats and of wild fauna and flora, *Official Journal* 22 July 1992, L 206, consolidated version of 2007 on ec.europa.eu/environment/nature/legislation/habitatsdirective/index_en.htm.

[18] Directive 2001/18/EC of the European Parliament and of the Council of 12 March 2001on the deliberate release into the environment of genetically modified organisms and repealing Council Directive 90/220/EEC, *Official Journal* 17 April 2001, L 106.

[19] Florida Native Plant Society, *Policy statement on Arundo donax*, p.1 http://www.weedcenter.org/management/docs/Arundo%20Policy%20Statement.pdf.

[20] www.ncbi.n1m.nih.gov/pubmed/20363125.

[21] www.ncbi.n1m.nih.gov/pubmed/16110677.

[22] www.issg.org/database/species/ecology.asp?si=112.

In 1992 the Parties to the CBD committed to protecting ecosystems, habitats, and species against alien threats.[23] Thus, it is beneficial to check national and local rules relating to non-native, alien species. For example, the Habitats Directive in the EU obliges the member states to *ensure that the deliberate introduction into the wild of any species which is not native to their territory is regulated so as not to prejudice natural habitats within their natural range or the wild native fauna and flora and, if they consider it necessary, prohibit such introduction.*[24] Examples of plants that fall under the Habitat Directive are Miscanthus, Switchgrass, and *Arundo donax*.

If plants and animals are on the chosen site, checking for protected species is also good practice. The IPPC concerns plant selection through article VII, 2, (i) *[c] ontracting parties shall, to the best of their ability, establish and update lists of regulated pests, using scientific names, and make such lists available to the Secretary, to regional plant protection organizations of which they are members and, on request, to other contracting parties*. These lists are available on the IPPC,[25] Regional Plant Protection Organization (RPPO) website,[26] and national or state authorities.

Other regional or national legislation may apply, like the EU Directives on protective measures against harmful organisms for plants and plant products. Annex III is relevant for phytoremediation as it contains a list of banned plants either in general or in protected zones.

#### 8.2.2.3 Genetically Modified Plants (GM Plant) and Soil

As with any issue, opinions diverge on genetic modifications of organisms (GMOs). On the one hand, modification can improve phytoremediation by increasing plant resilience to abiotic stress and improving contaminant uptake or promoting the root's immobilization capacity. On the other hand, much uncertainty still surrounds certain aspects of genetic modification, particularly their safety, integrity, and sustainability. In fact, the monocultures of GMOs systematically dry out, exploit, and eradicate fertile soil at alarming rates worldwide.

Soils are some of the most complex habitats on earth, with one gram of agricultural or forest soil from temperate regions containing thousands of species.

[23] Article 8, h of the Convention on Biological Diversity (CBD).

[24] Article 22 (b) of the Habitat Directive.

[25] Refer to https://www.ippc.int/countries/regulatedpests/, last visited 6 January 2015.

[26] The RPPOs are inter-governmental organizations functioning as coordinating bodies for National Plant Protection Organizations (NPPO) on a regional level. There are currently ten RPPOs:

- Asia and Pacific Plant Protection Commission (APPPC)
- Comunidad Andina (CA)
- Comite de Sanidad Vegetal del Cono Sur (COSAVE)
- Caribbean Plant Protection Commission (CPPC)
- European and Mediterranean Plant Protection Organization (EPPO)
- Inter-African Phytosanitary Council (IAPSC)
- Near East Plant Protection Organization (NEPPO)
- North American Plant Protection Organization (NAPPO)
- Organismo Internacional Regional de Sanidad Agropecuaria (OIRSA)
- Pacific Plant Protection Organization (PPPO).

Given this complexity, the impact of genetically modified plants on soil systems is unknown.[27] Moreover, a plant modified to be better suited to the environment can become invasive and potentially reduce biodiversity. Careful consideration of these risks is necessary.[28]

Genetic modification's considerable costs and implications need careful examination before selecting a species for a phytoremediation project. Besides the impact of regulation, the educated public's opinion of GMOs is rarely positive. GMO regulation in the US, for example, favors their development. GMOs are economically significant to the biotechnology industry and to the US in general. As a result, the US is the world's leading producer of genetically modified crops.[29] However, in the EU, GMO plants can only go to market if precautionary measures by member states guarantee their safety for humans and the environment. Thus, the precautionary principle should be a decision criterion, as the authorization of a GM plant happens individually and requires testing of its environmental effects.[30] Under European law, member states can ban an authorized GMO if the state can justify the prohibition. This was demonstrated when the European Environmental Ministers voted against forcing Austria and Hungary to allow Monsanto's MON810 GM maize to be grown in their countries. The ban was upheld, notwithstanding the full authorization of maize for food and feed, plus processing and import.[31] Thus continues the debate on the accidental spread and adverse effect on nature.[32] EU Member states can ban GMOs grown for food from their territory; however, the European Commission has the final say.[33]

## 8.2.3 Soil Management

### 8.2.3.1 Legal Framework

National and local rules govern soil, so it is essential to check this legal framework before implementing a phytoremediation project. Legislation varies in scope, from a broad focus, covering soil protection in general, to a narrow focus specific to soil remediation. The Netherlands, for example, passed the Soil Protection Act of 1987, with provisions on soil protection and remediation. This legislation concerns historical pollution, soil contamination, and sustainable soil management, with a focus that closely links soil and water.

In other countries, soil falls under the same regulation as groundwater or a water area. In Sweden, for example, the main criterion of the Swedish Environmental Code

---

27 The potential environmental, cultural and socio-economic impacts of genetically modified trees, *Conference on Biological Diversity*, March 2008, UNEP/CBD/COP/9/INF/27, p 20, www.cbd.int/doc/meetings/cop/cop-09/information/cop-09-inf-27-en.pdf, last visited on 6 January 2015

28 See also the paragraph on soil management.

29 Refer to http://www.loc.gov/law/help/restrictions-on-gmos/usa.php#_ftn2, last visited on 6 January 2015.

30 Article 4, GMO Directive.

31 http://www.gmo-compass.org/eng/gmo/db.

32 Austria, Hungary Allowed to Keep Ban on Genetically Modified Crops, *Deutsche Welle*, http://www.dw.de/austria-hungary-allowed-to-keep-ban-on-genetically-modified-crops/a-4068097, last visited 6 January 2015.

33 https://eur-lex.europa.eu/legal-content/EN/TXT/?uri=CELEX:32019R1381, Last accessed Dec 14, 2021

is the presence of potential harm to human health or the environment.[34] A supervisory authority may require a responsible party to remediate the damage caused, regardless of when the pollution occurred.

On the international level, soil protection plays an increasingly important role. The final document from the United Nations Rio +20 Conference aims for land degradation neutrality, defined as *a state whereby the amount of healthy and productive land resources, necessary to support vital ecosystem services, remains stable or increases within specified temporal and spatial scales. LDN can occur as the result of natural regeneration or improved land management practices and ecosystem restoration.*[35] To this end, a Working Group was created to report on the understanding and advancement of the common goal and could lead to the inclusion of LDN obligations in legal instruments soon.[36]

Even where no specific soil regulation exists, legislation relating to groundwater, chemicals, waste, and agriculture impacts soil protection and production. Furthermore, the legal national/regional regulation of land use or spatial planning impacts soil use. Once the site owner approves a project, local regulation should be consulted for further action.

Necessary approvals may differ between phytostabilization and phytoextraction. The following paragraphs further elaborate on these techniques.

## 8.3 PHYTOREMEDIATION: THE DIFFERENCE BETWEEN STABILIZATION AND EXTRACTION

The most common phytoremediation techniques are stabilization and extraction. Since there are differences in technique, impact, and legislation, the following analyzes each option.

Phytostablization focuses on immobilizing contaminants to prevent uptake or leaching into the environment. The contaminants do, however, remain in the soil. The technique aims to decrease bioavailability and can be used on-site with heavy metal contaminants through plants and soil amendments.[37] The amendments prevent uptake and diminish leaching.

Notably, amending substances is de facto changing soil composition. Furthermore, by immobilizing some contaminants, it is possible to mobilize others.[38] This process can lead to leaching into groundwater, at which point groundwater regulation takes effect. Additionally, amendments can change soil structure or immobilize essential elements.[39] When it results in a loss of biodiversity, this process will trigger related legislation.

---

[34] Ministry of Environment, *The Swedish Environmental Code, A résumé of the text of the Code and related Ordinances, Genetic Engineering,* Danagrards Grafiska, Stockholm, Sweden, p. 27.

[35] UNCCD Intergovernmental Working Group (IWG) on the follow up to the outcomes of Rio +20, Task1-Science-based definition of land degradation neutrality, 28 May 2014.

[36] For the latest report, refer to http://www.unccd.int/Lists/SiteDocumentLibrary/Publications/V2_201309-UNCCD-BRO_WEB_final.pdf?HighlightID=329.

[37] Vangronsveld et al. (2009), p. 766.

[38] Vangronsveld et al. (2009, p. 767 and 700).

[39] Ibid p.767

Finally, chemical legislation applies when soil is chemically treated, as is the case with phytostabilization. However, since the goal is not to excavate, sell, or use the soil in articles, this obligation remains theoretical.

On the other hand, phytoextraction reduces contaminants in the soil through uptake in harvestable plants. For the plan to succeed, contaminants must be bioavailable. The degree of uptake varies according to the contaminant and capability of the crop.

Many metal contaminants are essential plant micronutrients. In common non-accumulator plants, the accumulation of metal contaminants does not exceed their needs.[40] Some plants can accumulate more than they need. As a result, the plants accumulate high amounts of contaminants, storing them in foliage. Further processing and disposal of the plants then become necessary.

Contrary to stabilization, extraction amendments are used to increase the extraction bioavailability of contaminants. Legislation often includes requirements to prevent and limit leaching resulting from the increased mobility of contaminants.

To increase the uptake of lead (Pb) by crops, ethylene-diamine-tetraacetic acid (EDTA) is commonly added to the soil in phytoremediation projects.[41] EDTA is considered cytotoxic and weakly genotoxic, and it causes reproductive and developmental issues in animals.[42] Though the substance is currently unregulated, its input into groundwater should be limited.

Growing crops for biofuel in an economically viable way requires the harvest of enough biomass. Intensive cultivation necessitates frequent nutrient inputs through heavy fertilization, often done with water/sludge to limit costs.[43]

Weeds pose another hindrance to growth, especially during a plant's first year. Consequently, biomass production can heavily diminish in the following years. Certain situations necessitate weed control. An example is the problem of broomrape, a parasitic plant that destroyed up to 70 percent of sunflowers.[44] Pesticides and herbicides are used to control weeds.[45] Pesticides harm human health, so legislation on plant protection generally applies strict rules to these products. Residue limits also typically apply to food and feedstuffs.[46] Pesticide legislation must be checked before every phytoremediation project.[47]

---

40 Lasat (2000), p. 5.

41 Yuan and VanBriesen (2006), p. 533.

42 Lanigan and Yamarik (2002).

43 BIOPROS, Solutions for the safe application of wastewater and sludge for high efficient biomass production in Short-Rotation-Plantations, *D4-Report on ongoing research and gaps in SRP knowledge*, 14 February 2005, Sweden, February 2006, p. 7

44 Refer to http://www.depi.vic.au.agriculture-and-food/pests-diseases-and-weeds/weeds/state-prohibited-weeds/branched-broomrape; Refer to http://www3.interscience.wiley.com/journal/119212373/abstract?CRETRY=1&SRETRY=0.

45 Pesticides are usually divided into two major groups (plant protection products (PPPs) and pesticides for non-agricultural uses). PPPs contribute to high agricultural yields and help to ensure that good quality food is available at reasonable prices. Pesticides for non-agricultural uses are, e.g., important for public health protection and for preservation of materials.

46 Refer to e.g. in the EU the Plant Protection Directive.

47 Refer to e.g. in the EU Directive 2009/128/EC of the European Parliament and of the Council of 21 October 2009 establishing a framework for Community action to achieve the sustainable use of pesticides, *Official Journal* 24 November 2009, L 309.

Furthermore, post-treatment sewage sludge may be applied to land. Pertinent rules vary depending on the use of the sludge and the classification of the cultivated crops. These classifications vary by country. A separate framework typically regulates sludge use in agriculture. Its use outside of agriculture generally falls under waste law.

In general, countries should strive for integrated pest management systems, carefully considering all available plant protection methods. Subsequently, measures should discourage the use of harmful organisms, limit the use of plant protection products to economically and ecologically justified levels, and minimize risks to human and environmental health.[48] Thereby, measures taken under other regulations to protect sensitive groups and biodiversity should be respected.

## 8.4 GROWING AND HARVESTING THE CROPS

To analyze the legal use of crops grown on contaminated soil, one must consider principles of sustainability, environmental impact, and the optimal closed lifecycle of the crops. These considerations are crucial in case alternatives are available or interpretation of rules is necessary. The lack of specific reference to contaminated biomass means situations arise where legislation is applicable per analogy, based on the intention and goal of that legislation. For example, it is unclear whether legislation on agriculture applies to contaminated biomass. One of the most important objectives of a phytoremediation project is sufficient crop production, such as for conversion into biofuel. Soil fertilization and plant protection actions are often necessary to ensure adequate yields. The following sections respectively discuss the legal implications of plant fertilization and examine the rules concerning plant protection.

### Planting

Is planting crops for phytoremediation an agricultural activity? Agriculture generally connotes feeding humans and animals. Presently, however, farmers are pushed to produce high-quality foods and seek development, such as *renewable environmentally friendly energy sources.*[49]

One can conclude that good agricultural practices and legislation principles can apply to contaminated crops/biomass. We will thus approach other aspects of growing crops considering phytoremediation as an agricultural activity.

Since countries have different interpretations of agriculture, one must analyze local regulation and its relevance for growing crops on non-agricultural contaminated land to understand the local definition.

Agricultural definitions are often broad. Sweden, for example, defines it as "including forestry, hunting, and fishing," as well as "cultivation of crops and livestock production."[50] It remains unclear whether specific legislation applies to cultivation. However, one can choose the most environmentally friendly approach since the objective is to improve the environment and not reduce it by inappropriate remediation of soils.

---

48 Refer also to Article 3, 6 of Directive 2009/128/EC.

49 European Commission (2012).

50 Refer to www.indexmundi.com/facts/Sweden/agriculture.

### Fertilizing

Fertilizers are substances used to enhance the growth of vegetation and can include manure, sewage sludge, chemical fertilizer, and composts. Fertilizers contain nitrogen; thus, excessive use is an environmental concern, especially the eutrophication of water. It causes the growth of algae and plants, changing water quality and the balance of organisms.

Countries usually establish codes or legislation for good agricultural practices such as situations including inappropriate use of fertilization or water-saturated, flooded, frozen, or snow-covered grounds. Additionally, these rules include guidelines for crop rotations and minimum quantity of vegetation coverage. Fertilization should be limited to certain times of the year based on environmental factors. Furthermore, since nutrient losses to water should be limited, maintaining minimal vegetation coverage during the no-grow season is important.

Compost has unique characteristics as a fertilizer. Compost is the sanitized output from composting. Compost generally improves the structure of the soil and its chemical and biological properties. However, its impact depends on its input material, which can include biodegradable waste, commercial food wastes, forestry residues, waste from agriculture, and waste from food and beverage. Generally, compost slowly releases nitrogen and supplies essential quantities of phosphate, potassium, and micronutrients.

Consequently, compost can partially replace other fertilizers. However, its drawbacks are that compost can contain heavy metals and persistent organic pollutants, which can exacerbate conditions in soil with an elevated level of these pollutants. However, no specific legislation regulates the quality and use of compost. One study concludes that limiting negative impacts will require quality checks on the nature and origin of input material.[51]

Various countries regulate compost with statutory standards and agencies. For example, in the Netherlands, an independent third party certifies compost meets legal requirements under a quality assurance system. However, participation in the certification process is voluntary.

### Protecting

Plant protection regulation once focused on the marketing of plant products. For example, pesticide residue limits existed only for food and feedstuffs. Recently, countries have started regulating the use of plant protection products.[52] Given the adverse effects of pesticides, widening this focus from food safety to protecting the environment and human safety is an important step forward.

As a result, commercial plant protection products may need to be approved and labeled with the best use practices and intended purposes. The labels should also include safety measures, but one should still check local legislation before using commercial plant protection products.

Fortunately, countries should adopt plans to reduce the risk and impact of pesticides by having alternative and integrated pest management that use non-chemical

---

[51] Delgado et al. (2009), p. 53.

[52] E.g. the European Directive on the sustainable use of pesticides.

methods. However, importing foreign predators for pest control requires strict caution since they can negatively impact local biodiversity.

**Harvesting**

The ability to determine levels of contamination is essential when harvesting crops. As mentioned, phytoremediation is a more attractive option if plant contamination stays below permitted levels so that the plants are usable as food or feed. This is more common with phytostabilization than with phytoextraction. However, even if not valorized for food or feed, contaminated plants can be converted into biofuel. As is the case with maize, a plant can also be used as both food/feed and biofuel. While the green parts of the plant containing the contaminants can be converted into biofuel, the grain does not uptake any contaminants and can be used as food/feed. The valor of the crop will depend on its classification, which also dictates its legal regulation.

Within this framework, the impact of contamination on plants and their classification is further analyzed in the following paragraphs, focusing on contaminated plants that cannot be used as food or feed.

**Contaminating**

The idea behind phytoextraction is that the plants take up part of the contamination and remove it from the soil when harvested. This raises questions about the nature of the resulting biomass. For example, heavy metals, which occur naturally in plants, are taken by plants during phytoextraction, changing the natural and safe levels in the plant and soil.

There are no international guidelines to determine at what level contaminants in harvested crops are considered toxic and hazardous.[53] The process typically results in an increased concentration of toxic substances in the plant than in typical growing conditions. If this difference is considerable, plants may be classified as ecotoxic. This conclusion is based on arguments from the European Waste Framework Directive[54] and the European REACH Regulation.[55] According to the Waste Framework Directive, waste is ecotoxic when it *presents or may present immediate or delayed risks for one or more sectors of the environment.*[56] The contaminants in the soil are considered to have a toxic, negative impact, which is precisely the reason for phytoremediation at that site. The uptake of those same contaminants into plants could classify those plants as hazardous too. However, the concentration of

---

[53] It is important to note that the meaning of the term "contaminant" is not similar to the meaning of "hazardous substance" used in regulation and in particular chemical legislation. A contaminant is a substance which is in, on or under the land and has the potential to cause harm (or to cause pollution of controlled waters). A hazardous substance is classified as such on the basis of certain characteristics, i.e., carcinogenic, mutagenic, toxic for reproduction, PBT or vPvBvT ((very) persistent, (very) bioaccumulative and (very) toxic).

[54] Directive 2008/98/EC of 19 November 2008 on waste and repealing certain Directives, *Official Journal* 22 November 2008, L 312/3.

[55] Regulation (EC) No 1907/2006 of 18 December 2006 concerning the Registration, Evaluation, Authorisation and Restriction of Chemicals (REACH), establishing a European Chemicals Agency, amending Directive 1999/45/EC and repealing Council Regulation (EEC) No 793/93 and Commission Regulation (EC) No 1488/94 as well as Council Directive 76/769/EEC and Commission Directives 91/155/EEC, 93/67/EEC, 93/105/EC and 2000/21/EC, *Official Journal* 25 May 2007, L 136/3.

[56] Directive 2008/98/EC, Annex III, H14.

contaminants may differ significantly between the soil and the plant, which could influence the classification of the biomass.

One must consider two parameters to assess the toxicological effect: the number of contaminants and the nature of the substances present in the biomass. The latter should be compared with local legislation since many soil contaminants are listed as hazardous substances, often subject to authorization or restrictions. Thus, contaminated biomass may also be considered dangerous and ecotoxic. Regardless of its classification as waste or not, contaminated biomass can reasonably be considered toxic to the environment.

### 8.4.1 Waste or Not?

Crops grown for remediation can generate biofuel, which benefits farmers by adding to their income and interesting investors. Indeed, turning contaminated biomass into profitable biofuel would incentivize private parties to invest in phytoremediation, in turn helping the local population and environment.

Local legislation determines if biomass is waste. The European Waste Framework Directive defines waste as "any substance or object which the holder discards or intends or is required to discard."[57] The exception is *straw and other natural non-hazardous agricultural or forestry material used in farming, forestry, or for the production of energy from such biomass through process or methods which do not harm the environment or endanger human health.* Consequently, biomass must be natural and non-hazardous, which is not the case for contaminated biomass. In the EU this biomass is not excluded from the Waste Framework Directive.

By comparing the harvesting of contaminated crops with the definition of waste and court decisions, the conclusion can be drawn that the biomass of a phytoremediation site is not waste. The following arguments led to this conclusion:

- The economic feasibility of using contaminated biomass to produce biofuel is an important criterion for the viability and success of the projects.
- The suitability of the crops is considered during the selection process.
- The crops are grown with the explicit goal of creating biofuel and remediation contamination.
- Biofuel is a commercial product that must be manufactured. Its production is not just a deactivation of harmful biomass but actual processing.

Phytoremediation projects for biofuel production will be decided based on their economic viability. Biomass is not recycled or reused but grown with the intention to produce biofuel and is thus the first processing. Overall the grower or holder did not discard nor intend to discard the contaminated biomass. Consequently, its classification as waste would be challenging to defend.

[57] Ibid., Article 3, 1.

## 8.5 CONCLUSION

Soils are under increasing pressure around the globe from agriculture, forestry, industrial activities, tourism, and urban development. However, soil is of crucial importance, so its degradation should be prevented and reversed. Through phytoremediation, the use of plants to remove pollutants, degraded soil can be restored, and agriculture cultivated. Unfortunately, a sound legal framework for phytoremediation does not exist yet. This chapter follows the chronology of a phytoremediation project to analyze any potential legal bottlenecks.

Altogether, many rules must be checked prior to a project's implementation, including regulations on the use of certain plant species, GMOs, soil management, fertilization, plant protection products, and the use of harvested plants. Ideally, contamination levels remain low enough for the plant's acceptable use for food or feed, but if not, the biomass can be used as biofuel. Hopefully, these legal bottlenecks will not inhibit phytoremediation projects, and both agricultural cultivation and soil remediation can become part of sustainable land management.

## REFERENCES

Previous chapter by Barnard Vanheusden in *International Farm, Animal, Wildlife and Food Safety Law*. https://eur-lex.europa.eu/legal-content/EN/TXT/?uri=CELEX:32019R1381 (last accessed Dec 14, 2021).

Chaney, R.L., Malik, M., Li, Y.M., Brown, S.L., Brewer, E.P., Angle, J.S. and Baker, A.J.M. (1997) Phytoremediation of soil metals. *Current Opinion in Biotechnology*, 8, pp. 279–284.

Delgado, L., Catarino, A.S., Eder, P., Litten, D., Luo, Z. and Villanueva, A. (2009) End-of-Waste Criteria, final report, European Commission, Joint Research Centre, p. 379, available at http://ftp.jrc.es/EURdoc/JRC53238.pdf (last visited on 12 January 2015).

De Vries, B. (2000) Multifunctional agriculture in the international context: A review. *The Land Stewordship Project*, p. 15.

Dornburg, V. and Faaij, A.P.C. (2005) Cost and $CO_2$ emission reduction of biomass cascading: methodological aspects and case study of SRF poplar. *Climatic Change*, 71, pp. 373–408.

European Commission (2006) Communication from the Commission to the Council, the European Parliament, the European Economic and Social Committee and the Committee of the Regions: Thematic Strategy for Soil Protection, COM(2006) 231 final, p. 12.

European Commission (2012) The Common Agricultural Policy. A Partnership between Europe and Farmers, 16 p., available at http://ec.europa.eu/agriculture/capexplained/index_en.htm (last visited on 12 January 2015).

Firbank, L.G. (2008) Assessing the ecological impacts of bioenergy projects. *Bioenergy Research*, 1, pp. 12–19.

Ghosh, M. and Singh, S.P. (2005) A review on phytoremediation of heavy metals and utilization of its byproducts. *Applied Ecology and Environmental Research*, 3, pp. 1–18.

Koopmans, G.F., Römkens, P.F.A.M., Fokkema, M.J., Song, J., Luo, Y.M., Japenga, J. and Zhao, F.J. (2008) Feasibility of phytoextraction to remediate cadmium and zinc contaminated soils. *Environmental Pollution*, 156, pp. 905–914.

Lanigan, R.S. and Yamarik, T.A. (2002) Final report on the safety assessment of EDTA, Calcium Disodium EDTA, Diammonium EDTA, Dipotassium EDTA, Disodium EDTA, TEA-EDTA, Tetrasodium EDTA, Tripotassium EDTA, Trisodium EDTA, HEDTA, and Trisodium HEDTA. *International Journal of Toxicology*, 21(2), pp. 95–142.

Lasat, M. (2000) Phytoextraction of metals from contaminated soil: a review of plant/soil/metal interaction and assessment of pertinent agronomic issues. *Journal of Hazardous Substance Research*, 2, p. 5, available at citeseerx.ist.psu.edu/viewdoc/download?doi=10.1.1.15.7236&rep=rep1&type=pdf, last visited on 6 January 2015.

Meers, E., Ruttens, A., Hopgood, M., Lesage, E. and Tack, F.M.G. (2005) Potential of Brassica rapa, Cannabis sativa, Helianthus annuus and Zea mays for phytoextraction of heavy metals from calcareous dredged sediment derived soils. *Chemosphere*, 61, pp. 561–572.

Mench, M., Lepp, N., Bert, V., Schwitzguébel, J.P., Gawronski, S.W., Schröder, P. and Vangronsveld, J. (2010) Successes and limitations of phytotechnologies at field scale: outcomes, assessment and outlook from COST Action 859. *Journal of Soils Sediments*, 10, pp. 1039–1070.

Robinson, B.H., Leblanc, M., Petit, D., Brooks, R.R., Kirkman, J.H. and Gregg, P.E.H. (1998) The potential of Thlaspi caerulescenss for phytoremediation of contaminated soils. *Plant Soil*, 203, pp. 47–56.

Van der Grift, B. and Griffioen, J. (2008) Modelling assessment of regional groundwater contamination due to historic smelter emissions of heavy metals. *Journal of Contaminant Hydrology*, 96, pp. 48–68.

Vangronsveld, J., Herzig, R., Weyens, N., Boulet, J., Adriaensen, K., Ruttens, A., Thewys, T., Vassilev, A., Meers, E., Nehnevajova, E., van der Lelie, D. and Mench, M. (2009) Phytoremediation of contaminated soils and groundwater: lessons from the field, COST Action 859, *Environmental Science and Pollution Research*, 16, pp. 765–794.

Vanheusden, B., Hoppenbrouwers, M., Witters, N., Vangronsveld, J., Thewys, T. and Van Passel, S. (2011) Legal and economic aspects of crops selection for phytoremediation purposes and the production of biofuel, 104 p., available at http://projects.swedgeo.se/r2/wp-content/uploads/2012/04/REJUVENATE-2-Final-report-Legal-and-economic-aspects.pdf (last visited on 3 January 2015).

Van Nevel, L., Mertens, J., Oorts, K. and Verheyen, K. (2007). Phytoextraction of metals from soils: How far from practice? *Environmental Pollution*, 150, pp. 34–40.

Weyens, N., van der Lelie, D., Taghavi, S. and Vangronsveld, J. (2009) Phytoremediation: plant-endophyte partnerships take the challenge. *Current Opinion in Biotechnology*, 20, pp. 248–254.

Weyens, N., Truyens, S., Saenen, E., Boulet, J., Dupae, J., van der Lelie, D., Carleer, R. and Vangronsveld, J. (2010) Endophytes and their potential to deal with co-contamination of organic contaminants (toluene) and toxic metals (nickel) during phytoremediation. *International Journal of Phytoremediation*, 13, pp. 244–255.

Witters, N., Van Slycken, S., Ruttens, A., Adriaensen, K., Meers, E., Meiresonne, L., Tack, F.M.G., Thewys, T., Laes, E. and Vangronsveld, J. (2009) Short rotation coppice for phytoremediation of a Cd-contaminated agricultural area: A sustainability assessment. *BioEnergy Research*, 2(3), pp. 144–152.

Yuan, Z. and VanBriesen, J. (2006) The formation of intermediates in EDTA and NTA biodegradation. *Environmental Engineering Science*, 23(3), pp. 533–544.

# 9 Invasivorism as a Sustainable Strategy to Confront Animal and Resource Exploitation

*Morgan Boutilier*
University of California, Los Angeles School of Law

## CONTENTS



DOI: 10.1201/9781003137733-17

## 9.1 INTRODUCTION

Through hunting, overfishing, and farming, the human appetite drives plant and animal species to extinction. At the same time, human-introduced species are invading and deteriorating ecosystems throughout the world, with negative impacts on human health, livelihoods, and economies. Behind "invasivorism" is the premise that we could use our insatiable appetite for good by redirecting it toward these invasive alien species (IAS). Increasingly, governments and private ventures are adopting commercial markets as a tool for controlling IAS. And given current international concerns over IAS, biodiversity loss, and the sustainability of global food systems, one would expect policies and programs that promote the consumption of IAS to be thoroughly investigated. But despite the potential for invasivorism to cost-effectively manage IAS while providing an environmentally sound and high-welfare food source, little research closely examines the issue. There is a particular gap in the analysis of regulatory frameworks as they apply to invasivorist management efforts. However, implementing such a strategy requires intersection with and consideration of several legal areas, including but not limited to:

- Environmental Protection
- Wildlife Management, including hunting regulations
- Food Safety
- Labeling
- International Trade
- Animal Welfare and Humane Slaughter
- Biosecurity
- Private Property

The aim of this chapter is to begin to fill this gap. The next section sets the stage by describing the global threat posed by invasive species and animal farming that invasivorism seeks to address. The following section summarizes the current research analyzing invasivorist management strategies, which, while limited, suggests that a carefully implemented and monitored program could be an effective tool for IAS management. Some studies offer guidance for natural resource managers considering such an approach, noting certain regulatory hurdles to the successful implementation of such programs. Later sections explore legal and policy frameworks that could be used to address defects in current invasivorist management efforts. This examination begins with international legal instruments, which are well-poised for assisting countries' implementation of invasivorist programs and facilitating the safe trade in IAS food products. However, the context-specific nature of IAS management constrains the utility of international tools, requiring examination of local regulatory contexts. A comparative overview of the regulatory systems of the United States (US) and the European Union (EU) as they pertain to invasivorism reveals defects in the US's environmental and food safety regimes that impede the successful implementation of a large-scale invasivorist project. But despite the absence of these regulatory hurdles in EU legislation, the current lack of implemented invasivorist projects thwarts any assessment of their potential success. A more complete overview of the existing literature on invasivorism is provided by the table at the end of this chapter, followed by observations proposing areas for further research. Ultimately, this emerging area of research and policy will require the collaboration of a variety of experts around the world, including ecologists, animal welfare and food scientists, economists, and market researchers, to study all sides of the issue.

The scope of this chapter is limited to efforts promoting the consumption of invasive animal species, due to the greater potential to displace environmentally harmful forms of meat production. However, further research should examine strategies that could promote the harvest of invasive edible plant species like kudzu, "the vine that ate the south," as well as those promoting harvest of IAS for purposes other than food, such as for fertilizer or biofuel. This chapter also does not attempt to provide a comprehensive analysis of all legal instruments that apply to invasivorism, but serves to highlight selected regulatory bodies and instruments in the IAS management and food safety regulatory regimes that provide opportunities or hurdles to the commercial sale of invasive animal food products. The US and EU are compared as examples of regulatory landscapes that are less and more compatible, respectively, with an effective invasivorist strategy. However, public-private partnerships have launched invasivorist programs throughout the world, and scholarship from other jurisdictions should continue this comparative analysis. IAS and the environmental threat they pose are not confined by borders. As a result, managing and mitigating their effects will require a collaborative effort at the international, national, and local levels.

## 9.2 BACKGROUND

### 9.2.1 Scope of the Biodiversity Crisis

There is a current global biodiversity crisis that will necessitate a swift and coordinated international effort to ensure the continued functioning of vital ecosystems. At the United Nations' third international conference on sustainable development, the parties recognized the critical role that biodiversity plays in maintaining ecosystems that provide essential services for human well-being.[1] They also recognized the severity and gravity of global biodiversity loss, which affects food security and nutrition, provision and access to water, and the health of present and future generations. The way humans source food has been identified by the international community as a significant driver of the current biodiversity crisis. The Convention on Biological Diversity Overexploitation through hunting and overfishing is threatening the extinction of wild populations.[2] Additionally, farmed food serves as another key player in the biodiversity crisis. With its 2006 seminal report, Livestock's Long

---

1 G.A. Res. 66/288, 197 (July 27, 2012).

2 *See* Convention on Biological Diversity, *Global Biodiversity Outlook 5.* (2020), https://www.cbd.int/gbo5.

Shadow, the UN described the livestock industry as perhaps "the leading player in the reduction of biodiversity, since it is the major driver of deforestation, as well as one of the leading drivers of land degradation, pollution, climate change, overfishing [to produce fish meal for livestock feed], sedimentation of coastal areas and facilitation of invasions by alien species."[3] Further, welfare concerns over aspects of modern methods of animal farming, like intensive confinement, unnatural living conditions, and untreated injuries, draw criticism from international welfare groups.[4] Environmental, ethical, and health concerns with meat production are turning many consumers to look for alternative sources of protein.[5] But current FAO predictions of the rising rate of global meat consumption indicate the effects of this industry are likely only to worsen.[6]

As the human diet drives species to extinction, humans continue to introduce new species into areas where they too threaten global biodiversity. The International Union for Conservation of Nature (IUCN) defines IAS as "animals, plants or other organisms introduced by man into places out of their natural range of distribution, where they become established and disperse, generating a negative impact on the local ecosystem and species."[7] The Convention on Biological Diversity (CBD) points to growing global trade, travel, and tourism as factors facilitating the intentional and unintentional introductions of nonnative species.[8] According to the CBD, invasive species are "main direct drivers of biodiversity loss across the globe."[9] It also estimates IAS damages and control costs at over hundreds of billions USD per year, and possibly more than one trillion.[10] And this figure does not include the cost to human health when chemicals used to manage IAS infiltrate water and soil.[11] The environmental and economic damages caused by IAS are further exacerbated by climate change, pollution, and habitat alteration.[12] Optimizing cost-effective IAS management strategies is thus necessary for protecting global ecosystems, economic welfare, and human health. However, IAS and their impacts are not confined by political boundaries, meaning the successful control of an IAS population is highly dependent on the collective action taken by all resource managers within an invaded

3 Food & Agricultural Organization of the U.N. (FAO), *Livestock's Long Shadow: Environmental Issues and Options*, at xxiiii (2006), https://www.fao.org/3/a0701e/a0701e.pdf.

4 *See Stopping the spread of intensive confinement,* Humane Society International, https://www.hsi.org/news-media/stop_confinement (last visited Dec 17, 2021); *The Mission*, Compassion in World Farming, https://www.ciwf.com/about/the-mission (last visited Dec 17, 2021). ("Pigs reared for meat are often kept in barren, crowded conditions; on slatted or concrete floors without straw for bedding or rooting. They have no access to outdoors and will never experience fresh air or daylight. They are unable to behave naturally and become bored and frustrated. They tend to fight and to bite each other, sometimes causing severe injury, particularly to their tails.").

5 Organisation for Economic Co-operation Development (OECD) and FAO, *Agricultural Outlook 2021–2030,* 6 (July 5, 2021), https://doi.org/10.1787/19991142.

6 *Id.*

7 Invasive Species Specialist Group (ISSG) of the International Union for Conservation of Nature (IUCN), http://www.issg.org/is_what_are_they.htm (last visited Dec. 17, 2021).

8 *Invasive Alien Species,* Convention on Biological Diversity, https://www.cbd.int/invasive/problem.shtm (last visited Dec 17, 2021).

9 *Id.*

10 *Id.*

11 *Id.*

12 *Id.*

zone.[13] The parties to the CBD recognize that IAS are "a global issue that requires international cooperation and actions."[14] And the threat posed by IAS has prompted the establishment of international bodies like the Global Invasive Species Program (GISP) and Invasive Species Specialist Group (ISSG) to provide guidance for the implementation of frameworks that prevent IAS introduction and promote IAS eradication or control.[15]

### 9.2.2 Introduction to Invasivorism

Common IAS management efforts, if implemented, can be costly, ineffective, and may themselves risk harm to the health of humans and wildlife.[16] Since human predation was effective at extinguishing animals like wooly mammoths and giant sloths in the past, why not direct our appetites instead toward ecologically harmful species?[17] This question has given rise to the invasivore movement, which aims to lessen the harmful impacts of invasives by eating them.[18] Among the first scientists to promote the idea was Joe Roman, a conservation biologist at the University of Vermont.[19] His website, eattheinvaders.org, posts harvest tips and recipes for dishes like wild boar tenderloin, smoked pulled nutria, and sautéed garden snails.[20] Chefs like Bun Lai of Miya's sushi restaurant have started serving IAS to the public. His menu replaces traditional yet unsustainable sushi ingredients with IAS, for example, swapping out bluefin tuna for feral boar meat.[21] But beyond these grassroots efforts, interest in the logic of invasivorism has spread to government agencies, now increasingly looking to use the consumer market in their toolbox of IAS management strategies.[22] For example, In 2010, the National Oceanic and Atmospheric Administration (NOAA) began a new campaign that encourages consumers to eat invasive lionfish causing damage to the Atlantic Ocean, with the catchphrase: "If we can't beat them, let's eat them!"[23] And such efforts seem promising. The Jamaican government has reported a 66 percent reduction in lionfish sightings after the launch of its "Eat it to Beat it"

---

[13] Rebecca S. Epanchin-Niell et al., *Controlling invasive species in complex social landscapes*, 8 Frontiers in Ecology and the Environment 210 216 (2009).

[14] Convention on Biological Diversity, *supra* note 8.

[15] See Global Invasive Species Program, https://www.gisp.org/ (last visited Dec. 17, 2021); Invasive Species Specialist Group, http://www.issg.org/ (last visited Dec. 17, 2021).

[16] See *infra*.

[17] Joshua Ulan Galperin & Sara E. Kuebbing, *Eating invaders. Managing biological invasions with fork and knife?* 28 Natural Resources & Environment 41 (2013).

[18] *Id.*

[19] Michael Snyder, *Can we really eat invasive species into submission?*, Scientific American (May 19, 2017), https://www.scientificamerican.com/article/can-we-really-eat-invasive-species-into-submission/.

[20] *Fighting invasive species, one bite at a time*! Eat The Invaders, http://eattheinvaders.org/ (last visited Dec 17, 2021).

[21] Miya's Sushi New Haven, CT Menu, http://miyassushi.com/assets/Miyas-Menu.pdf (last visited Dec. 14, 2021).

[22] Martin A. Nuñez et al., *Invasive species: to eat or not to eat, that is the question.* 5.5 Conservation Letters 334 (2012).

[23] *Filleting the lion*, NOAA's National Ocean Service (2016), https://oceanservice.noaa.gov/news/lionfish/eatlionfish.html (last visited Dec 17, 2021).

campaign.[24] Despite IAS's potential as an environmentally sound food source, there is a notable lack of peer-reviewed sources examining the use of human consumption to manage IAS or providing guidance on how to effectively implement a strategy. However, implementation of a successful strategy will require careful consideration of factors that include population ecology, market dynamics, and a health assessment. The next section provides a summary of the existing literature on the topic.

## 9.3 SUMMARY OF LITERATURE

The general consensus finds prediction, prevention, early detection, and other rapid responses are more effective and less costly than reactive responses once an IAS has become established and widespread.[25] But once an IAS becomes widespread, control through regular intervention is required in order to maintain the population at a level limiting the disturbances and damage caused. However, several obstacles are posed by the complexity in implementing this type of management, including increasing expenditures, and the requirement that technical measures are systematically adapted to the specific site of invasion. Though invasivorism presents an effective way to implement cost-effective regular intervention, it cannot entirely remove, and may introduce, complexities inherent to remedial control programs. Most research on invasivorism raises, without answering, questions on the possible consequences of the approach. Little research has examined the effectiveness of programs already using human consumption to control IAS. Some studies analyze the potential of commercial markets to control certain species of IAS, though most focus on aquatic species. Overall, there are gaps in the scientific literature that prevent a holistic understanding of the viability of invasivorism as an IAS management strategy. The table at the end of the chapter provides a more comprehensive overview of the existing literature, which is summarized in this section.

### 9.3.1 Invasivorism as an Effective Tool Against IAS

The few studies that do examine the issue suggest that if carefully managed, a strategy that encourages the commercial harvest and consumption of invasive species could be an important tool for managing IAS. Two papers by Martin Nunez and his coauthors[26] and by Susan Pasko and Jason Goldberg[27] conduct the most thorough analyses of the benefits and risks of the approach while providing guidance for natural resource managers in its implementation. Pasko and Goldberg review the use of harvest incentives as a means to control IAS populations, including those that

---

[24] Athaliah Reynolds-Baker, *Lionfish population down*, Jamaica Information Service, https://jis.gov.jm/lionfish-population/ (last visited Dec 17, 2021).

[25] Center for International Environmental Law & Defenders of Wildlife, *Invasive alien species and trade: integrating prevention measures and international trade rules,* at 6, (January 2006), https://cleantrade.typepad.com/clean_trade/files/invasives_trade_paper_0106.pdf.

[26] Nuñez, *supra* note 22, at 334–341.

[27] Susan Pasko & Jason Goldberg, *Review of harvest incentives to control invasive species*, 5 Management of Biological Invasions 263–277 (2014).

incentivize the harvest of IAS for food.[28] They discuss the biological, ecological, health, and socioeconomic factors involved and offer guidance in the form of 11 recommendations to resource managers considering this type of strategy.[29] The authors first note that typical IAS control activities often require financial resources, time, and effort that are not always available.[30] This is exemplified by an IUCN report from France, which found that a 4-year project to eradicate a population of bullfrogs through monitoring, destruction of eggs, traps, hunting, nightly shooting, and dewatering of sites, while successful, cost over 385,000 USD.[31] The cost and technical difficulties involved in IAS management have turned attention toward the use of incentive programs using human consumption as a management tool.[32] Though few studies have critically examined the success of these programs, commercial and government efforts to develop and implement such programs have already begun.[33] In the US, for instance, private industry is developing products and markets that use invasive species of Asian Carp, and the State of Illinois' Target Hunger Now! campaign encourages hunters and anglers to donate invasive carp and other "nuisance animals" to be processed and distributed to needy families.[34] Pasko and Goldberg underline the significant opportunity for these programs to support ecosystem and natural resource management in a cost-effective way while simultaneously boosting economic development and environmental awareness.[35]

Moreover, promoting IAS consumption as a replacement for traditionally farmed meat and fish could have the added benefit of reducing demand on, and thus, the environmental impact of these industries. A consumer perception study from the Czech Republic concludes that consumers can accept invasive wild boar as a substitute for pork.[36] Its authors note that if harvested under strictly regulated hunting practices, meat from animals hunted in the wild can have the highest level of animal welfare among meat products, as they are not subject to extreme confinement, transport stress, or other common practices of intensive animal farming.[37] Meat harvested from the wild may even pose a lower health risk for consumers. Unlike domestic pigs, wild boars have not been administered antibiotics or hormones and have not been selectively bred to commonly suffer diseases like transmissible gastroenteritis, circovirosis, vesicular rash, Aujezky's disease, and Swine erysipelas.[38] Beyond the beneficial effects to the ecosystem brought by removing IAS, they can also serve as a healthy, more ethical, and environmentally sound source of food reducing pres-

---

28 *Id.*
29 *Id.*
30 *Id.* at 263.
31 IUCN French Committee, *Making use of invasive alien species settled in natural environments. An effective approach to management? An initial analysis and discussion of points requiring attention*, at 7, (June 2019).
32 Pasko & Goldberg, *supra* note 27, at 263.
33 *Id.*
34 *Id.*
35 *Id.*
36 Karolina Macháčková et al., *Wild boar meat as a sustainable substitute for pork: A mixed methods approach*, 13 Sustainability 2490 (2021).
37 *Id.* at 2492.
38 *Id.* at 2493.

sure on factory farming. However, to ensure that an invasivorist program or policy achieves a net positive outcome for the environment first requires careful consideration of a variety of factors.

### 9.3.2 Policy Considerations

Pasko and Goldberg's paper makes clear that there is no "one-size-fits-all" approach to IAS and encourages natural resource managers not to rely on invasivorism as a panacea to problems posed by IAS.[39] Managers must first contemplate whether a harvest strategy is appropriate for a given species and population of IAS. Some IAS are simply inedible or pose a health hazard when ingested.[40] The authors recommend managers first conduct a health assessment like those conducted to assess fisheries or commercial markets to investigate if the IAS is fit for human consumption.[41] Determining whether harvest is an appropriate form of invasive species management also requires understanding the population dynamics of the targeted species, including the species' density-dependent processes, demographic structure, and specific vital rates.[42] A 2021 field experiment by Grosholz and coauthors demonstrated why this understanding is necessary.[43] The researchers instituted traps that reduced the population of an invasive crab species by over 90 percent. However, the traps only removed the larger members of the population that had previously held the population in check by cannibalizing the juveniles. As a result, the crab population nearly tripled in size from before removal efforts began.[44] Pasko and Goldberg warn that such biological overcompensation may obstruct the ability of an IAS control method, including a harvest program, to deliver a long-term reduction of the target population.[45] Lastly, the success of commercial markets or an incentive program to control an IAS will depend in part on the value of the harvested commodity, the cost of harvest, and the minimum profit acceptable to the harvester.[46] In a 2012 paper, Nunez and his coauthors discuss if, when, and how an approach of controlling IAS via gastronomy should be used. The authors call attention to the fact that people are inherently conservative in food preferences and harbor a general tendency to dislike new foods. This can pose a formidable barrier to the ability to significantly affect a population through consumption. Nuñez and coauthors identify the fundamental role played by marketing in a successful program.[47] But successfully marketing certain invasive species, like nutria (also known as river rat), could be challenging.[48] Assessing the market potential of a species is therefore crucial to a cost-benefit analysis of the project. Even with a viable target species, market forces alone may not reliably control or

---

39 Pasko & Goldberg, *supra* note 27, at 273.
40 *Id.* at 267–268.
41 *Id.* at 268.
42 *Id.* at 265.
43 Edwin Grosholz et al., *Stage-specific overcompensation, the hydra effect, and the failure to eradicate an invasive predator*, 118 Proceedings of the National Academy of Sciences, 1–8 (2021).
44 *Id.* at 119.
45 Pasko & Goldberg, *supra* note 27, at 265.
46 *Id.* at 269.
47 Nuñez, *supra* note 22, at 339.
48 *Id.*

eradicate a species. When IAS are at high population densities, financial incentives from resource managers may not be needed as market prices alone may support active harvest.[49] But Pasko and Goldberg emphasize that as density decreases, population removal requires greater resources. Often, removal of the last 1 percent of an IAS population is more difficult than removal of the first 99 percent.[50] The cost and effort needed to capture remaining individuals may outweigh the benefits for commercial harvesters, who may also be reluctant to deplete the last of their source of income. However, removing the remaining individuals is necessary for eradication as populations can quickly rebound.[51] The authors recommend either combining a commercial harvest program with other IAS control strategies or weighing the costs of increasing financial incentives for harvesters as population densities decrease.[52]

Possible unintended consequences associated with an invasivorist strategy could result in it backfiring. Pasko and Golberg caution managers about the possibility for harvest programs to inflict further damage to native species through bycatch or increased human disturbance.[53] But the primary concern with promoting the commercial harvesting of IAS is the potential to generate perverse incentives that further spread the target species.[54] The notorious Cobra Effect is based on an anecdote in which the British government set out to rectify a venomous snake problem in Delhi by offering a bounty for dead snakes.[55] When Delhi residents discovered that the bounty exceeded the cost of breeding, many began to raise snakes for income. The government discontinued the bounty program, prompting breeders to release their snakes which exacerbated the snake problem.[56] Similarly, demand for IAS or incentive programs that encourage their harvest may unintentionally increase the population size of the target species by providing a financial incentive to breed IAS, which could result in their intentional or unintentional release. For example, illegal fish stocking is now a global issue in which fish are intentionally released into a body of water to facilitate a fishery, frequently at the expense of native species.[57] Pasko and Goldberg advise that natural resource managers implement a plan only after careful consideration of human behavior and potential responses, including fraud by breeding the IAS, intentionally introducing the species into new areas, or other means that result in additional spread of the target species in order to reap the benefits.[58] An "exit strategy" may be required to identify other profitable activities to compensate for the reduction of the IAS resource.[59]

Incorporating an IAS into local cuisine additionally risks creating cultural attachments to the species that make its eradication or control impossible.[60] Wild boar, for

---

[49] Pasko & Goldberg, *supra* note 27, at 270.
[50] *Id.*
[51] *Id.*
[52] *Id.* at 266–267.
[53] *Id.* at 270.
[54] *Id.*
[55] *Id.*
[56] *Id.*
[57] Nuñez, *supra* note 22, at 337.
[58] Pasko & Goldberg, *supra* note 27, at 274–275.
[59] IUCN French Committee, *supra* note 29, at 45.
[60] Nuñez, *supra* note 22, at 337.

example, was initially introduced to the Hawaiian Islands as a food source. Despite the significant damage it inflicts to native ecosystems, strong cultural ties to the animal as a hunting resource obstruct any eradication and control efforts.[61] Nuñez and his coauthors warn there may be a delicate line between promoting the consumption of a species and incorporating it into local culture.[62] Marketing campaigns included as part of the strategy should thus remind consumers of the end goal to eradicate the species.[63] Overall, the studies emphasize the need for close monitoring of unintended consequences, including periodic population and damage assessments, and adaptive management that responds to unintended consequences and the need for additional incentives.

Properly monitoring a commercial project will require the cooperation of a variety of different actors throughout its life. A 2019 survey of commercial projects by the IUCN French Committee warns that a project promoted exclusively by a commercial entity makes it likely that ecological objectives will fall secondary or not be taken into account at all.[64] But including all the stakeholders in IAS management, like state services, public agencies, researchers, and local governments, makes it possible to monitor the benefits, risks, and progress toward the ecological objectives of the project. Though the studies express concerns about the risks of using invasivorism as a management strategy, they agree that if properly planned, monitored, managed, and implemented, a strategy that promotes human consumption of IAS could serve as a cost-effective tool against IAS. However, the more comprehensive overview of the literature provided in the table below reveals gaps in the scientific literature as to the market potential and efficacy of programs promoting IAS consumption. There is also a gap in the analysis of regulatory frameworks as they apply to invasivoristy strategies; however, some studies touched on certain regulatory hurdles which are summarized below.

### 9.3.3 Administrative and Legislative Issues

Regulatory limitations on the harvest or transportation of IAS may need to be amended to allow an invasivorist strategy to reach intended markets. In *Eating Invaders*, authors Joshua Galperin and Sara Kuebbing describe the legal barriers to the invasivore movement in the US, contending that Federal and state law act to prevent its success.[65] Federal environmental laws, such as the Lacey Act (16 U.S.C. §§ 3371–3378) and the Plant Protection Act (7 U.S.C. §§ 7701–7786), prohibit the possession, transportation, and sale of certain invasive species on the premise that these activities encourage further invasions.[66] They note that federal and state environmental laws place restrictions on the sale, transportation, and possession of most of the IAS most commonly recommended as targets for the invasivore movement.[67] Additionally, food safety laws make the commercial sale of invasive animals exceedingly difficult by imposing strict requirements on the slaughter and inspection of

61 *Id.*
62 *Id.* at 338.
63 *Id.*
64 IUCN French Committee, *supra* note 29, at 73.
65 Galperin & Kuebbing, *supra* note 16.
66 *Id.* at 43.
67 *Id.*

invasive species. Under both the Food and Drug Act (21 U.S.C.A § 301 et. seq) and the Meat Inspection Act (21 U.S.C.A. § 601), the sale of adulterated meat is prohibited.[68] Meat that is not inspected before slaughter or is not slaughtered pursuant to USDA regulations is deemed adulterated, which effectively prevents the interstate sale of any meat (but not fish) harvested by hunting.[69] These restrictions make the significant ingestion of IAS in the US unrealistic, and state laws only solidify these regulatory hurdles with additional restrictions.[70]

Pasko and Goldberg point to private property rights as another important consideration for incentive programs and propose that legislation may be needed to encourage or allow access into private lands to ensure the removal of all targeted individuals in a population.[71] As an example, they cite regulations promulgated by The Wisconsin Department of Natural Resources that allow state inspectors to enter private lands to ensure prohibited species are not present.[72] Legislation may also be required to ease state regulation of incentive programs, which are currently either unlawful or heavily restricted in California, Maryland, Massachusetts, New Jersey, New Mexico, New York, and Oregon.[73] Lastly, the authors note that license requirements or bag limits that hinder the capture of IAS and therefore may need to be eased prior to a project's implementation.[74] Some countries already ease hunting requirements for IAS, like in France, which allows the trapping of nutria year-round without a permit.[75]

There is, however, a lack of any further analysis of regulatory systems as they would pertain to commercial invasivorist projects, and no research offers a comparative analysis between regulatory systems as they apply to this topic. The following sections aim to address this gap, beginning with an examination of international legal instruments that offer unique tools for promoting or facilitating IAS consumption.

## 9.4 INTERNATIONAL LAW

Strong international guidance is important for dealing with IAS, which pose a global threat to food security and livelihoods and compound the effects of climate change.[76] Nations vary in their capacity to respond to IAS threats. An uneven distribution of resources (or concern) toward IAS or a lack of coordination between nations can undermine the management efforts of any one nation.[77] Thus, guidance, clear direction, and common goals must be implemented at an international level for the development of effective strategies to manage IAS. Additionally, food safety laws

---

[68] *Id.* at 44.
[69] *Id.*
[70] *Id.*
[71] Pasko & Goldberg, *supra* note 27, at 271.
[72] *Id.*
[73] *Id.*
[74] *Id.*
[75] IUCN French Committee, *supra* note 14, at 73.
[76] *Invasive alien species and climate change*, IUCN (Feb. 2021), https://www.iucn.org/resources/issues-briefs/invasive-alien-species-and-climate-change.
[77] Clare Shine, *Overview of the management of invasive alien species from the environmental perspective*, IUCN Commission on Environmental Law & IUCN Invasive Species Specialist Group (2003), https://www.fao.org/3/y5968e/y5968e07.htm.

are important to minimize the transmission of pathogens, zoonoses, or other animal diseases between countries. And, since IAS are spread through trade, legislation governing international trade is also important for managing pathways of IAS introduction or re-introduction. However, these laws should not be overly restrictive so as to prevent IAS products from reaching desired markets. For instance, the US has struggled to find a market for invasive Asian Carp in the US, so much of the Asian Carp currently harvested from US waterways are shipped to China for consumption.[78] (Though, increasing the food miles of the species may need to be factored into an environmental cost-benefit analysis of a program.)

The CBD provides a comprehensive but general framework for managing IAS, setting goals for sustainable food production and IAS management. However, there is potential to offer greater guidance to nations home to edible IAS that are ripe for invasivorist efforts. The World Organization for Animal Health (OIE) assists the safe trade in IAS by tracking and reporting data on wildlife disease surveillance which can ensure proper safeguards are in place when collecting IAS to prevent the spread of zoonoses or other animal diseases. The OIE also regulates the safety of IAS products through its codes applying sanitary measures to terrestrial and aquatic species of animals. The World Health Organization (WHO) and Food and Agriculture Organization (FAO)'s Codex Alimentarius ensures a baseline of food safety standards for the trade in meat harvested from the wild. The FAO also supports invasivorist initiatives by providing financial and technical assistance for projects in the Mediterranean seeking to use invasive crabs as an export commodity, which will hopefully serve as a model for future projects. Overall, the international regulatory framework offers unique tools pertaining to invasivorist efforts.

### 9.4.1 The Convention on Biological Diversity (CBD)[79] and International IAS Management

The CBD, effective since 1993, is a multilateral treaty with three objectives: Conservation of biological diversity, the sustainable use of the components of biological diversity, and the fair and equitable sharing of the benefits arising out of the utilization of genetic resources.[80] The convention is a legally binding instrument, requiring its 196 signing parties to implement its provisions.[81] In 2010, the parties to the CBD adopted the Strategic Plan for Biodiversity 2011–2020 which provided a 10-year framework for action by all countries and stakeholders to safeguard biodiversity and its benefits.[82] This plan incorporated 20 Aichi Biodiversity Targets, which

[78] Sarah Varble & Silvia Secchi, *Human consumption as an invasive species management strategy. A preliminary assessment of the marketing potential of invasive Asian carp in the US*, 65 Appetite 58–67 (2013).

[79] Convention on Biological Diversity, 1760 U.N.T.S. 143, June 5, 1992, [hereinafter CBD], available at http://www.cbd.int/doc/legal/cbd-en.pdf.

[80] *Id.*

[81] *Id.*

[82] Conference of the Parties to the CBD, The Strategic Plan for Biodiversity 2011–2020 and the Aichi Biodiversity Targets, UNEP/CBD/COP/DEC/X/2 (Oct. 29, 2010).

included the goal to identify, control, or eradicate priority invasive species by 2020.[83] To help countries meet this target, the IUCN published a Toolkit of Best Prevention and Management Practices.[84] Though this toolkit recognized the commercial hunting of IAS for meat as a possible control strategy, it warns that due to the danger of providing a perverse incentive, stating "[t]his issue has to be evaluated on case-by-case basis in order to estimate the potential danger and benefits."[85] However, further guidance for evaluating this strategy is entirely absent from the toolkit.

The CBD recently reported that none of its Aichi targets were fully achieved by the 2020 goal.[86] The report finds that "food and agricultural production remains among the main drivers of global biodiversity loss," "a third of marine fish stocks are overfished," and that though governments and businesses are developing plans for more sustainable production and consumption, these plants "are not being implemented on a scale that eliminates the negative impact of unsustainable human activities on biodiversity."[87] However, the CBD is currently preparing for the Post-2020 Biodiversity Framework, which presents the opportunity to promote the research and development of strategies that encourage the commercial harvesting of IAS as a food source as a way to meet its new targets.[88]

### 9.4.2 OIE Animal Health Codes

The OIE is the intergovernmental organization responsible for improving animal health worldwide.[89] The OIE develops animal health codes and manages the World Animal Health Information Systems for domesticated and wild animals, which disseminates timely information on 117 listed animal diseases.[90] The Aquatic Animal Health Code and Terrestrial Animal Health Code provide standards with the goal of promoting the "safe international trade of animals and animal products while avoiding unnecessary impediments to trade."[91] These codes are technically voluntary, but influential as many nations fulfill their Phytosanitary Measures (SPS) and Technical Barriers to Trade (TBT) agreements through adherence to the OIE's codes.[92] Of primary relevance to the trade in IAS are the OIE's codes pertaining to wildlife trade and animal-derived food production. For example, the Terrestrial Health Code requires that feral boar meat must derive from boars killed in a country or zone free from ASF or subjected with favorable results after a post-mortem

---

83 *Id.*

84 Global Invasive Species Programme, Invasive alien species: A toolkit of best prevention and management practices (Rudiger Wittenberg & Matthew J. W. Cock eds., 2003).

85 *Id.* at 156.

86 CBD, *supra* note 2 at 2.

87 *Id.* at 13.

88 *Preparations for the Post-2020 Biodiversity Framework*, CBD, https://www.cbd.int/conferences/post2020 (last visited Dec. 17, 2021).

89 *Who We Are*, World Organization for Animal Health [hereinafter OIE], https://www.oie.int/en/who-we-are/ (last visited Dec. 17, 2021).

90 *What We Do*, OIE, https://www.oie.int/en/what-we-do/ (last visited Dec. 17, 2021).

91 OIE, *supra* note 76.

92 World Trade Organization [WTO], *Sanitary and Phytosanitary Measures* at 19, the WTO Agreements Series (2010).

inspection approved by the relevant veterinary authority.[93] The OIE also provides guidelines for members to implement a wildlife health surveillance system, which involves hunters, trappers, and wildlife meat markets as sources of health data.[94] However, the OIE has not specifically encouraged or addressed trade in invasive animal products. Understandably, the OIE is primarily concerned with the potential for trade, especially in wildlife products, to spread zoonotic and emerging infectious diseases.[95] Growing evidence implicates the global wildlife trade in human disease outbreaks as well as outbreaks impacting livestock, native wildlife, and ecosystems.[96] Recognizing a need to support legal, sustainable, and responsible wildlife use, the OIE has recently announced that it is developing new guidelines for trade in wildlife. The OIE will base these guidelines "on sound governance that reduces health risks, and supports animal welfare and biodiversity conservation."[97] In fitting with these principles, the OIE could seek to research and develop additional safety guidelines that contemplate the safe harvest and trade in terrestrial IAS meat products. Nonetheless, the Terrestrial and Aquatic Animal Health Codes and Animal Health Information Systems can serve as important tools for implementing invasivorist strategies with international markets that minimize the spread of disease.

### 9.4.3 FAO/WHO Codex Alimentarius

The FAO was created in 1945, with the goal to "achieve food security for all" and ensure that people "have regular access to enough high-quality food to lead active, healthy lives."[98] The FAO has over 194 member states and works in over 130 countries worldwide.[99] Recognizing that "the conflicting nature of [national] food regulations may be an obstacle to trade and may therefore affect the distribution of nutritionally available food." The FAO and WHO established the Codex Alimentarius Commission. The Commission serves as the international food standards-setting body with the primary concern of protecting the health of consumers and ensuring fair practices in the food trade.[100] The commission publishes the Codex Alimentarius (Food Code), which includes provisions on food hygiene, methods of analysis and sampling, and import and export inspection and certification.[101] A number of codes are relevant to trade in IAS products. These include the general principles of Food

93 World Organization for Animal Health [hereinafter OIE], Terrestrial Animal Health Code (2006), Chapter 15.1: Infection with African Swine Fever Virus.

94 OIE, *supra* note 76.

95 OIE, *A Rapid Review of Evidence on Managing the Risk of Disease Emergence in the Wildlife Trade*, at 16, (February 18, 2021).

96 OIE Wildlife Working Group, *Wildlife Trade and Emerging Zoonotic Diseases* (April 2020), https://www.oie.int/app/uploads/2021/03/a-oiewildlifetradestatement-april2020.pdf.

97 *Id.*

98 *About FAO*, FAO, https://www.fao.org/about/en/ (last visited Dec 17, 2021).

99 *Id.*

100 *Codex and the international food trade*, FAO, https://www.fao.org/3/w9114e/W9114e06.htm (last visited Dec 17, 2021).

101 *About Codex*, FAO, https://www.fao.org/fao-who-codexalimentarius/about-codex/en/ (last visited Dec 17, 2021).

Hygiene, and the Code of Hygienic Practice for Meat.[102] Even though the FAO/WHO code is not binding for member countries, like the OIE codes, these standards have become a baseline for compliance with SPS and TBT Agreements.[103] The Code of Hygienic Practice for Meat governs the hygiene, transport, inspection, and processing of wild game killed in the field.[104] Additionally, the Code imposes origin requirements, for example, requiring that the game animal originates from an area that is not subject to chemical pest control programs or active quarantine measures.[105] The Code also requires that primary production includes officially recognized programs for monitoring zoonoses in animal populations and reporting notifiable zoonotic diseases.[106] Overall then, the current landscape of international trade agreements imposes safety and hygiene restrictions that can promote the safe trade in IAS products, important for their ability to reach new markets.

The FAO, along with the CBD, OIE, IUCN, and other international groups, is a member of the Inter-agency Liaison Group on Invasive Alien Species (IALG-IAS).[107] The IALG-IAS "facilitate[s] cooperation among relevant organizations to support measures to prevent the introduction and mitigate the impacts of invasive alien species."[108] And the FAO indicated its awareness of the potential of using IAS as a food source to meet this goal. Indeed, in 2021, the FAO reported on its project that assisted small-scale fishers in Tunisia to turn an invasive blue crab that was harming their industry into a source of income.[109] The blue crab was introduced to the Tunisian coast in 1993 and by 2014 began proliferating massively, causing damage to native species. The crab also threatened the livelihood of local fishermen by damaging nets and traps and feeding on the trapped fish. This species of crab is the fifth most popular crab in the world, with an especially large demand in Asia, the US, and Australia. Seeing this opportunity, the FAO partnered with the Tunisian government to provide local fishermen with crab traps and host training sessions for catching and selling crabs. In collaboration with the Tunisian marine research center, the FAO produced a manual on best practices in blue crab fishing.[110] The FAO has also assisted the government to develop a complete value chain to facilitate the local and international marketing of the blue crab. As a result, exports of blue crab rose significantly, to a value of USD 7.2 million during the month of May 2021, where exports for the same period in 2020 were valued at USD 3.1 million. The FAO will continue to support similar efforts through a newly established research program that will monitor the dispersion, impacts, and income potential of

---

[102] Codex Alimentarius Comm'n, *General Principles of Food Hygiene*, CXC 1-1969.; Codex Alimentarius Comm'n, *The Code of Hygienic Practice for Meat*, CXC 58-2005.

[103] FAO *supra* note 87.

[104] Codex Alimentarius Comm'n, *The Code of Hygienic Practice for Meat*, CXC 58-2005.

[105] *Id.*

[106] *Id.*

[107] Inter-agency Liaison Group on Invasive Alien Species, CBD, https://www.cbd.int/invasive/lg/ (last visited Dec. 17, 2021).

[108] *Id.*

[109] *From invasive species to prized export*, FAO (Oct. 19, 2021), https://www.fao.org/fao-stories/article/en/c/1443166/.

[110] FAO, *Guide des bonnes pratique de pêche des crabes bleus aux nasses polyvantes*, https://www.fao.org/3/cb1493fr/CB1493FR.pdf. (2020).

two other crab species throughout the region.[111] If this program proves successful, it can serve as a model for the research, collaboration, and implementation of future projects aimed at edible IAS, for the FAO as well as for other national and local governments throughout the world.

While this paper offers a preliminary assessment of the international regulatory landscape, there are additional tools not covered here that can be explored for their potential to promote invasivorist strategies. The Kyoto Protocol, for example, commits its signing parties to the reduction of greenhouse gases. Article 12 of the protocol defines its Clean Development Mechanism, which allows a signing nation to meet its targets by implementing an emission-reduction project in developing countries.[112] Since feral pigs around the world release the equivalent of the greenhouse gas emissions of 1.1m cars annually,[113] a country could potentially help meet its targets by implementing a commercial project in a developing country that harvests feral swine for their meat. This mechanism could therefore allow countries to stimulate sustainable development in a country that would otherwise have inadequate resources to devote to feral pig control.

International frameworks are important for setting targets, imposing safety standards, and providing guidance for invasivorist strategies. But, as the literature points out, there cannot be a one-size-fits-all approach to IAS management. Targeting an IAS population requires a highly context-specific approach, close supervision and management throughout the project, as well as the cooperation of a variety of local actors. More local regulatory bodies will ultimately need to be involved in the implementation of an effective IAS management strategy. Local food safety legislation is also important for its ability to establish requirements that ensure the safety of IAS meats, where the entire life cycle of the animal cannot be controlled like in farming. However, as is seen in the US, some food safety laws may impose hurdles that make an invasivorist project prohibitively difficult or expensive. The following sections will compare the regulatory landscapes of the US and the EU to highlight the potential hurdles and opportunities for invasivorist management efforts.

## 9.5 US REGULATORY FRAMEWORK

IAS are a growing concern for the US government. The annual economic costs associated with the roughly 50,000 nonnative species introduced to the US are estimated at more than $100 billion USD.[114] In an effort to prevent, control, and eradicate IAS, the federal government spent around $2.3 billion in the 2016 fiscal year.[115] The invasivore movement in the US is currently a disjointed effort by a handful of local and

111 *Research programme on blue crabs*, General Fisheries Commission for the Mediterranean [GFCM], FAO, https://www.fao.org/gfcm/researchprogramme-bluecrabs (last visited Dec. 17 2021).

112 Kyoto Protocol to the United Nations Framework Convention on Climate Change, Dec. 10, 1997, 2303 U.N.T.S. 162.

113 Christopher J. O'Bryan et al., *Unrecognized threat to global soil carbon by a widespread invasive species*, 28 Global Change Biology 877–882 (2021).

114 U.S. Congressional Research Service. Invasive Species: Major Laws and the Role of Selected Federal Agencies (R43258; Jan. 17, 2017).

115 *Id.*

state governments, environmental groups, and private entities. These efforts can take the form of public-private partnerships like in Maryland, where its Department of Natural Resources sends snakeheads to a private wholesaler for processing and distribution to the commercial market and food banks.[116] There are also private efforts, like those by Maui Nui Venison, a company that sells venison harvested from Hawai'i's invasive Axis deer population.[117] Though this may soon change, there is currently no concerted national effort controlling IAS through invasivorism.[118] Moreover, a patchwork of laws regulating invasive species, food safety, and food labeling create financial and legal hurdles for governments and enterprises interested in promoting invasivorism. However, there may be a trend toward removing these hurdles, like in Maryland's neighboring state of Virginia. In 2018, the Virginia governor signed a bill creating an exception to a state law banning the commercial sale of snakeheads, to allow restaurants and retailers to sell processed snakeheads from certified dealers.[119] However, overall, a high degree of fragmentation in IAS laws and management efforts has hindered the implementation of an effective invasivorist project. The following section provides an overview of the relevant national IAS and food safety management frameworks.

### 9.5.1 Invasive Species Management

Today, many levels of government and non-governmental organizations, private industry, and academic institutions are involved in the patchwork that makes up the US's IAS response efforts.[120] In addition to federal laws, many state laws regulate IAS, with potential impacts for invasivorist efforts. Indeed, differing hunting laws in a target population's range will likely affect the feasibility of harvest. For instance, there are approximately 6 million feral swine distributed over more than 31 states that are causing damage to native ecosystems and are linked to the decline of nearly 300 native species.[121] While some states like Texas and Oklahoma have laws that encourage hunting these species as a method of control, other states like Kansas and Nebraska have restrictions or bans on feral swine hunting altogether to avoid creating incentives to repopulate an area for recreational hunting opportunities.[122] Their statutes do not explicitly contemplate feral swine harvest for commercial sale; however,

---

116 *Invasive Fish from Conowingo Dam Donated to Food Banks, Science*, Md. Dept. of Nat. Res. (May 13, 2021), https://news.maryland.gov/dnr/2021/05/13/invasive-fish-from-conowingo-dam-donated-to-food-banks-science/.

117 *About Us,* Maui Nui Venison, https://mauinuivenison.com/pages/about-us (last visited Dec. 17, 2021).

118 Asian Carp Regional Coordinating Committee, *2021 Asian Carp Action Plan* (March 22, 2021). https://invasivecarp.us/Documents/2021-Action-Plan.pdf (the Asian Carp Regional Coordinating Committee announced its plan to launch a new brand to promote the human consumption of invasive species of Asian carp.)

119 Va. Code Ann. § 29.1-574 (West).

120 Invasive Species Advisory Comm. [hereinafter ISAC], *Strengthening Federal-State Coordination* 1, (2017), https://www.doi.gov/sites/doi.gov/files/uploads/isac_federal-state_white_paper.pdf.

121 US Dep't of Agriculture [USDA], *FERAL SWINE: Impacts on Threatened and Endangered Species*, Program Aid 2195f (May 2020).

122 Terence J. Centner & Rebecca M. Shuman, *Governmental provisions to manage and eradicate feral swine in areas of the United States*, 44 AMBIO 121–130 (2014).

the laws could potentially obstruct the capture of all the individuals of a population distributed across state lines. This type of variability in IAS law can frustrate effective control by harvest, especially for interstate populations of rapidly reproducing species like swine, whose females can birth up to 24 piglets per year.[123] The harmonization of certain state laws may therefore be necessary for an invasivorist project to succeed in controlling a population.

At the federal level, efforts to control invasive species include both administrative and legislative actions.[124] Several executive orders have provided an overarching federal framework to address invasive species. Broad statutes like the Lacey Act,[125] Endangered Species Act,[126] and National Environmental Policy Act[127] (NEPA) also provide authority to address certain IAS, depending on the nature of its impacts. Some laws are tailored to individual species, and others peripherally address invasive species in pursuit of their broader purpose.[128] Several federal and interagency efforts share responsibility regarding IAS, but activities at the USDA account for the bulk of federal funding on IAS at around $1.6 billion.[129] However, comprehensive legislation on the treatment of invasive species has never been enacted, and no single law provides for coordination among federal agencies.[130] The primary leadership responsibility for coordinating IAS management lies with the National Invasive Species Council (NISC).

#### 9.5.1.1 National Invasive Species Council

In 1999, then-president Clinton signed EO 13112, which established the NISC.[131] NISC is "an interdepartmental body that helps to coordinate and ensure complementary, cost-effective Federal activities regarding invasive species."[132] Its overarching duty is to "provide the high-level vision and leadership necessary to sustain and expand Federal efforts to safeguard the interests of the U.S. by preventing, eradicating, and controlling invasive species, as well as restoring ecosystems and other assets impacted by invasive species."[133] The council is co-chaired by secretaries of the Interior, Agriculture and Commerce, and its members include the heads of 13 federal agencies, including the administrator of the Environmental Protection Agency.[134] The NISC is thus well positioned to deliver a coordinated response to IAS.

Among the NISC's duties is to issue the National Invasive Species Management Plan. The purpose of this plan is to identify the high-priority, interdepartmental

---

123 USDA, *supra* note 106.

124 Flynn Boonstra, *Leading by example: A comparison of New Zealand's and the United States' invasive species policies*, 43 Connecticut Law Review. 1185 (2011).

125 16 U.S.C. §§ 3371-3378.

126 16 U.S.C. §§1531-1544.

127 42 U.S.C. §§ 4321 et seq.

128 U.S. Congressional Research Service *supra* note 114.

129 *Id.*

130 U.S. Congressional Research Service *supra* note 114.

131 Exec. Order No. 13112, 3 CFR 13112 (199).

132 *Federal Government's Response for Invasive Species*, USDA, https://www.invasivespeciesinfo.gov/federal-governments-response-invasive-species (last visited Dec. 17, 2021).

133 *Id.*

134 *About the Council*, National Invasive Species Council, US Dep't of the Interior, https://www.doi.gov/invasivespecies/about-nisc (last visited Dec. 17, 2021).

actions that the Federal government and its partners can take to prevent, eradicate, and control invasive species.[135] In the most recent version of the plan, NISC noted that "[t]here is increasing interest in the harvest and use of invasive species as a means of eradicating or controlling invasive populations."[136] It recognized that "If undertaken in a well-informed manner, incentivizing and encouraging public or commercial harvest represents a significant opportunity to support ecosystem restoration and natural resource management while simultaneously boosting economic development and environmental awareness." However, the Plan acknowledges that "little guidance is available on: (a) when and how to use incentivized harvest as an effective management tool for invasive species and (b) strategies for overcoming the logistical barriers to establishing harvest programs in the market place (e.g., availability of start-up funding for companies that intend to deplete the commodity on which revenues are based)."

In a white paper using Pasko and Golberg's article as a primary resource, the NISC approved recommendations for member agencies to consider and prepare management strategies harvesting aquatic invasive aquatic species.[137] Although many of these recommendations are potentially applicable, the NISC has yet to issue any guidance for implementing programs that promote the harvest or consumption of terrestrial IAS, likely due to the restrictions imposed by the meat regulatory regime, as discussed below. Among the paper's recommendations are for member agencies to[138]:

- Develop a management plan that incorporates program goals, measures of success, cost analysis, target species biology, health risks to humans, and humane treatment of target species.
- Ensure monitoring for unintended consequences like perverse incentives or ecological disturbances and incorporate adaptive management and active enforcement disincentivizing the release of the target species.
- Incorporate outreach to stakeholders and the public to promote understanding of the harm caused by the target species and the end goal of eradication.

The paper also notes that managing aquatic IAS with harvest incentives becomes complex when the population is spread between multiple jurisdictions, advising managers to consult federal, state, and local laws.[139] It mentions the possibility that alternative control methods or new legislation may be required to ensure the greatest reduction in the target population.[140]

A recent Invasive Species Advisory Committee (ISAC) report identified weaknesses in the NISC's ability to coordinate with state governments, finding that the lack of a national or regional framework to address IAS consistently across political

135 *National Invasive Species Management Plan,* USDA, https://www.invasivespeciesinfo.gov/national-invasive-species-management-plan (last visited Dec. 17, 2021).

136 National Invasive Species Council. *Management Plan: 2016–2018.* (July 11, 2016).

137 ISAC, *Harvest Incentives: A Tool for Managing Aquatic Invasive Species.* (May 15, 2014).

138 *Id.* at 4.

139 *Id.* at 3.

140 *Id.*

boundaries hampers the NISC's ability to target IAS threats on a large scale.[141] ISAC's recommendation for the implementation of a Federal-State Invasive Species Coordinating Committee may therefore improve capacity to implement large-scale invasivorist programs that incorporate the necessary research, monitoring, and adaptive management on the ground.

#### 9.5.1.2 Lacey Act

The NISC's guidance document also identified the Lacey Act as an obstacle to reaching markets where live IAS are preferred.[142] When the Lacey Act was passed in 1900, it became the first federal law protecting wildlife.[143] Today, the Lacey Act prohibits imports and exports of certain fish, wildlife, or plants, including those listed as "injurious species." The Act enforces civil and criminal penalties against those who violate its provisions. The Secretary of the Interior is vested with the authority to declare species of mammals, birds, fish, amphibians, and reptiles to be "injurious to human beings, to the interests of agriculture, horticulture, forestry, or to wildlife…" Listing an IAS as "injurious wildlife," under the Lacey Act prohibits its live cross-border movement. However, as was pointed out by Galperin and Kuebbing's article, prohibiting cross-border movement of an edible IAS may be at odds with an invasivorist strategy. For example, Asian Carp are a delicacy in China where consumers prefer to eat the fish fresh. However, the species' listing as injurious wildlife under Title 18 of the Lacey Act (18 U.S.C. 42) makes it impossible to reach these markets without first processing the fish, limiting their income potential.[144] States can only issue permits to transport or purchase live Asian carp for scientific, medical, or educational purposes.[145] If the Lacey Act was amended to allow for the secure transport of IAS for commercial purposes when part of a management strategy, it could allow fresh fish to reach a wider market. However, other regulatory safeguards would need to be put in place to prevent the species' intentional and unintentional spread, which could include the requirement of stricter transport protocols, as required by the EU's IAS legislation. Still, the interstate sale of products from processed Asian Carp, including canned fish, fish meal, fish oil, and frozen fish, remains an option for producers. But the US food safety regime makes the sale of invasive terrestrial animals much more difficult.

### 9.5.2 Food Safety Management

The US food regulatory system is a patchwork of state and federal laws and involves several agencies including the USDA, FDA, EPA, and CDC. The Federal Food, Drug, and Cosmetics Act (21 U.S.C. ch. 9 § 301 et seq) (FDCA) vests the FDA with regulatory

---

[141] ISAC, *supra* note 105.

[142] ISAC, supra note 117 at 3 ("Certain markets may prefer live Asian carp, but their listing as injurious wildlife under Title 18 of the Lacey Act (18 U.S.C. 42) prohibits live in- terstate and cross-border movement.").

[143] Lacey Act, US Fish & Wildlife Service, https://www.fws.gov/international/laws-treaties-agreements/us-conservation-laws/lacey-act.html (last visited Dec. 17, 2019)

[144] Dapeng Li et al., *Asian carp: A threat to American lakes, a feast on Chinese tables*, 20 Comprehensive Reviews in Food Science and Food Safety 2968–2990 (2021).

[145] 18 U.S.C. 42.

authority over most food products, including fish. The Federal Meat and Inspection Act of 1906 (21 U.S.C. ch. 12 § 601 et seq) (FMIA) and Poultry Products Inspection Act of 1957 (21 U.S.C. ch. 10 § 451 et seq) (PPIA) place responsibility with the USDA to regulate most meats. However, since most IAS are not named by these acts as "amenable" species, they are left in a regulatory gray zone. Further, the current prohibition on the commercial sale of meat from animals hunted in the wild places places additional limits on the ability to control terrestrial IAS through commercial consumption. Food labeling regulations promulgated by the FDA may also harm the commercial potential for IAS by limiting how their products can be marketed.

#### 9.5.2.1 Federal Meat Inspection Act of 1906 (FMIA) and Poultry Products Inspection Act of 1957 (PPIA)

The FMIA requires pre-slaughter and post-mortem inspection of amenable species of livestock, implements sanitary standards, and authorizes the USDA to monitor and inspect slaughter and processing facilities.[146] Livestock Products amenable to the FMIA are those prepared from cattle, sheep, swine, goats, horses, mules, and other equines. The PPIA implements similar requirements for the facilities and processing of domesticated birds, which include chickens, turkeys, ducks, geese, guinea fowl, and ratites (ostriches).[147] Invasive feral swine (*Sus scrofa fera*) are the product of domestic pigs (*Sus scrofa domestica*) used for pork production that have been released or escaped into the wild.[148] This means that the meat from feral swine is covered as amenable by the FMIA; however, the meat from most other IAS falls outside of this classification. The Food Safety and Inspection Service (FSIS) is the agency of the USDA responsible for enforcing the two acts.[149] Federal inspection personnel must be present at all times during slaughter operations. Federal inspectors must ensure live animals are fit for slaughter and observe slaughter to verify the humane handling of animals in accordance with the Humane Methods of Slaughter Act[150] (7U.S.C. ch. 48 § 1901 et seq), which applies to cattle, calves, horses, mules, sheep, and swine (including feral swine). In addition to overseeing processing, FSIS personnel must inspect the facilities and equipment to ensure sanitary conditions are maintained.

About half of the states have their own meat-inspection programs, which can only approve meat products for intra-state commerce, and must enforce requirements "at least equal to" those imposed under the Federal Acts.[151] These regulations mean that feral swine must be live-caught and held until they can be transported to processing facilities where they receive a pre- and post-mortem inspection by FSIS personnel. However, this may not be feasible if the nearest processing facility is in a state that prohibits the transport or import of feral swine. Transport over long distances additionally causes stress for the animals. To process animals "in the field" requires

---

[146] Federal Meat Inspection Act, Pub.L. 59–382, 34 Stat. 669 (1906).

[147] Poultry Products Inspection Act, Pub.L. 85–172, 71 Stat. 441.

[148] *askFSIS Public Q&A: Amenable Swine*, USDA, https://ask.usda.gov/s/article/Which-species-of-swine-are-amenable-to-the-Federal-Meat-Inspection-Act (last visited Dec 17, 2021).

[149] Food Safety and Inspection Service, USDA, https://www.fsis.usda.gov/ (last visited Dec. 17, 2021).

[150] Humane Methods of Slaughter Act, Pub.L. 85–765, 72 Stat. 862 (1958).

[151] *State Inspection Programs*, USDA, https://www.fsis.usda.gov/inspection/apply-grant-inspection/state-inspection-programs (last visited Dec. 17, 2021).

additional expense for voluntary inspection, as discussed below. The requirements become more expensive to comply with for terrestrial IAS not covered by these acts, which are technically governed by the FDA.

### 9.5.2.2 Federal Food, Drug, and Cosmetics Act (FDCA)

Under the FDCA,[152] the FDA regulates all food products except those specifically covered by the FMIA and PPIA. The FDA promulgates regulations applying to wild-caught fish which include requirements for producers to conduct hazard analyses and implement HACCP plans and implement sanitation control procedures.[153] For non-amenable species of terrestrial animals, the FDA provides general guidelines in the form of a model Food Code.[154] These guidelines require that the processing of meat from live-caught wild animals occurs under a voluntary inspection program "conducted by a regulatory agency such as the agency that has animal health jurisdiction."[155] Animals that are hunted and dressed in the field must receive a post-mortem examination by an approved veterinarian or veterinarian's designee and must be processed according to laws governing meat and poultry "as determined by the agency that has animal health jurisdiction and the agency that conducts the inspection program."[156] Since the FDA does not have meat inspection capabilities, this means that meat from IAS must undergo a voluntary inspection by the FSIS before it can be sold in interstate commerce.

### 9.5.2.3 Agricultural Marketing Act of 1946

The production of amenable meats is eligible for mandatory federal inspection, meaning inspections are funded by the government. The Agricultural Marketing Act of 1946 (7 U.S.C. 1621 et seq.) gives the FSIS authority to provide voluntary inspections for non-amenable meats; however, producers must pay for these inspections out of pocket.[157] This can come at a high cost for producers of IAS products, as the 2021 rate for an inspector was $66.56 per hour.[158] Slaughterhouses also require FSIS approval to process each species, which comes with additional costs for processors. As a result, few slaughterhouses take non-amenable species.[159] And reaching these slaughterhouses can be made difficult or impossible in states where there are restrictions on the cross-border movement of live IAS.

Regulations allow producers the option to hunt and process animals in the field, but these also come at a high cost. Producers are not only responsible for paying the

---

[152] Federal Food, Drug, and Cosmetic Act, 21 U.S.C. §§ 301-399f (2012).

[153] 21 C.F.R. § 123 Fish and Fishery Products.

[154] *Model Food Code*, U.S. Department of Health and Human Services. (2017), https://www.fda.gov/media/110822/download.

[155] *Id.* at 3-201.17 (3)(a).

[156] *Id.* at 3-201.17 (4).

[157] 9 C.F.R. § 352.5.

[158] 2021 Rate Changes for the Basetime, Overtime, Holiday, Laboratory Services, and Export Application Fees (Notice by the Food Safety and Inspection Service). Federal Register 85:79992 (Dec. 11, 2020) p. 79992.

[159] Elizabeth S. Byrd, Nicole J. Olynk Widmar & John G. Lee, *Non-amenable meat consumption, sale, and regulation: bison, beef and Bambi, oh my! all meats are not created equal*, 20 Drake Journal of Agricultural Law 337 (2015).

hourly rate for the ante-mortem and post-mortem inspection but are also responsible for the inspectors' travel and per diem costs.[160] Producers unable to reach processing facilities could implement mobile processing facilities, which would contribute to high start-up costs for potential producers of IAS.

It should also be noted that labeling requirements will play a large role in the marketability of IAS. The purpose of labeling laws is to facilitate consumers' informed choices, prevent deception, educate consumers, assure food safety, and promote honest and fair dealing in the marketplace.[161] The USDA has jurisdiction over the labeling of food products from amenable species, including feral pigs, but the FDA regulates the labeling of meat from all other species including fish. Concerned about fraud, the FDA published guidance for the seafood industry that limits the available acceptable market names.[162] However, connotations with certain IAS names, like the "snakehead" fish, may make them less appealing to consumers. Meat from other non-amenable animals must still adhere to the requirements of the Fair Packaging and Labeling Act (15 U.S.C. ch. 39 § 1451 et seq), but there is little guidance as to the marketable names for these IAS species. This could offer greater flexibility to market products under names more desirable for consumers, however, may open liability for consumer claims of deception. Further issues may emerge as to how to inform consumers when an IAS product has been harvested from the wild while assuring consumers of its safety.

Overall, the requirements imposed by the meat regulatory system create financial and logistical obstacles that result in prohibitively expensive start-up and inspection costs for any producer considering harvesting terrestrial IAS for meat. However, amending the FMIA or PPIA to list edible IAS as amenable under their provisions or offering federal subsidies through IAS control funds to offset start-up costs for IAS producers that are operating under an IAS management program could remove some of these obstacles and thereby encourage the harvest of IAS meat products. The Farm Bill may serve as another potential tool to encourage producers to target IAS. The Farm Bill, an omnibus bill that is revised and renewed every 5 years, governs an array of agricultural and food programs.[163] The most recent version, signed into law in 2018, has potentially applicable provisions that create funds for wildlife services to address feral swine populations, as well as an Environmental Quality Incentives Program that provides financial and technical assistance to "non-industrial forest managers to address natural resource concerns and deliver environmental benefits such as...improved or created wildlife habitat..."[164] These funds could potentially be used to fund research as well as start-up and inspection costs for IAS meat producers. Still, depending on consumer demand, government subsidies for the ante and post-mortem inspection of each animal may negatively impact the cost-to-benefit ratio of an invasivorist management program. Thus, a food regulatory

---

160 9 CFR § 352.4.

161 Michael T. Roberts, § 4.02 *Regulation of Food Labelling, in* Food Law in the United States 209 (2016).

162 The Seafood List FDA Guide to Acceptable Market Names for Seafood Sold in Interstate Commerce; Availability (Notice of Availability). Federal Register (Sept. 14, 1994).

163 *Farm Bill*, USDA, https://www.usda.gov/farmbill (last visited Dec. 17, 2021).

164 *Environmental Quality Incentives Program*, USDA Natural Resources Conservation Series, https://www.nrcs.usda.gov/wps/portal/nrcs/main/national/programs/financial/eqip/ (last visited Dec. 17, 2021).

system more like the EU's may be more conducive to an invasivorist strategy while still ensuring the safety of IAS products.

## 9.6 EU REGULATORY FRAMEWORK

The EU's current regulatory system is more conducive to an invasivorist management strategy, so long as one is formalized and implemented by a Member State. And while there is some effort to promote invasivorism at the grassroots level,[165] so far, member states have yet to demonstrate a harmonized approach that controls IAS through commercial sale.

### 9.6.1 Invasive Species Management

#### 9.6.1.1 Regulation (EU) No. 1143/2014

Invasive species are regulated by Regulation (EU) No 1143/2014 on the prevention and management of the introduction and spread of invasive alien species (the IAS Regulation).[166] The regulation entered into force in 2015 and was fully implemented by member states in 2019. It aims to: (1) prevent, minimize, and mitigate the adverse impacts of IAS on biodiversity and ecosystem services; and (2) limit social and economic damage. The IAS Regulation was adopted to fulfill targets of the EU Biodiversity Strategy to 2020[167] and the strategic plan for biodiversity 2011–2022 under the CBD.[168] The IAS regulation calls for the establishment of a committee, which is assisted by a scientific forum, consisting of members of the scientific community appointed by the member states for providing advice on the application of the regulation.[169]

At the regulation's core is the list of IAS of Union Concern (The Union list), which lists species whose adverse impact has been deemed sufficiently severe to require concerted action at Union level.[170] Member states are required to take action on pathways of unintentional introduction, to take measures for the early detection and rapid eradication of these species, and to manage species that are already widely spread in their territory. The Union list currently contains 66 species. Listed species are subject to the restrictions and measures set out in the Regulation, which include restrictions on keeping, importing, selling, breeding, and growing.[171] Unlike under the Lacey Act, member states may issue permits for the temporary commercial use of established IAS. The stipulation is that commercial use of IAS must form part of the member states' management measures aimed at the species' eradication, control or

[165] *Could invasive species, harming the ecosystem, be our next source of food?* The Dutch Institute of Food & Design, https://thedifd.com/articles/could-invasive-species-harming-the-ecosystem-be-our-next-source-of-food/ (last visited Dec. 17, 2021). (Article from the Netherlands describing "menu from the new wild" a culinary concept by industrial designer Alexandra Fruhstorfer that sees invasive species as a food resource.).

[166] OJ L 317, 4.11.2014, pp. 35–55.

[167] Our life insurance, our natural capital: an EU biodiversity strategy to 2020 (COM(2011) 244).

[168] Conference of the Parties to the CBD *supra* note 68.

[169] OJ L 317, 4.11.2014, pp. 35–55.

[170] *Id.*

[171] OJ L 189, 14.7.2016, pp. 4–8.

containment, and must be "under strict justification and provided that all appropriate controls are in place to avoid any further spread." [172]

Although member states have not yet taken advantage of the exception, there are opportunities to do so. For example, there is a demand for Chinese mitten crab in China, and prior to the IAS regulation, a Dutch seafood company exported 95% of harvested mitten crab to Hong Kong.[173] However, the IAS regulation now entirely prevents the trade of mitten crabs caught in European waters until Holland implements a national management plan.[174] The European Commission has noted the commercial potential for the species, and even suggested management measures that include the crabs' consumption as a source of food or animal feed, "subject to the control of their pollutant concentrations and provided that this is integrated into the management measures of the Member State."[175]

Though it temporarily halts individual commercial efforts targeting IAS, this system can ultimately help ensure that the commercial use of IAS is closely monitored and adaptively managed by local governments. But, until member states have implemented and studied the effects of invasivorist strategies, it is too soon to tell what will be the effect of this exception. If successful, the IAS regulation could provide a blueprint for amendments to the Lacey Act that would allow US IAS products to reach broader markets without promoting their spread.

## 9.6.2 Food Safety Laws

### 9.6.2.1 Regulation (EC) No. 852/2004 and Regulation (EC) No. 854/2004

An extensive body of EU-wide law covers the entire food production and processing chain.[176] The EU's food regulations are guided by the advice from the European Food Safety Authority (EFSA), which is responsible for ensuring science-based risk management.[177] The EFSA's work involves gathering scientific data and expertise, providing independent, up-to-date scientific advice on food safety issues, communicating its scientific work to the public, and cooperating with EU countries, international bodies, and other stakeholders.

In the EU, the regulations applicable to the sale of invasive animal meat are Council Regulation (EC) No. 852/2004 on the hygiene of foodstuffs,[178] Council Regulation (EC) No. 853/2004, laying down specific hygiene rules for the hygiene of foodstuffs,[179] and Council Regulation (EC) No. 854/2004 laying down specific

172 *Supra* note 141.

173 Mark Godfrey, Dutch firm loses China crab export business with EU ruling, Seafood Source (August 8, 2016), https://www.seafoodsource.com/news/supply-trade/dutch-firm-loses-china-crab-export-business-with-eu-ruling.

174 *Id.*

175 European Union, *Invasive Alien Species of Union concern*, (2017).

176 *Food safety in the EU*, European Union, https://european-union.europa.eu/priorities-and-actions/actions-topic/food-safety_en (last visited Dec. 17, 2021).

177 European Food Safety Authority (EFSA), European Union, https://european-union.europa.eu/institutions-law-budget/institutions-and-bodies/institutions-and-bodies-profiles/efsa_en#what-does-efsa-do (last visited Dec. 17, 2021).

178 OJ L 139, 30.4.2004, pp. 1–54.

179 OJ L 139, 30.4.2004, pp. 55–205.

rules for the organization of official controls on products of animal origin intended for human consumption.[180] The safety strategies for handling game meat laid out in these regulations include establishing efficient surveillance, providing continuing training for hunters regarding hygiene and zoonoses, assuring traceability of products through proper documentation, and the provision of well-equipped facilities.[181]

These regulations require meat inspection, but unlike in the US, there are exceptions for ante-mortem inspection of wild game. Trained hunters can conduct an initial examination of game on the spot to identify anomalies or hazards, and official post-mortem inspection can be done at a game-handling establishment. Hunters are also responsible for ensuring traceability of wild game meat by labeling each carcass. A veterinarian then must examine the carcass to test for trichinella.[182] Member states can establish stricter rules to account for specific risks.

The EU's regulations allow the hunting of IAS from the wild without imposing costly additional inspection requirements on hunters or processors. However, removing veterinary supervision over the entire game meat chain means the microbiological quality of the meat depends on hunters' operating procedures, which may not be adequate for inspection and control.[183] One alternative approach could impose stricter training requirements for hunters, including awareness of disease risks, proper slaughter methods, and meat hygiene.

Overall, while the EU's IAS and food safety regulatory systems are more conducive to an invasivorist approach, especially for terrestrial species of IAS, it is too soon to tell whether these changes will lead to the implementation of a successful invasivorist management program. However, IAS pose a dire international risk that will require action from all angles. Swift enactment of regulatory changes, stakeholder engagement, and removal of perverse incentives are essential to reducing the negative effects of IAS.

## 9.7 CONCLUSION

IAS and unsustainable food production pose growing and solemn threats to biodiversity. To address these impacts, governments around the world are beginning to promote IAS consumption as a control method. Invasivorism offers the potential to cost-effectively manage IAS while providing an environmentally sound and high-welfare food source. However, this is a new area of research drawing from several complex areas of science and policy. Regulations at a local, national, and international level are critical tools for developing a successful strategy. Given the lack of research on the issue, stakeholder engagement may be necessary to spark more public awareness and debate on the issue. This chapter provides only a glimpse into relevant regulatory systems, and further research should continue this comparative analysis in other countries with opportunities to control IAS through consumption.

---

[180] OJ L 139, 30.4.2004, pp. 206–320.

[181] Niels Bandick & Andreas Hensel, *Zoonotic diseases and direct marketing of game meat: Aspects of consumer safety in Germany*, 1, Game Meat Hygiene in Focus 93–100 (2011).

[182] European Union Commission Implementing Regulation, 2015. Law 1375/2015 of 10 August 2015 laying down specific rules of official controls for Trichinella in meat

[183] Germana Giuggioli et al., *The hygiene-sanitary control in the Wild Game Meats*, 6 Italian Journal of Food Safety, 222–224 (2018).

## 9.8 LITERATURE REVIEW

| Authors | Country, Region | Purpose | Type of Source | Summary Points |
|---|---|---|---|---|
| Nuñez, Kuebbing, Dimarco, and Simberloff (2012) | US | To analyze the pros and cons of attempting to control invasive species by eating them, focusing on the general issues of this approach, and discussing if, when, and how the approach of controlling nonnative species via gastronomy should be used. | Policy Perspective | Government agencies, academic institutions, independent groups, citizen scientists, and modern hunter-gatherers are already partaking in the consumptive use of invasive species. The potential benefits associated with the attempt to manage invasive species by eating them are increasing awareness of IAS, boosting local economy.<br>However, decisions to start or maintain a program based on human consumption of an invasive species as a way to control it should arise by carefully considering the potential benefits and problems that the program could produce. These risks include failing to affect population size, creating a market for IAS that will need to be maintained, promoting further invasion, or promoting incorporation of invasives into local culture. |
| Galperin and Kuebbing (2013) | US | To examine the legal barriers to invasivore movement. | Law article | The government plays a dissonant role in the eating invaders' movement. Invasive species laws pose a challenge to invasivore movement by preventing their spread. The Lacey Act and Plant Protection Act restrict the import, export, and movement of certain species. However, for an invasivore effort to succeed would require broad uptake and commercialization, which may be impossible if interstate commerce is restricted. Further, food safety laws prevent the sale of hunted game. |
| Orth, Schmitt, and Hilling (2020) | US, Chesapeake Bay | To explore the narrative surrounding the introduction and expansion of Blue Catfish in mid-Atlantic waters. | Media analysis | Invasivorism is the only actively pursued management solution to the problem of Invasive Blue Catfish in the Chesapeake Bay region; however, wildlife managers disagree as to the effectiveness of the strategy. |

*(Continued)*

| Authors | Country, Region | Purpose | Type of Source | Summary Points |
|---|---|---|---|---|
| Sim (2013) | US | To propose a tactic for the management of invasive species. | PhD dissertation | Launched web-blog invasivore.org which shares information about invasive species, harvesting techniques, and recipes. Invasivory's merits are that it mitigates IAS impacts and can increase awareness of invasive species, thus potentially preventing the introduction and entrenchment of new invasive species. The drawbacks of the approach are the "invasivore's dilemma," which refers to the potential to create an economic incentive for harvesters to purposely propagate invasive species. |
| Varble and Secchi (2013) | US | To investigate the potential of creating a US consumer market for Asian carp meat. | Marketing study based on a national survey | Creating demand for Asian carp could be a market-based, cost-effective solution for the problem of invasive carp, if coupled with appropriate policies and safeguards to ensure the fish is eventually eradicated and not cultivated for profit after removal from US rivers and lakes. |
| Weis (2011) | US | To review aquatic biological invasions in which predation (or its absence) plays a major role in the success of the invader. | Article/review | Human consumption of invasive species is a relatively new management approach in the US. The approach is being utilized in Florida and throughout the Caribbean against the lionfish, but could be extended to other fishes, as well as to various invasive crustaceans and mollusks. This idea is controversial because some environmentalists argue that invasivory could promote the development of a sustainable fishery instead of eliminating the target species. Further, current regulations prohibiting the possession of individuals of the invasive species (e.g., mitten crabs or snakefish) would preclude the development of a fishery for them. Ultimately, eating invasive species will not solve the problem by eradicating them, but it will lessen their impacts and at the same time provide a new food source for a world in which many traditional food fishes are in short supply because of overfishing, habitat loss and pollution. |
| Hayward (2014) | Australia | To introduce the concept of eco-culinary activism and explore the potential for cane toad meat in local cuisine as an export commodity. | Article/Analysis of media coverage | Eco-culinary activism can be defined as the strategic targeting and promotion of particular species for either consumption or avoidance of consumption on ecological grounds. Factors for considering invasive cane toads as a potential mass-market product include their nutritional value, environmental benefits of reducing the population, and plentiful supply. However, the challenges to this approach are the brand image of the toads and their safety as a food source. |

*(Continued)*

| Authors | Country, Region | Purpose | Type of Source | Summary Points |
|---|---|---|---|---|
| Ninon (2018) | Pacific Island countries and Territories | To assess the commercial markets' potential to control exotic invasive species of the PICTs. | Project Note Invasive species listing, table construction, general species data collection, and the economic potential assessment | With the exception of isolated initiatives, economic valorization of IAS remains relatively new in the Pacific with very few documentations existing on the subject. Very few studies have critically assessed the success of these market-based programs. Market-based harvest incentives are therefore a promising but delicate management strategy. With this approach, each implementation case will involve its own economic and ecological risks. Its success implies careful considerations and relies on the quality of the economic and environmental feasibility studies conducted and on the effectiveness of the monitoring and evaluation strategy. With the exception of the RESCCUE project actions on the commercial valorization of the Rusa Deer in New Caledonia, there are no regional or country-specific programs working on the development of this management approach. |
| Hayward (2013) | Tasmania | To address the role of the sea urchin in 20th–21st century cuisine, fisheries and aquaculture in Tasmania, and the manner in which consumption of the invasive sea urchin has been promoted as a strategy to control its spread in coastal waters. | Article | Hard data on the extent to which invasive urchin products have been a success with fishers, wholesalers, retailers, and the public has yet to emerge; anecdotal evidence of the product's popularity on local menus in Caribbean locations such as Bonaire island indicate that it has become well-established as a culinary item; and preliminary scientific surveys indicate that harvesting is having a significant impact on fish numbers in focal areas.<br>This integration of traditional ecological activism and culinary promotion adopted to combat environmental menaces offers a model for other creatures and regions. |

(*Continued*)

| Authors | Country, Region | Purpose | Type of Source | Summary Points |
| --- | --- | --- | --- | --- |
| | | | | However, the volume of harvesting activity is not likely to have a drastic impact on the local sea urchin invasives since the cost-effectiveness of dive harvesting means that no comprehensive clearance will be achieved unless the demand/price for sea urchins increases exponentially. Further, if a large and relatively stable market for sea urchins can be created, reliable supply of the product is likely to be better served by aquacultural systems, thereby leaving the invasive to flourish unchecked outside of these. Thus eco-culinary activism of the type discussed in this article needs to develop and adapt in a strategic and programmatic manner in order to have significant success in its enterprise. |
| Galperin, Killiam, and Kuebbing (2016) | US | To describe the rationale and breadth of the eating invaders movement and challenge the efficacy and presumptions of the movement. | Research Report | Legal restrictions do not make it impossible to eat invasive species, but by limiting possession, transportation, and sale of many invaders, the restrictions could make it impossible to establish effective commercial markets for these species.<br>A critique of this movement is that consumers may begin to recognize invasive species as a valuable part of their milieu and then protest declining populations that tend to remove a cultural icon. Although not food-related, the American public has shown deep attachment to wild mustangs (*Equus ferus caballus*), despite their invasiveness, and object to government management plans. An eat the invaders plan is a tool for educating the public about an important issue. However, there are substantial challenges to its feasibility, and the more popular it becomes, the more likely it is to exacerbate ecological problems. It may be, in other words, a self-defeating effort. A more critical and public debate of the idea is necessary. |

(*Continued*)

| Authors | Country, Region | Purpose | Type of Source | Summary Points |
|---|---|---|---|---|
| Cerveira (2021) | Europe | To evaluate the acceptance of weakfish by a panel of Portuguese fish consumers and create outreach actions – partnerships with local Chefs and press releases – to explain to a broader public what invasive species are and promote the consumption of edible aquatic invasive species. | Study based on Survey and Outreach | Consumers showed that weakfish has great chances of being well accepted by the public – 90% of consumers would buy weakfish because they appreciate its appearance, flavor, and texture, besides being a wild fish. The outreach actions reached a few million people because 46 online articles were published, and three news pieces broadcasted on national television. Overall, the strategy greatly increased the public's awareness about invasive species, which can be replicated elsewhere in the world. Outreach campaigns focusing on the conservation of native fisheries should simultaneously be encouraged, as well as the commercialization of native species that are not being valued to shift the focus from perpetuating the fishery of an invasive species and introducing it somewhere else. |
| Carballo-Cárdenas (2015) | Caribbean | To investigate how the lionfish (*Pterois* sp.) invasion of the Western Atlantic Ocean has been socially constructed by natural scientists, the media, and stakeholders associated with various marine protected areas in the Caribbean. | Media analysis | Promotion of invasive species consumption to control populations is not new, has intensified in recent years, and has been picked up by the media, using the term "invasivores" to refer to those who eat nonnative (invasive) species. The Mexican government is promoting the harvesting of lionfish for local and international seafood markets. Respondents from Curaçao and Bonaire contend that some individuals purposefully leave the small lionfish in the reef to grow larger. Among ecologists, the dominant discourse is that lionfish consumption is a promising control strategy. Only one article was found where a marine biologist questioned the effectiveness of lionfish commercialization, in particular if regulatory aspects are not addressed simultaneously. For some respondents, harvesting lionfish manually for food is not worth the trouble because of the effort, costs, and risk involved in capturing and processing lionfish, and because the flesh yield is limited. Within the eat-lionfish coalition, development of a specialized lionfish trap is advocated, but the counterarguments are that, besides being expensive to develop, traps are not as species-specific (hence sustainable) as manual removals. A spear lionfish versus trap-lionfish debate has emerged on the grounds of cost-effectiveness and sustainability of removal methods, but thus far no data had been brought forward to substantiate claims. |

(*Continued*)

| Authors | Country, Region | Purpose | Type of Source | Summary Points |
|---|---|---|---|---|
| IUCN France (2018) | France | To conduct a study to identify the challenges and risks of socioeconomic valuation of IAS established in natural environments as a possibility of regulating these species. | Analysis of case studies of commercial harvest initiatives | There is little quantitative data to support possible successes in reducing IAS populations. Few scientific evaluation methods are proposed and implemented, and few projects are accompanied by recommendations or warnings regarding certain negative consequences. At the same time, many experts report risks and difficulties in implementation and control. Projects consisting of incentives should be differentiated from those involving only commercial exploitation. Indeed, for the former, the main project promoters are public institutions, in partnership with research organizations, user associations, and communities. These projects are integrated into global management strategies, display exclusively ecological objectives, are all accompanied by awareness-raising actions, and are the subject to monitoring. There are thus presumably fewer environmental risks than for commercial exploitation projects. |
| Pasko and Goldberg (2014) | US | To review the biological, ecological, human health, and socioeconomic factors involved in invasive species incentive programs, and offer recommendations to assist in the development and implementation of a successful harvest program. | Review | Despite the widespread attention harvest has received, there are limited case studies where its value has been successfully demonstrated for control or eradication, with relatively few analyses or recommendations available. Programs that encourage incentivized harvest may be an effective management tool in targeting small, distinct populations or they may play a supplementary role within larger control or eradication programs. In many cases, there will be a need for partnerships between government, natural resource managers, and private industry to develop and implement efficient programs to the control invasive species and to mitigate the threats to native ecosystems. The authors offer recommendations for natural resource managers considering a plan that incentivizes harvest of an IAS. The authors recommend that natural resource managers:<br>1. Define the management plan and objectives, which should include an exit strategy to avoid complications with ending a program that generates revenue for the private sector. |

*(Continued)*

| Authors | Country, Region | Purpose | Type of Source | Summary Points |
|---|---|---|---|---|
| | | | | 2. Compare the anticipated benefits of control or eradication to the costs of control.<br>3. Understand the target species' population dynamics, including its demographic structure, specific vital rates, and minimum viable population size.<br>4. Consider risks to human health and safety and conduct an appropriate health assessment as they might do for assessing any fishery.<br>5. Evaluate potential ecological outcomes, to evaluate species interactions and effects or removing an invasive species from the ecosystem.<br>6. Monitor for unintended outcomes like breeding or intentional introductions of the target species.<br>7. Prevent re-introduction through prevention, monitoring, and rapid response programs,<br>8. Incorporate adaptive management to determine when and if incentives should be implemented.<br>9. Conduct outreach so that the public understands and dislikes the long-term harm the target species can cause to native species and habitats.<br>10. Restore impacted areas to ensure long-term benefits of native restoration outweigh the short-term benefits accrued by harvest.<br>11. Determine appropriate points for government intervention, as financial support from state and federal agencies to encourage harvest may be more necessary when population levels of the invasive species are low. Agencies and legislatures may also need to review and consider new authorities or regulations that may be needed to effectively manage and control the target IAS. |

(Continued)

| Authors | Country, Region | Purpose | Type of Source | Summary Points |
|---|---|---|---|---|
| Street (2014) | US | To discuss the biological, ecological, and socioeconomic considerations involved in programs that utilize harvest incentives to manage aquatic invasive species and to provide recommendations for consideration in the development, implementation, or support of incentive or harvest efforts that target aquatic invasive species. | White Paper for the Department of the Interior | The success of any harvest incentive program to address aquatic invasive species will depend upon numerous biological, socioeconomic, and legal considerations. Programs that encourage harvest may be a successful management tool in targeting small, distinct populations; in high-priority areas within a larger invasion; or they may play a supplementary role within larger control programs. Their use, however, will require careful review, planning, and monitoring to ensure success and that they do not unintentionally lead to further spread of invasive species, cause additional harm to native species, or waste valuable resources.<br>Resource managers are recommended to develop a management plan prior to undertaking a harvest incentive program. The plan should incorporate each of the following: a. Program goals and measures of success b. Cost analysis c. Target species' biology d. Address humane treatment e. Human and wildlife health risks f. Potential ecological outcomes g. Risk of creating perverse incentives. After developing a plan, managers should: a. Monitor for unintended consequences b. Monitor for ecological disturbances c. Incorporate adaptive management d. Encourage active enforcement to help mitigate perverse incentives by creating a disincentive to release the target species back into the control area or previously non-invaded areas. Lastly, a plan should incorporate outreach that is clear about the goals of the program and to ensure the public does not desire the targeted species. Success is more likely if the public understands the long-term harm the species can cause. |
| Bouska et al. (2020) | US | To explore the efficacy of a consumer-based market (i.e., invasivorism) to manage Invasive bigheaded carps | Experimental Study | Contracted removal was effective for suppressing small populations at the edge of the range but cannot support a market. "Fisher-side" removals totaled 225,372 kg in 1 year. However, participation was low, likely due to reporting requirements for fishers. The "market-side," set-quota approach removed about 1.3 million kg of bigheaded carp in less than 6 months. Larger, older fish were disproportionately harvested, which may hinder the ability to suppress |

(*Continued*)

| Authors | Country, Region | Purpose | Type of Source | Summary Points |
|---|---|---|---|---|
| | | | | population growth. Total density declined in one river reach, and harvest may reduce upstream movement toward the invasion fronts. With sufficient market demand, harvest may control bigheaded carp. However, lack of processing infrastructure and supply chain bottlenecks could constrain harvest, particularly at low commodity prices. Given the geographical scale of this invasion and complicated harvest logistics, concerns about economic dependence on invasivorism that encourage stock enhancement are likely unmerited. |
| Gressier (2016) | Australia | To analyze Australians' relationship with feral meat. | Ethnography | While wild meat previously elicited disgust owing to the lack of regulating controls, the increasingly widespread critiques of the industrial food system and the excessive human interventions therein have resulted in greater aversion toward factory-farmed meats. On a structural level, legislation currently prohibits the harvesting of wild meat for human consumption on a commercial scale in several Australian states. Australians are generally enthusiastic meat eaters, consuming almost twice that of the British in an average year. This proclivity, in conjunction with the increasingly mainstream concerns with environmental sustainability, animal welfare, and human health, suggests we should consider alternatives to conventionally produced meat. The growing market for organic, free-range, and high welfare meat demonstrates that many people are prepared to pay more for that which is ethically produced. |
| Keane, Mundy, Porteus, Johnson, IMAS-FA, University of Tasmania, (2019) | Tasmania | To analyze whether the commercial harvest of long-spined sea urchins can reduce the impact of urchin grazing on abalone and lobster fisheries. | Survey Program | There are positive benefits for abalone habitat at when invasive urchins are exploited at low levels, with the extent and spatial magnitude of benefits expected to rapidly increase with increasing urchin exploitation. It is unlikely that commercial harvesting will lead to eradication of the long-spined sea urchin, but there is potential for commercial fishing to be a primary contributor to mitigating the destructive grazing of this species. |
| Macháčková, Zelený, Lang, and Vinš (2021) | Czech Republic | To identify consumers' attitudes toward plant and animal products from wild and conventional foods and put these findings into a broader social context. | Sensory analysis and qualitative research with focus groups | Statistical testing in blind evaluations indicates that wild boar meat could represent a partial substitute for pork. Despite the mostly indifferent sensory evaluation, focus group responses showed barriers to wild food, due to consumer perceptions of food safety and of hunted meat as less humane. The paper concludes that possible educational and popularizing procedures are presented, including forest pedagogy, which could eliminate consumers' prejudices. |

### 9.8.1 Observations

Invasivorism is an emerging area of research, and much of the literature remarked on the lack of scientific research on the issue. Many articles cited to or noted the same observations made by Pasko and Goldberg and Nuñez et al. Often, the articles only mention invasivorism as a potential strategy for managing certain IAS, without providing additional insight. Though some studies analyze the potential of an invasivorist strategy against certain species, these mainly focused on aquatic species of IAS. However, the ecology and potential markets for terrestrial IAS like feral boar and invasive deer species should be more thoroughly studied to determine their potential as a source of food. There is also a gap in scientific studies that examine the effectiveness of programs already using human consumption to control IAS. This could examine the ability to reduce the target population, whether incentives to spread IAS are effectively curtailed, and the program's effect on consumer awareness of the IAS. Overall, more research and guidance is needed, especially in areas outside of the US, Europe, and Australia.

# 10 Managed Bees vs. Pollinator Welfare[1]

*Sahana Ramdas*

## CONTENTS



## 10.1 THE SYNERGISTIC IMPACT OF BEE FARMING ON HONEYBEES

In the industrial agriculture setup, monoculture crops are produced by the extensive use of harmful pesticides and the pollination services of exploited honeybees.[2] In the utilitarian conservation setup, honeybees are treated as the brand ambassadors of the pollinator crisis. However, unlike honeybees that are surplus in number and abundant in anthropocentric landscapes, wild bees, and other animal pollinators are threatened with extinction.[3] The protection of wild bees and pollinator biodiversity is mainly overlooked, as wild animal pollinators are not subjugated to human confinement, control, and oppression.[4] However, the abundance of managed honeybees does not mean that individual bees' welfare and the protection of managed bee's health being taken care of. Instead, to replace the lost animal pollinators' ecological

[1] This Chapter by Sahana Ramdas is a precursor to the book shortly to be published on protecting animal pollinators. The Book being authored by Sahana Ramdas is titled, "Saving Pollinator Biodiversity: *Examinations of Environmental Conundrums and Looming Ecological Crises*" and it is forthcoming in Essentials in Animal Welfare Series at CRC Press, Taylor &Francis Publications. The author retains full copyrights to this chapter.

[2] Buchmann, S. L., and Nabhan, G. P., 1996, *The Forgotten Pollinators*. Island Press, Washington, D.C.; Allan Stromfeldt Christensen, *Honeybee Collapse is the Result of Their Enslavement in Industrial Monocultures*, Resilience (May 26, 2015). https://www.resilience.org/stories/2015-05-26/honeybee-collapse-is-the-result-of-their-enslavement-in-industrial-monocultures/; Kremen, C., et al., Pollination and other ecosystem services produced by mobile organisms: a conceptual framework for the effects of land-use change. *Ecology Letter*, 10, 299 (2007).

[3] Buchmann, S. L., and Nabhan, G. P., 1996, *The Forgotten Pollinators*. Island Press, Washington, D.C.; National Research Council, 2007, *Status of pollinators in North America*. National Academies Press, Washington, D.C., Chapter 2, p. 289; Myth: Beekeeping is needed to conserve pollinator populations. Truth or Drought. https://www.truthordrought.com/beekeeping-for-conservation-myths.

[4] Myth: Beekeeping is needed to conserve pollinator populations. Truth or Drought. https://www.truthordrought.com/beekeeping-for-conservation-myths.

DOI: 10.1201/9781003137733-18

services, generalist bees are being reared and managed intensively as fungible commodities in industrial agriculture.[5]

Managed honeybee populations are being wiped out at alarming rates. Due to industrial farming, the production of pollinated crops and honey has steadily increased, but in the United States alone, almost 50% of honeybee colonies are lost annually.[6] Since the last decade, beekeepers have normalized losing at least 30% of their bees every year.[7] In 2015, over half of the managed honeybee colonies were wiped out.[8] Over one-third of commercial bee colonies with around 50 billion honeybees were lost during the winter of 2018–2019 alone.[9] Commercial beekeepers lost over four in ten of their honeybee colonies during the 2020 winter period.[10]

Honeybee declines are the cumulated effect of their exposure to harmful pesticides, poor beekeeping practices, loss of habitats, and nutritional food sources.[11] Despite the persistent loss of bees, no federal law or regulation aims to protect pollinators from these synergistic stressors. Nonetheless, the United States heavily depends on honeybee colonies as crop pollination by bees generates as much revenue as honey production does.[12] Consequentially, honeybee colonies are unscrupulously transported across the country and commercially exploited for crop pollination , their produced wax, and honey.[13]

Livestock animals are unnaturally kept, bred, and harvested under abhorrent conditions in the industrial animal agriculture system.[14] Managed honeybees too, are meted out the same inhumane treatment. Beekeeping is an integral but less noticed

---

5 Buchmann, S. L., and Nabhan, G. P., 1996, *The Forgotten Pollinators*. Island Press, Washington, D.C.; Pat Hurley. 1st place Rezendes ethics essay 2020. AT 5. https://honors.umaine.edu/wp-content/uploads/sites/184/2020/10/Pat-Hurley-1st-Place-Rezendes-Ethics-Essay-2020.pdf.

6 US beekeepers continue to report high colony loss rates, no clear progression toward improvement. June 25th 2021. https://ocm.auburn.edu/newsroom/news_articles/2021/06/241121-honey-bee-annual-loss-survey-results.php.

7 McGivney, A., 'Like sending bees to war': the deadly truth behind your almond milk obsession. Guardian, January 8th 2020. https://www.theguardian.com/environment/2020/jan/07/honeybees-deaths-almonds-hives-aoe.

8 Hellerstein, D., Hitaj, C., Smith, D., and Davis, A. (2017), Land use, land cover, and pollinator health: A review and trend analysis. USDA Economic Research Service. Research Report #232, p. 1. Retrieved from: https://www.ers.usda.gov/webdocs/publications/84035/err-232.pdf?v=42908.

9 https://www.ifis.org/blog/californias-almond-trade-exploiting-bee-population.

10 https://www.theguardian.com/environment/2020/jan/07/honeybees-deaths-almonds-hives-aoe.

11 Knobbe, E., Honeybee and the law, protecting our pollinators, *Journal of Environmental Law and Litigation*, 30, 228–234. (2015); https://scholarsbank.uoregon.edu/xmlui/bitstream/handle/1794/18888/Knobbe.pdf?sequence=1; Pat Hurley, 1st Place Rezendes Ethics Essay 2020. AT 5–11. https://honors.umaine.edu/wp-content/uploads/sites/184/2020/10/Pat-Hurley-1st-Place-Rezendes-Ethics-Essay-2020.pdf.

12 See: Ramanujan, K., Insect pollinators contribute $29 billion to U.S. farm income, *Cornell Chronicle* May 22nd, 2012. https://news.cornell.edu/stories/2012/05/insect-pollinators-contribute-29b-us-farm-income; Helping agriculture's helpful honey bees, U.S.FDA. https://www.fda.gov/animal-veterinary/animal-health-literacy/helping-agricultures-helpful-honey-bees.

13 See: Exploitation of bees by humans, Animal Ethics. https://www.animal-ethics.org/exploitation-of-bees-by-humans/.

14 See: Noah Harari, Y., Industrial farming is one of the worst crimes in history. https://www.theguardian.com/books/2015/sep/25/industrial-farming-one-worst-crimes-history-ethical-question; Also see: Animal cruelty is the price we pay for cheap meat – Rolling stones. https://www.rollingstone.com/interactive/feature-belly-beast-meat-factory-farms-animal-activists/.

part of industrial animal agriculture in America. Industrial farming in the United States treats honeybees as livestock.[15] Accordingly, the Internal Revenue Service classifies commercial beekeeping as farming,[16] and the United States Department of Agriculture (USDA) includes honeybees in the "livestock" category.[17] Since bees are made to live under constant stressful conditions, the industrial farming of honeybees is resulting in the loss of billions of bees every year.[18] The *Guardian* coverage in the *Age of extinction* series points out the impact of treating honeybees as livestock; writer Annette McGivenney highlights in the article "Like sending bees to war" that "more bees die every year in the US than all other animals raised for slaughter combined."[19]

International animal welfare standards require human caretakers to provide animals with Five Freedoms from cruelty: *Freedom from Hunger and Thirst; Freedom from Discomfort; Freedom from Pain, Injury, and Disease; Freedom to Express Natural and Expected Behavior; and Freedom from Fear and Distress.*[20] Each one of these rights is violated in animal agriculture.[21] Bees, too are subjected to poor nutrition, pain, injury, disease, and discomfort, and their natural behavior patterns are constantly oppressed to control their movement and extract their services.[22] Despite the lack of bee welfare, in the United States, rules and regulations on bees and beekeepers are limited to protecting commercial operations involving honeybees.[23]

---

15 Think of Honeybees as 'livestock,' not Wildlife, University of Cambridge (Jan. 25, 2018). https://phys.org/news/2018-01-honeybees-livestock-wildlife-experts.html#nRlv (accessed Oct. 21, 2020); Annette McGivney, 'Like sending bees to war': the deadly truth behind your almond milk obsession. Guardian, January 8th 2020.

16 Farmer's Tax Guide, Publication 225. IRS. Chapter 13, Page 81. October 15th 2021. https://www.irs.gov/pub/irs-pdf/p225.pdf; When honey bees hit the road: The role of federal, state, and local laws in regulating honey bee transportation. https://law.uoregon.edu/sites/law1.uoregon.edu/files/when_honey_bees_hit_the_road.pdf.

17 USDA, Emergency Assistance for Livestock, Honey Bees, and Farm-raised Fish (ELAP). https://www.fsa.usda.gov/programs-and-services/disaster-assistance-program/emergency-assist-for-livestock-honey-bees-fish/index; McGivney, A., 'Like sending bees. to war': the deadly truth behind your almond milk obsession. Guardian, January 8th 2020. https://www.theguardian.com/environment/2020/jan/07/honeybees-deaths-almonds-hives-aoe.

18 McGivney, A., 'Like sending bees to war': the deadly truth behind your almond milk obsession. Guardian, January 8th 2020. https://www.theguardian.com/environment/2020/jan/07/honeybees-deaths-almonds-hives-aoe.

19 McGivney, A., 'Like sending bees to war': the deadly truth behind your almond milk obsession. Guardian, January 8th 2020. https://www.theguardian.com/environment/2020/jan/07/honeybees-deaths-almonds-hives-aoe.

20 Animal Welfare, OIE World Organization for Animal Health. https://www.oie.int/en/what-we-do/animal-health-and-welfare/animal-welfare/.

21 See: Noah Harari, Y., Industrial farming is one of the worst crimes in history. https://www.theguardian.com/books/2015/sep/25/industrial-farming-one-worst-crimes-history-ethical-question; Also see: Animal cruelty is the price we pay for cheap meat – Rolling stones. https://www.rollingstone.com/interactive/feature-belly-beast-meat-factory-farms-animal-activists/.

22 Exploitation of bees by humans, Animal Ethics, https://www.animal-ethics.org/exploitation-of-bees-by-humans/; When honey bees hit the road: The role of federal, state, and local laws in regulating honey bee transportation. https://law.uoregon.edu/sites/law1.uoregon.edu/files/when_honey_bees_hit_the_road.pdf; Also see: Myth: Beekeeping is needed to conserve pollinator populations, https://www.truthordrought.com/beekeeping-for-conservation-myths.

23 When honey bees hit the road: The role of federal, state, and local laws in regulating honey bee transportation. https://law.uoregon.edu/sites/law1.uoregon.edu/files/when_honey_bees_hit_the_road.pdf

The *Emergency Assistance for Livestock, Honeybees, Farm-Raised Fish Program* (ELAP) declares that managed bees used for pollination, honey production, and bee breeding are eligible under the initiative.[24] However, that does not mean the ELAP offers preventative measures or remediation to bees succumbing to weather conditions and disease outbreaks. Instead, it only compensates beekeepers who lose bee colonies.[25] The ELAP primarily provides financial assistance to livestock producers and commercial honeybee keepers when they face losses due to bees succumbing to diseases and weather conditions.[26]

Honeybees are regulated as though they are agricultural commodities at both the national and state level.[27] The USDA Animal and Plant Health Inspection Service (APHIS) and other agencies watch over the effectiveness of bee colonies as they are considered instrumental for pollinating commercial crops and honey production.[28] The Federal Government only regulates the import of bees and bee products into the United States to prevent the spread of disease outbreaks from imported bees.[29] For instance, the Honeybee Act of 1922 (7 USC § 281–286) lays down the federal shipping specifications for importing honeybees and federal rules that the exporting countries need to follow like the kind of beekeeping equipment and the kinds of bee species that can be imported.[30]

The state governments in the United States are primarily responsible for laying down regulations to protect bee health and regulating the interstate and intrastate movement of beehives.[31] Except for a few states establishing and enforcing well-funded programs of inspections and transportation controls,[32] most states have inadequate enforcement mechanisms of apiary laws,[33] and some states have no beekeeping regulations to control the spread of diseases and pests.[34] Only a couple of states have mandated hive registrations, inspections, and certification programs to maintain healthy bee colonies and bee products.[35] The majority of the states have either made registrations non-mandatory or rolled back registration requirements—that were once in place.[36] Even states with solid beekeeping regulations cannot effectively

---

24 USDA, Emergency Assistance for Livestock, Honey Bees, and Farm-raised Fish (ELAP). https://www.fsa.usda.gov/programs-and-services/disaster-assistance-program/emergency-assist-for-livestock-honey-bees-fish/index.

25 See: https://ask.usda.gov/s/article/What-types-of-bees-are-covered-by-the-Emergency-Assistance-for-Livestock-Honey-Bees-and-Farm-raised.

26 See: Id.

27 When honey bees hit the road: The role of federal, state, and local laws in regulating honey bee transportation. At 7. https://law.uoregon.edu/sites/law1.uoregon.edu/files/when_honey_bees_hit_the_road.pdf.

28 Id. AT 2.

29 Id. AT 8–10, 16.

30 Id. AT 8–10, 16; See: 7 USC Ch. 11: Honeybees from Title 7—Agriculture. https://uscode.house.gov/view.xhtml?path=/prelim@title7/chapter11&edition=prelim.

31 Id. AT 1.

32 When honey bees hit the road: The role of federal, state, and local laws in regulating honey bee transportation. At 7. https://law.uoregon.edu/sites/law1.uoregon.edu/files/when_honey_bees_hit_the_road.pdf (AT 1, 3).

33 Id. AT 1, 3, 20.

34 Id. AT 1.

35 Id. AT 3.

36 Id. AT 17, 21, 25.

implement the existing rules due to inadequate funding for regulatory programs.[37] Some states have also rescinded funding for inspection programs or altogether deregulated the beekeeping industry.[38] Thus, the inconsistent laws combined with gaps in bee regulation create an open space for the ill-treatment of managed bees.[39]

The irreplaceable value of pollinators and honeybee colonies is not translated into securing bee colonies' welfare and individual bee's health through the current Apiary Law Framework or beekeeping regulations. For instance, the *Bees, Beekeeping ByProducts, and Beekeeping Equipment Regulations APHIS* do not regulate the conditions in which honeybees are reared, kept, or used.[40] The result of such legal vacuums is the unregulated and unscrupulous use of honeybees in industrial agriculture.

Managed honeybees treated as livestock are extensively used to pollinate commercial crops in monocultures, where honeybees' food sources are not only limited but severely lacking in nutrition and doused with toxic pesticides.[41] Unlike polycultures, monocultures are devoid of local diversity, depriving honeybees of essential nutrition.[42] Moving bee colonies to pollinate monocultures lowers their immunity to devastating varroa mites, pests, and pathogen attacks.[43] Mainly, poor apiculture practices have significantly degraded bees' health and hampered genetic diversity, making them susceptible to large-scale colony collapses.[44] Thus, honeybee colonies are succumbing to management stressors, exposure to harmful pesticides, and other environmental stressors.[45]

To meet consumer demands and rake profits, commercial crop growers harvest their fields at the earliest time they can, causing food scarcity and leaving honeybees to survive in unnaturally poor conditions.[46] By expediting the pollination process of

---

[37] Id. AT 3.

[38] Id. AT 29.

[39] Id. AT 1, 6, 9, 20, 22.

[40] USDA APHIS, honeybees and other bees. https://www.aphis.usda.gov/aphis/ourfocus/planthealth/import-information/permits/plant-pests/bees/honeybees-other; Title 7, Chapter III, Part 322 – Bees, Beekeeping Byproducts, and Beekeeping Equipment. Authority: 7 U.S.C. 281; 7 U.S.C. 7701–7772 and 7781–7786; 7 CFR 2.22, 2.80, and 371.3. https://www.ecfr.gov/current/title-7/subtitle-B/chapter-III/part-322.

[41] Pettis, J. S., Lichtenberg, E. M., Andree, M., Stitzinger, J., and Rose, R., Crop pollination exposes honey bees to pesticides which alters their susceptibility to the gut pathogen Nosema Ceranae. *PLoS One*, 8, e70182 (2013); Ricketts, T. H., et al., Landscape effects on crop pollination services: Are there general patterns? *Ecology Letters*, 11, 499–515 (2008); Knobbe, E., Honeybee and the law, protecting our pollinators. *Journal of Environmental Law and Litigation*, 30, 228–234 (2015).

[42] Armstrong, L., California's almond trade is exploiting one of nature's most essential workers. IFIS, 5th June 2020. https://www.ifis.org/blog/californias-almond-trade-exploiting-bee-population.

[43] Id; Knobbe, E., Honeybee and the law, protecting our pollinators. *Journal of Environmental Law and Litigation*, 30, 228–234 (2015); Allen Christensen, Bee collapse is the result of their enslavement in industrial monocultures (May 2015). https://theecologist.org/2015/may/02/bee-collapse-result-their-enslavement-industrial-monocultures (accessed Oct. 21, 2020).

[44] Armstrong, L., California's almond trade is exploiting one of nature's most essential workers. IFIS, 5th June 2020. https://www.ifis.org/blog/californias-almond-trade-exploiting-bee-population; When Honey Bees Hit the Road. AT 219–234.

[45] Knobbe, E., Honeybee and the law, protecting our pollinators. *Journal of Environmental Law and Litigation*, 30, 228–234 (2015).

[46] Armstrong, L., California's almond trade is exploiting one of nature's most essential workers. IFIS, 5th June 2020. https://www.ifis.org/blog/californias-almond-trade-exploiting-bee-population.

bees, commercial beekeepers disrupt the symbiotic relationship between bees and their dependent plants.[47]

Throughout a single bee's life, it ends up producing around one and a half tablespoons of honey.[48] Ethical beekeeping requires beekeepers to collect the surplus produce of honey only.[49] Instead, commercial beekeepers are known to extract excess honey from bees' produce and leave bees to feed on sugary substitutes.[50] Humans often engage in this form of Interspecific Kleptoparasitism by harvesting food that animals produce for their own needs or the food needs of their offspring and other animal species.[51] Honeybees are reared and transported in artificial and taxing nesting sites where unnatural feeding regimes are practiced.[52] By compelling honeybees to intake artificial sugars mixed with antibiotics, honeybees are made to work tirelessly. Further, this entrapped labor is sustained by the excessive use of antibiotics to fight off pests, pathogens, and diseases inherent in monocultures.[53] Thus, the consumption of honey is an extravagant food choice that is directly causing the unnecessary stealing of an individual bee's lifelong hard work. Moreover, the nutrients derived from consuming honey are easily replaceable by other sugary syrups.[54]

Though commercial beekeepers move packed bee colonies across the nation under unhygienic conditions, there is still no interstate regulation of beekeepers' operations.[55] Consequentially, bees are kept under distress conditions and transported for thousands of miles before arriving at the destination of labor—where poor nutrition, pesticide exposure, and poor environmental quality combine to severely compromise bees' immune systems, making them susceptible to diseases.[56] Since there is no federal law that requires best beekeeping practices, beekeepers' mismanagement includes stuffing honeybee colonies into cramped trucks and handling bee colonies

---

47 Id.

48 Grad, P., 10 of the most interesting facts about bees & honey. May 19th 2010. https://bigislandbees.com/blogs/bee-blog/14137349-interesting-bee-honey-facts; Cristina Diaconu, Catch the buzz- how do honey bees make honey? October 19th 2020, https://www.beeculture.com/catch-the-buzz-how-do-honey-bees-make-honey/

49 Ethical consumers guide to honey. AT 3. https://cdn.friendsoftheearth.uk/sites/default/files/downloads/Ethical%20consumers%20guide%20to%20honey.pdf.

50 Id; Hurley, P., Zombie Fields: Ethical Concerns of Pollination in Industrial Agriculture. The John M. Rezendes Annual Ethics Essay Competition 2020. AT 5.

51 Myth: Beekeeping is needed to conserve pollinator populations. Truth or Drought. https://www.truthordrought.com/beekeeping-for-conservation-myths

52 Ethical consumers guide to honey. AT 8. https://cdn.friendsoftheearth.uk/sites/default/files/downloads/Ethical%20consumers%20guide%20to%20honey.pdf

53 Ethical consumers guide to honey. AT 8. https://cdn.friendsoftheearth.uk/sites/default/files/downloads/Ethical%20consumers%20guide%20to%20honey.pdf; Also See: Recent regulatory changes, Helping agriculture's helpful Honey Bees, U.S.FDA. https://www.fda.gov/animal-veterinary/animal-health-literacy/helping-agricultures-helpful-honey-bees

54 https://www.truthordrought.com/beekeeping-for-conservation-myths.

55 When honey bees hit the road: the role of federal, state, and local laws in regulating honey bee transportation.

56 Knobbe, E., Honeybee and the law, protecting our pollinators. *Journal of Environmental Law and Litigation*, 30, 228–234 (2015).

with no regard to individual bees' welfare or bee colonies' health.[57] Thus, millions of honeybees are transported in overpacked trucks.[58] Once they reach new places, adjusting to non-native landscapes, fluctuating temperatures, and adapting to bees from different colonies and regions puts them under severe stress.[59]

The reproduction process among the bees is thwarted to keep enough bees pro ducing honey for monetary gains.[60] Commercial beekeepers wreck the natural thriving of honeybee colonies and disrupt healthy brood patterns.[61] Queen bees' natural behavior to lead worker bees to new wild hives is curbed by either culling the queen bee or clipping her wings.[62] When the productivity of queen bees declines, older queen bees are often killed and replaced with new queens.[63] Queen bees are also artificially inseminated using drones that are often crushed to extract their eversion.[64]

The forced labor does not spare honeybees to rest even during winters. Industrial-scale beekeeping practices during winter involve practices like culling entire beehives or making bees survive harsh winters by feeding them sugar, corn syrup, and pollen substitutes.[65] Such bee farming practices allow beekeepers to cut costs involved in the care, upkeep, and maintenance of honeybees.[66] Thus, honeybees get no time to hibernate or recuperate and are constantly moved from one crop to another under deplorable conditions.[67]

These stressors have a compounded effect on bees, weakening their immunity and increasing mortality rates amongst bee populations.[68] Hence, in the absence of bee protection policies, the intensive management of honeybees involves the poorest working environment and inhumane transportation of honeybees from one region to the other.

---

[57] Also see: Knobbe, E., Honeybee and the law, protecting our pollinators. *Journal of Environmental Law and Litigation*, 30, 228–234 (2015); Mailander, D., and Grant, Z., 2019, When honey bees hit the road: The role of federal, state, and local laws in regulating honey bee transportation, pp. 1–9. (The only exception is Hawaii, as restrictions are imposed on honeybee imports into the State, p. 8, footnote 35). https://law.uoregon.edu/sites/law1.uoregon.edu/files/when_honey_bees_hit_the_road.pdf.

[58] Factory farmed bees, Bee welfare issues in honey production, Transportation. AT 8. https://cdn.friendsoftheearth.uk/sites/default/files/downloads/Ethical%20consumers%20guide%20to%20honey.pdf.

[59] When honey bees hit the road. AT 2, 3, 6.

[60] Ethical consumers guide to honey. AT 8. https://cdn.friendsoftheearth.uk/sites/default/files/downloads/Ethical%20consumers%20guide%20to%20honey.pdf.

[61] Lewis Armstrong, California's almond trade is exploiting one of nature's most essential workers. IFIS, 5th June 2020. https://www.ifis.org/blog/californias-almond-trade-exploiting-bee-population.

[62] https://cdn.friendsoftheearth.uk/sites/default/files/downloads/Ethical%20consumers%20guide%20to%20honey.pdf.

[63] https://www.animal-ethics.org/exploitation-of-bees-by-humans/.

[64] https://www.animal-ethics.org/exploitation-of-bees-by-humans/.

[65] Id at 8; Ethical concerns in commercial beekeeping. AT 5. https://honors.umaine.edu/wp-content/uploads/sites/184/2020/10/Pat-Hurley-1st-Place-Rezendes-Ethics-Essay-2020.pdf

[66] Ethical consumers guide to honey. AT 8. https://cdn.friendsoftheearth.uk/sites/default/files/downloads/Ethical%20consumers%20guide%20to%20honey.pdf.

[67] Ethical consumers guide to honey. AT 8. https://cdn.friendsoftheearth.uk/sites/default/files/downloads/Ethical%20consumers%20guide%20to%20honey.pdf.

[68] Doeke, M. A., Frazier, M., and Grozinger, C. M., Overwintering honey bees: Biology and management. *Current Opinion in Insect Science*, 10, 185–193 (2015). http://dx.doi.org/10.1016/j.cois.2015.05.014; Knobbe, E., Honeybee and the law, protecting our pollinators. *Journal of Environmental Law and Litigation*, 30, 228–234 (2015).

As seen, despite the importance of bees on American farms, there is a severe dearth of bee welfare laws. Due to these issues, inexperienced and ill-equipped beekeepers often mismanage bees and flagrantly disregard bee welfare.[69] In the end, it is largely left to beekeepers to self-regulate their business operations and maintain healthy colonies and prevent pest and disease outbreaks.[70] Thus, the piecemeal bee management laws and enforcement of state beekeeping laws harbor commercial beekeeping practices in an unrestricted manner.

## 10.2 THE IMPERILED POLLINATOR BIODIVERSITY

Monocultures aided by agrochemicals and managed bees have jeopardized the diversity and health of surrounding wild pollinators.[71] Although many wild animal pollinators are threatened or endangered within the United States,[72] the federal government has not taken any legal action to buffer pollinator biodiversity from the detrimental impact industrial agriculture has on them. Permitting the interstate movement of managed bees without federal oversight not only leads to the ill-treatment of honeybees as agricultural commodities but also compromises the role of pollinator biodiversity in natural ecosystems.[73] For instance, numerous studies indicate that managed bees are susceptible to becoming carriers of pests, pathogens, and parasites due to the conditions in which they are bred, reared, and transported. Further, scientific findings indicate that such managed bees consequentially pass on the pests, pathogens, and parasites to native pollinators and cause disease outbreaks amongst native bees like the endangered wild bumblebees.[74]

---

[69] When honey bees hit the road: The role of federal, state, and local laws in regulating honey bee transportation. At 7. https://law.uoregon.edu/sites/law1.uoregon.edu/files/when_honey_bees_hit_the_road.pdf.

[70] Id.AT 1, 10, 20, 21, 25, 26, 28–30.

[71] **See: Impact of managed honeybees on pollinator biodiversity** - Katy- Bee Missionary, Honeybees and Monocultures, Save the Bees, (Aug. 20, 2019). https://beemission.com/blogs/news/honeybees-and-monocultures (accessed Oct. 16, 2020); Geslin, B., Gauzens, B., Baude, M., Dajoz, I., Fontaine, C., et al., Massively introduced managed species and their consequences for plant–pollinator interactions. *Advances in Ecological Research*, Elsevier, 57, 147–199 (2017). https://doi.org/10.1016/bs.aecr.2016.10.007ff; Angelella, G. M., McCullough, C. T., and O'Rourke, M. E., Honey bee hives decrease wild bee abundance, species richness, and fruit count on farms regardless of wildflower strips. *Scientific Reports*, 11, 3202 (2021). https://doi.org/10.1038/s41598-021-81967-1; **See: Impact of industrial agriculture's practices on pollinators** - Dan Charles, *Wild Bees Are Good For Crops, But Crops Are Bad For Bees* https://www.npr.org/sections/thesalt/2013/03/01/173167125/wild-bees-are-good-for-crops-but-crops-are-bad-for-bees (accessed Oct. 21, 2020); Kremen, C., Williams, N. M., and Thorp, R. W., Crop pollination from native bees at risk from agricultural intensification. *National Academy of Sciences*, 99, 16816–16821 (2002).

[72] Buchmann, S. L., and Nabhan, G. P., 1996, *The Forgotten Pollinators*. Island Press, Washington, D.C.; National Research Council, 2007, *Status of pollinators in North America*. National Academies Press, Washington, D.C., p. 289; Myth: Beekeeping is needed to conserve pollinator populations. Truth or Drought. https://www.truthordrought.com/beekeeping-for-conservation-myths.

[73] Buchmann, S. L., and Nabhan, G. P., 1996, New bee on the block. In *The Forgotten Pollinators*, pp. 169–183; Valido, A., Rodríguez-Rodríguez, M. C., and Jordano, P., Honeybees disrupt the structure and functionality of plant-pollinator networks. *Scientific Reports*, 9, 4711 (2019). https://doi.org/10.1038/s41598-019-41271-5; https://www.truthordrought.com/beekeeping-for-conservation-myths.

[74] https://sgp.fas.org/crs/misc/R43191.pdf.

Often bee conservation movements hyper-focus on protecting honeybees.[75] However, protecting, domesticated, managed, and non-native bee populations is not equivalent to the conservation of imperiled pollinator biodiversity.[76] Though honeybees display a tendency to be highly invasive, they are still not treated as invasive animals in native landscapes. Instead, the subspecies are regarded as the global ambassadors for the pollinator conservation movement and poster children for the pollinator crisis. This is because the domesticated, managed honeybees sustain business-as-usual cash crop production practices, though the foraging bees are lethally or sub-lethally exposed to monocultures, pesticides, and degraded habitats. And despite the subspecies failing health—, honeybee colonies are raised, maintained, and disposed of as fungible goods.[77] Hence, the invasive bees are recast as coveted pollinators.

Native and wild pollinator communities were once widespread in agricultural areas.[78] After the industrialization of crop production, intensive agricultural areas are only able to sustain generalist honeybee populations.[79] In these areas, native pollinators' visitation continues to plummet as their natural habitats hardly exist.[80]

Amidst the falling pollinator diversity, a small subset of animal pollinators, the honeybees have become the dominant taxon due to two main reasons. One is the honeybee colonies' relative resilience to agricultural intensification. Unlike the dominant taxon's ability to thrive or survive in anthropogenic landscapes, other pollinator communities are severely weakened by agricultural intensification.[81] Two, commercial beekeepers have extensively reared generalist pollinators like honeybees and

---

75 See: https://www.truthordrought.com/beekeeping-for-conservation-myths.

76 See: Pearson, G., WIRED, You're worrying about the wrong bees. https://www.wired.com/2015/04/youre-worrying-wrong-bees/; https://www.truthordrought.com/beekeeping-for-conservation-myths; Conserving honeybees does not help wildlife. https://www.science.org/doi/10.1126/science.aar2269; https://theconversation.com/keeping-honeybees-doesnt-save-bees-or-the-environment-102931?fbclid=IwAR3DlH_a5DWafEl7uMogp8hXPeR6csoNwVZSteEljpwExijxQOortuPwWxk; Chris Helzer, Focus on native bees, not honey bees. https://blog.nature.org/science/2019/08/19/focus-on-native-bees-not-honey-bees/?fbclid=IwAR1934GRlhsN4E-6iEX1cSpxAtlkfSP6zcCwqBKYtBGqM-BEHUJAq5wDRYCM.

77 See: Myth: Beekeeping is needed to conserve pollinator populations. Truth or Drought. https://www.truthordrought.com/beekeeping-for-conservation-myths.

78 See: Bommarco, R., et al., Drastic historic shifts in bumble-bee community composition in Sweden. *Proceedings of the Royal Society*, 279, 309 (2011); Kremen, C., Crop pollination services from wild bees. *Bee Pollination in Agricultural Ecosystems*, 10, 10, 15 (Rosalind R. James & Theresa L. Pitt-Singer eds. 2008); https://honors.umaine.edu/wp-content/uploads/sites/184/2020/10/Pat-Hurley-1st-Place-Rezendes-Ethics-Essay-2020.pdf (pp. 1–5).

79 Suding, K. N., Functional-and abundance-based mechanisms explain diversity loss due to N fertilization. *Proceeding of the National Academy of Sciences*, 102, 4387 (2005); Kleijn, D., Does conservation on farmland contribute to halting the biodiversity decline? *Trends in Ecology & Evolution*, 26, 474 (2011).

80 Kennedy, C. M., et al., A global quantitative synthesis of local and landscape effects on wild bee pollinators in agroecosystems. *Ecology Letters*, 16(5), 10–13 (2013); Dauber, J., et al., Landscape structure as an indicator of biodiversity: matrix effects on species richness. *Agriculture, Ecosystems & Environment*, 98, 321–329 (2003).

81 Fahrig, L., and Jonsen, I., Effect of habitat patch characteristics on abundance and diversity of insects in an agricultural landscape. *Ecosystems*, 1(2), 197–205 (1998); Kremen, C., Williams, N. M., and Thorp, R. W., Crop pollination from native bees at risk from agricultural intensification. *Proceedings of the National Academy of Sciences*. 99, 16812, 16815–16816 (2002).

made them to be managed bees as discussed above.[82] Hence, the conservation of pollinator biodiversity is not a priority as the management of the dominant bees' ecosystem services satisfies industrial agriculture's short-term goals of reaping quick economic returns with no thought for other considerations.[83]

Although honeybees significantly contribute to crop pollination, generating honeybee colonies in unsuitable conditions causes more harm than good.[84] The micro-management of bee populations has positioned managed bees as an immediate threat to the survival of remaining wild bees.[85] Empirical evidence points to show the adverse impact of generalist honeybees on wild bee populations.[86] Behavioral interactions between wild and honeybees are favorable to honeybees as it increases their efficiency.[87] Whereas such interspecific interactions are not so favorable toward wild bees' well-being or survival.[88]

---

[82] Vergara, C. H., 'Environmental impact of exotic bees introduced for crop pollination. In Rosalind, J., and Pitts-Singer, T. L. (eds), *Bee Pollination in Agricultural Ecosystems* (New York, 2008; online edn, Oxford Academic, 1 Sept. 2008), https://doi.org/10.1093/acprof:oso/9780195316957.003.0009, accessed 28 Feb. 2023; Buchmann, S. L. and G. P. Nabhan, 1996, New bee on the block. In *The Forgotten Pollinators.* Island Press, Washington, D.C. (pp. 169–183).

[83] Also see: Kleijn, D., et al., Delivery of crop pollination services is an insufficient argument for wild pollinator conservation. *Nature Communications*, 6, 3–4 (May 2015); Scheper, J., Environmental factors driving the effectiveness of European agri-environmental measures in mitigating pollinator loss – a meta-analysis. *Ecology Letters*, 16, 912, 918 (2013); Stanley, D. A. Pollinators and pollination of oilseed rape crops (*Brassica napus L.*) in Ireland: ecological and economic incentives for pollinator conservation. *Journal of Insect Conservation*, 17, 1181, 1187 (2013).

[84] Id. Buchmann, S. L., and Nabhan, G. P., (Shuffling bees from one area to the other also exposes wild pollinators to a host of infections and diseases and exacerbates the wild bee species declines in the background.) New bee on the block (pp. 172–178, 181–183), The little lives keeping crops fruitful (pp. 194–196).

[85] Buchmann, S. L., and Nabhan, G. P., New bee on the block (pp. 170–183), Fractured Fairy Tails, (pp. 124–125), Need nectar, will travel (pp. 105–116), The little lives keeping crops fruitful (pp. 193–196) (The impact of managed honeybees that act as exotic invaders in introduced environments); University of Illinois at Urbana-Champaign, Commercial bumble bee industry amplified a fungal pathogen of bees, study suggest. *Science Daily*, (Apr. 4, 2016). https://www.sciencedaily.com/releases/2016/04/160404152854.htm (accessed Oct. 4, 2020); https://phys.org/news/2018-01-honey-bees-livestock-wildlife-experts.html#nRlv.

[86] Buchmann, S. L., and Nabhan, G. P., 1996, *The Forgotten Pollinators*, Island Press, Washington, D.C. (Chapter 10 – New Bees on the Block honeybees adversely affecting wild bee networks by engaging in genetic introgressions, passing on disease pathogens, out-competing wild pollinators in seeking food and nesting sources etc.); Goulson, D., *Effects of introduced bees on native ecosystems. Annual Revolution of Ecology, Evolution and Systematics*, 34, 1, 2. (2003); Geslin, B., Gauzens, B., Baude, M., Dajoz, I., Fontaine, C., et al.. Massively introduced managed species and their consequences for plant–pollinator interactions. *Advances in Ecological Research*, Elsevier, 57, 147–199 (2017). https://doi.org/10.1016/bs.aecr.2016.10.007; Angelella, G. M., McCullough, C. T., and O'Rourke, M. E., Honey bee hives decrease wild bee abundance, species richness, and fruit count on farms regardless of wildflower strips. *Scientific Reports*, 11, 3202 (2021). https://doi.org/10.1038/s41598-021-81967-1.

[87] Greenleaf, S. S., and Kremen, C., Wild bees enhance honey bees' pollination of hybrid sunflower. *Proceedings of National Academy of Sciences, USA*, 103, 13890–13895.

[88] Id. Buchmann, S. L., Nonnative bees domesticated for their pollination services tend to become invasive pests in native lands (when they are accidentally or intentionally introduced.) New bee on the block (pp. 170–183), The little lives keeping crops fruitful (pp. 194–196); Daily, G. C., et.al., Nature's bounties: Reliance on pollinators for health (July 15, 2015). https://doi.org/10.1016/S0140-6736(15)61244-2 (accessed Oct. 21).

As pointed out in "Forgotten Pollinators," honeybees have become "*introduced invaders on a trajectory of ecological conquests*."[89] Honeybees' disproportionate numbers mismatch the native plants' needs, which require specialized pollinators' services; this leads to massive negative spillovers in native landscapes.[90] By their strength in numbers, western honeybees act as rapid foragers, harvesters, and hoarders of floral resources, depriving the wild pollinators of pollen and nectar.[91] Generalist honeybees are also said to have come between highly evolved, specialized interactions between native plants and native pollinators.[92] Cultivated bees significantly impact flowering crops, which only require pollination services for sporadic periods.[93] Since honey bees are active for most of the year and tend to travel far more than native pollinators, they end up extracting massive amounts of natural resources such as the pollen and nectar on whom wild pollinators are entirely dependent for their survival and diversity.[94] Despite removing excessive amounts of floral resources, non-native feral bees can leave pollinated plants unfertilized.[95] Wild pollinators threatened by anthropogenic activities are not in a position to cope with such demobilizing competition for scarce sources.[96] Such losses of nectar sources and the decline of legitimate pollinators go hand in hand in adversely impacting pollinator community interactions. Thus, the adverse competitive interactions between honeybees and wild pollinators have far-reaching consequences on pollinator biodiversity and the reproductive success of native plants.[97]

---

89 Id. Buchmann, S. L., New Bee on the block (pp. 170–183), The little lives keeping crops fruitful (pp. 194–196).

90 Id. Yuan, Y. W., et al., The genetic control of flower-pollinator specificity. *Current Opinion in Plant Biology*, 16, 422, 424, 428 (2013); James Hung, K. L., et al., Non-native honey bees disproportionately dominate the most abundant floral resources in a biodiversity hotspot. *Proceedings of the Royal Society B: Biological Sciences* (2019). DOI:10.1098/rspb.2018.2901.

91 Id. Buchmann, S. L. and Nabhan, G. P., New bee on the block. In *The Forgotten Pollinators* (pp. 170–183), The little lives keeping crops fruitful (Page 194–196); Lindstrom, S. A. M., et al., Experimental evidence that honeybees depress wild insect densities in a flowering crop. *Proceedings of Royal Society B*, 283, 1–2 (2016).

92 Id; Also see: Woodward, D. R., Monitoring for impact of the introduced leafcutting bee, *Megachile rotundata* (F.) (Hymenoptera: Megachilidae), near release sites in South Australia. *Australian Journal of Entomology*, 35, 187 (1996); Kato, M., et al., Plant-pollinator interactions in New Caledonia influenced by introduced honey bees. *American Journal of Botany*, 91, 1814, 1826 (2004); Morales, C. L., and Aizen, M. A., Does invasion of exotic plants promote invasion of exotic flower visitors? A case study from the temperate forests of the southern Andes. *Biological Invasions*, 4, 87, 97 (2002); Hanley, M. E., and Goulson, D., Introduced weeds pollinated by introduced bees: Cause or effect? *Weed Biology & Management*, 3, 204, 211 (2003).

93 Id. Buchmann, S. L., and Mendes do Carmo, R., et al., Introduced Honeybees (*Apis Mellifera*) Reduce pollination success without affecting the floral resource taken by native pollinators. *Biotropica*, 36, 371, 374 (2004).

94 Butz Huryn, V. M., Ecological impacts of introduced honey bees. *Q. R. Biology*, 72, 275, 277 (1997).

95 Id. Buchmann, S. L., and Nabhan, G. P., New bee on the block (pp. 170–183).

96 Id.

97 Id. Buchmann, S. L., and Nabhan, G. P., New bee on the block (pp. 172–173, 176,178, 181–183), The perils of matchmaking (pp. 73, 82); Wojcik, V. A., et al., Floral resource competition between honey bees and wild bees: Is there clear evidence and can we guide management and conservation? *Environmental Entomology*, 47, 822 (2018).

Overall, generalist bees diminish the vitality of biodiversity interactions and endanger native pollinators' survival.[98] When introduced honeybees enter nonnative areas, negative interbreeding occurs between nonnative bees and native wild bees.[99] The resulting hybridized honeybee strains have resulted in the growth of highly aggressive "Africanized" honeybee strains.[100] Also, the introduction of honeybee populations leads to inviable offspring and the spread of pathogens in native ranges.[101] In addition, honeybees' negative effects on the surrounding environment cause a tremendous population increase in the generalist population's numbers, compounding to bring down native pollinator biodiversity further.[102] Since introduced bees are also the carriers of regional areas' diseases, parasites, and pathogens, hostile interactions with invasive bees further push wild bees toward loss in numbers and extinction.[103] Hence, wild bees are languishing not just because of the most prominent anthropogenic factors like agrochemical-doused landscapes but also due to the invasiveness of nonnative honeybees and other feral bee populations.[104]

To add to the woes of wild pollinators, the loss of honeybee colonies made the public overattentive to the western bee's importance in food production—while

---

98 Id. Buchmann, S. L., Le Féon, V., et al., Intensification of agriculture, landscape composition and wild bee communities: a large-scale study in four European countries. *Agriculture, Ecosystems, & Environment*, 137, 148 (2010); Garibaldi, L. A., et al., Stability of pollination services decreases with isolation from natural areas despite honey bee visits. *Ecology Letters*, 14, 1062 (2011).

99 Joshi, N. K., Impact of biotic and abiotic stressors on managed and feral bees. *Insects*, 10(8), 233 (2019). DOI:10.3390/insects10080233.

100 Id.

101 Id.

102 Id; University of California - San Diego, Foreign bees monopolize prize resources in biodiversity hotspot: Immense numbers of introduced honey bees overwhelm native pollinators for access to pollen and nectar. *Science Daily* (Feb. 20, 2019). https://www.sciencedaily.com/releases/2019/02/190220103339.htm (accessed Oct. 4, 2020); James Hung, K. L. et al., Nonnative honey bees disproportionately dominate the most abundant floral resources in a biodiversity hotspot. *Proceedings of Royal Society B*, 286, 20182901 (2019).

103 Graystock, P., et al., Do managed bees drive parasite spread and emergence in wild bees? *International Journal of Parasitology: Parasites and Wildlife*, 5, 64, 70 (2015); Amdam, G. V., et al., Altered physiology in worker honey bees (Hymenoptera: Apidae) infested with the mite Varroa destructor (Acari: Varroidae): a factor in colony loss during overwintering? *Journal of Economic Entomology*, 97, 741–747; Highfield, A. C., et al., Deformed wing virus implicated in overwintering honeybee colony losses. *Applied & Environmental Microbiology*, 75, 7212 (2009); University of California – Riverside, Managed bees spread and intensify diseases in wild bees: Research shows wild bees are harmed even when managed bees are disease-free. *Science Daily* (Nov. 5, 2020). https://www.sciencedaily.com/releases/2015/11/151105152147.htm (accessed Oct. 17, 2020); University of Exeter, Wild pollinators at risk from diseased commercial species of bee. *Science Daily* (Jan. 16, 2015). https://www.sciencedaily.com/releases/2015/01/150116085604.htm (accessed Oct. 17, 2020).

104 Buchmann, S. L., and Nabhan, G. P., Silent springs and fruitless falls (pp. 124–125), Need nectar, will travel (pp. 105–116), Fractured fairy tales (pp. 24, 25, 42–45, 82, 88–89, 94–95, 105, 290). https://www.sustainability-times.com/environmental-protection/bees-are-livestock-not-wildlife/; Jennifer Oldham, Will Putting Honey Bees on Public Lands Threaten Native Bees? September 15th 2020, Yale Environment 360. https://e360.yale.edu/features/will-putting-honey-bees-on-public-lands-threaten-native-bees?fbclid=IwAR3n8OkggbMGcipJIdD7MAxw90WpY0ttvLfB80bPZz19iLww2QHIe4DshBU; Foreign bees monopolize prize resources in biodiversity hotspot, PHYSORG. https://phys.org/news/2019-02-foreign-bees-monopolize-prize-resources.html?fbclid=IwAR2Bu-CCu8MGuTYYXiOhtHcWoszeLJfkVBVunNapC5iBz2GQ4Tw5yIOA48M.

completely ignoring the multiple roles played by other animal pollinators.[105] The lack of familiarity with pollinator biodiversity and overfamiliarity with managed bees has undermined the role played by vital pollinators who chiefly maintain the overall health of natural ecosystems.[106] Contrary to common belief, honeybees are not the most efficient pollinators of crops.[107] Non-bee insects and other animal pollinators are as crucial as honeybees for crop production and are critical for pollinating wild plants.[108] Hence the loss of genetic diversity in bee pollinators mainly occurs from the misguided food production practices of humans.

## 10.3 CONCLUSION

There is an immediate need for robust pollinator protection laws as all animal pollinators have an inherent and immeasurable value in worldwide ecosystems. To this end, there is also a dire need for stricter pesticide regulation to prevent its exponential ecological harm to the environment and to the well-being of pollinator biodiversity. Thus, if the approach to protecting pollinators continues to be charted based on patent economic returns, the generalist bee populations will continue to diminish the vitality of biodiversity interactions and further endanger native pollinators' communities and survival. Upon this, pollinator biodiversity would be unable to buffer ecosystems from the impending crisis brought about by humans' mismanagement of the environment. Hence, in the longer run, sustainability of the ecosystem and persistence of plant diversity is only possible by maintaining the pollinator biodiversity and protecting individual pollinator's welfare.[109]

---

[105] Stanley, D. A., Don't forget about the (other) bees. *British Ecological Society* (Aug. 19, 2016). https://www.britishecologicalsociety.org/dont-forget-bees/.

[106] See: Id. Buchmann, S. L., and Nabhan, G. P., The forgotten pollinators (p. 5)

[107] Saunders, M., Flower Visitors vs. Pollinators: no evidence that honey bees are the most important pollinator worldwide, (Jan. 11, 2018). https://ecologyisnotadirtyword.com/2018/01/11/flower-visitors-vs-pollinators-no-evidence-that-honey-bees-are-the-most-important-pollinator-worldwide/ (accessed Oct. 21, 2020); Kremen, C., Crop pollination services from wild bees. *Bee Pollination in Agricultural Ecosystems*, 10, 10, 15 (Rosalind R. James & Theresa L. Pitt-Singer eds. 2008).

[108] See: Stein, K., et al., Bee pollination increases yield quantity and quality of cash crops in Burkina Faso, West Africa. *Scientific Reports*, 7, 17691 (2017); Garibaldi, L. A., et al., Stability of pollination services decreases with isolation from natural areas despite honey bee visits. *Science*, 344, 6183 at 1255213; Kremen, C., Williams, N. M., Bugg, R. L., Fay, J. P., and Thorp, R. W., The area requirement of an ecosystem service: Crop pollination by native bee communities in California. *Ecology Letters*, 7, 1109–1119 (2004); Mertz, L., Wild Pollinators Get the Job Done. https://entomologytoday.org/2019/12/13/wild-pollinators-get-the-job-done/.

[109] Also See: Fontaine, C., et al., Functional diversity of plant-pollinator interaction webs enhances the persistence of plant communities. *PLoS Biology*, 4, 0129, 0134 (2006); Fontaine, C., et al., Diverse pollination networks key to ecosystem sustainability. *PLoS Biology*, 4, 0005, 0006 (2006); North Carolina State University, Genetic diversity key to survival of honey bee colonies. *Science Daily* (June 17, 2013). https://www.sciencedaily.com/releases/2013/06/130617111341.htm (accessed Oct. 17, 2020).

# 11 Brazil's Role in Food Production – Food Security and Sustainability

*Gerardo Figueiredo Junior*
Milton Campos Colleges
Gama Filho University
International Food Law, Michigan State University
Founding partner in Zeigler Law Firm

*Paulo Bellonia*
Data Processing and Technology, FUMEC University
Founder of the Brazilian N.G.O. SaveCerrado

**CONTENTS**



## 11.1 INTRODUCTION

What is going to save the planet? Genetically modified food? Plant-based food? Organics? Combating food waste? This kind of discussion should never be subjugated to questions related to ideology, religious beliefs, or political and commercial

DOI: 10.1201/9781003137733-19

disputes, even though we recognize that human beings are both the result and the generators of these very influences. Therefore, every one of these themes is present when deciding the direction of society, and it would be utopian to believe that we would follow one path or another without taking everything that constitutes our intellectual and social development into consideration. However, one central truth emerges from all of this: we all need food to continue our daily activities and disputes, even if only to decide who is right and about what.

The world is demanding more and more food – of this, we can be certain. The Food and Agriculture Organization of the United Nations estimates that by 2050, it will be necessary to increase food production by up to 70 percent beyond what we are producing today.[1] The amount, quality, and access to food will all be vitally important in feeding the world. In this scenario, Brazil holds a privileged position as one of the world's largest food producers, likely reaching the top of the list within a few years. However, the country still needs to reach benchmarks imposed by international markets to ensure continuity and further increases in Brazilian exports. It is not just about the quality of what is produced. Buyers are demanding more from food producers, such as certifications of product origin, sustainability in production methods, respect for environmental issues, proof of product authenticity, and many other requirements, without which national producers will be pushed out of this highly competitive market.

There is no doubt that increased demand for food represents a gigantic commercial opportunity, while at the same time an enormous strategic responsibility toward the inhabitants of our planet, as it will be up to Brazil to guarantee the food security of many countries in the coming years. This will require a series of measures to be carefully undertaken in order to meet increasingly important commercial and health safety requirements. In the coming sections, this chapter will discuss major ideas in the push to protect the future of the food supply in Brazil, including confronting questions on genetically modified foods, investigating plant-based alternatives, preserving crucial natural areas, and reducing waste through donations, in order to reduce food insecurity globally (Figure 11.1).

## 11.2 BRAZIL AND GENETICALLY MODIFIED FOOD (GMO)

### 11.2.1 The GMO Safety Debate

When it comes to genetically modified foods, or GMOs,[2] there is a very real dispute regarding the monopolization of truth. Even so, after all the arguments have been

---

[1] "How to Feed the World in 2050" p. 8, 2.2. https://www.fao.org/fileadmin/templates/wsfs/docs/expert_paper/How_to_Feed_the_World_in_2050.pdf.

[2] Editor Note: Genetically Modified Organisms (GMOs) are organisms that have been modified through genetic modification technology. The industry and academia have adopted many terms and synonyms for foods produced from genetic technology that are used interchangeably, such as Genetically Engineered (GE), Genetically Modified (GM), Genetically modified crops or foods (GMOs), and Bioengineered (BE). The authors in this chapter use the term "GMO" to describe foods produced by genetic modification. However, the reader should be aware that there are various regulatory definitions and terms used by different entities and regulating bodies. Of importance, many also use these terms above interchangeability with transgenic, which the authors attempt to explain.

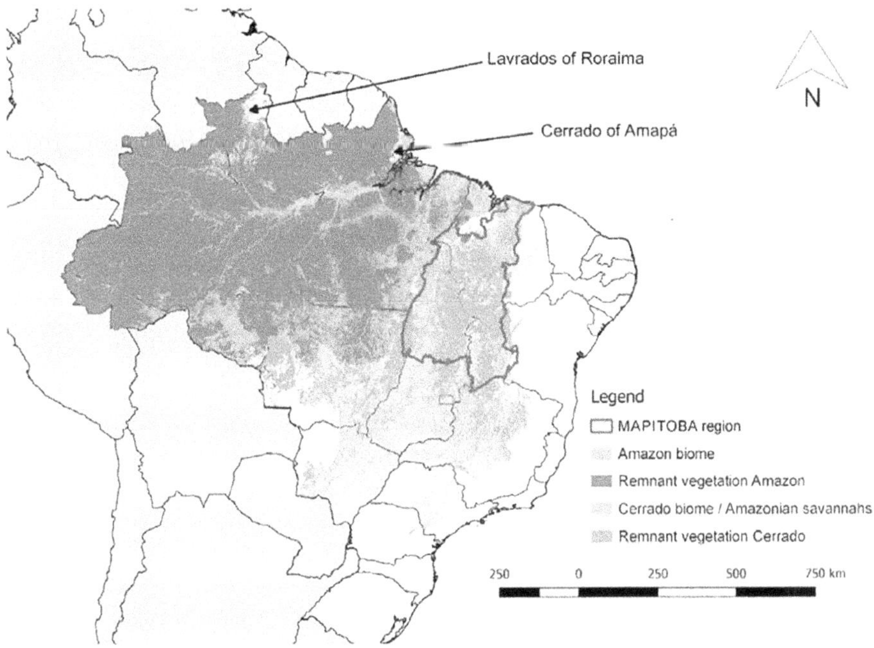


**FIGURE 11.1** Deforestation control in the Brazilian Amazon: A conservation struggle being lost as agreements and regulations are subverted and bypassed – Scientific Figure on ResearchGate. (Available from: https://www.researchgate.net/figure/Map-of-Brazil-showing-the-Amazon-and-Cerrado-biomes-with-remnant-vegetation-highlighted_fig1_334115444 [accessed 18 Dec, 2021].)

made, can one say with certainty that consuming products that contain GMOs is safe? Perhaps science is still not able to answer with a simple "yes" or "no". After all, we are talking about relatively recent technological advances, although the use of organisms to alter or create other substances has been around for a thousand years. And as long as it is not possible to reach definitive conclusions, countries are looking for alternatives to meet the growing demand for food, including the use of GMOs, which are widely touted by those who have the most to gain from their commercialization. For the time being, these nations are allowing the presence of genetically modified organisms in their production chain, while at the same time seeking to adopt measures that minimally guarantee the safety of living beings and the environment.

Divergences of ideas and interests are inevitable and, to a certain extent, necessary for the construction of a food system capable of meeting the growing needs of populations.

Thus, only by utilizing the law and legislation, in general, we can find a possible answer for alleviating the tensions between the different currents in favor of or against the use of GMOs and those that use all sorts of arguments to try to impose their own truth.

### 11.2.2 GMO Legislation in Brazil

Brazil's reality is not disassociated from that of the rest of the planet. There are those who defend the use of GMOs as a safe solution (if such a solution exists) for the development of food production, especially considering that Brazil is, today, one of the largest international suppliers of food products such as animal protein, soy, coffee, and other plant and animal product cultivation. But there are also those who would like to simply ban these organisms from Brazilian territory, because they consider them to be extremely risky to human and animal health and to only benefit large producers and corporations, contributing to the further elimination of smaller producers unable to compete in these markets. Considering that these organisms are already part of the daily lives of Brazilians, other questions have become central to the debates, such as what are the safe quantities of GMO raw material contained in commercialized products that can be consumed? And, also, the rights of consumers to know and to exercise choice as to whether or not to purchase products that contain GMOs.

In principle, the answer would seem to be quite simple and that which is stipulated by legislation: every product that contains GMOs exceeding 1 percent of its total volume must provide this information, clearly displayed on the label. However, this rule, initially established by the relevant authorities, has since been modified by a decision of the Supreme Court, as we will see later. There are many repercussions of this recent Supreme Court decision. For example, there is no longer any tolerance regarding the duty to inform consumers about the presence of GMOs among the ingredients used, recognizing that the food chain is composed of many suppliers with each one of them linked together and considered as a whole under Brazilian law, which results in even more responsibilities. The applicable legislation that is in force which must be analyzed to better understand how to deal with this issue starts with the Constitution of the Republic and continues on through Law No. 11.105, of March 24, 2005 (Biosafety Law), Decree No. 4.680, of April 24 of 2003, Decree No. 9013, of March 29, 2017 (Regulation of Industrial and Sanitary Inspection of Products of Animal Origin - RIISPOA), Law No. 8.078, of September 11, 1990 (Consumer Protection Code), Law 7889, of November 23, 1989 (Sanitary and Industrial Inspection of Products of Animal Origin), as well as international treaties and bilateral agreements for the importation and exportation of food products.

As was stated earlier, GMOs are not new but recent enough to warrant debate since the green revolution. In fact, it can be said that the combination of plant, animal, and microorganism genomes is a very old branch of biotechnology which, in turn, can be described as the technique involving the utilization of living organisms, or a part of them, in the preparation or modification of other products, resulting in "improvements" or in seeking to develop new substances for specific uses. A valid example for GMO debates is beer, a beverage that has existed since ancient times. Yeast is used and is an example of a cultivated organism, as it is in the production of yogurt from milk, among other foods that have been in our society for centuries or even millennia. However, with the evolution of genetic engineering, especially in the 1970s, a new phase of biotechnology was reached, in which methods capable of modifying the genetic material of living cells were developed and utilized.

Industry began to use this knowledge in the development of new substances or the improvement of existing ones. Then, the name "transgenic" appeared, which is a term referring to nothing more than transferring or inserting only a specific gene chosen from a donor organism in order for a new variety that is considered better, more developed, and more capable of achieving the results desired by the producer to emerge. It is important here to clarify that GMOs and transgenics, although different concepts are often confused and treated as synonyms. GMOs or GE foods are modified to do something or to react in a specific way that would not have been possible in nature. For example, the introduction of genetic material from another species, such as fish in a tomato, results in traits of the new GE plant or GMO that either thrive under certain conditions, such as drought, or high herbicide use, that make the plant produce its own pesticide, such as in corn, or some other "unnatural" effect.

Since the initial evolution of GMO technology in the 1970s, the applications have been innumerous, although accompanied by incessant monitoring by academics and authorities seeking to understand what are the risks involved in the use of GMOs for living beings and for the environment. The modification and utilization of microorganisms has become a tool in the constant striving for improvement of plants and animals and in various processes and products, generating, in turn, numerous studies seeking to clarify whether or not there are risks to health of humans, animals, and the environment. In sum, scientists around the globe have yet to prove the safety of GMOs as touted by the industry.

### 11.2.3 The Proliferation of GMOs in Brazil

The fact is that there already are millions of hectares of GMOs planted throughout the world today, in dozens of countries, which is the reality that society, authorities, and academics must face. The need to boost food production is one of the driving forces behind the expansion of the use of these organisms by producers around the world. The Food and Agriculture Organization of the United Nations itself considers biosafety to be a component of food security, recognizing that the healthy and sustainable use of this technology is one of the ways to guarantee population access to food, although it recommends caution regarding the practical results of its use. There is still much misinformation regarding the use of GMOs in production, and this is one of the reasons Brazilian authorities, as well as those from other countries, adopted the Precautionary Principle as a reference in the elaboration of policies and in the analysis of issues involving this theme.

It is important to note that Brazil stands out as having one of the highest total GMO areas planted. Though still behind the United States, GMO areas are growing rapidly, and part of this expansion is due to the approval process for these products in our country. Many people assert that Brazil is an extremely liberal country in terms of the approval and use of GMOs. However, the reality is that specialists from many countries who are familiar with our health regulatory system consider the quantity and quality of information required to be satisfactory. The fact is that the growth in total areas planted with GMOs represents a source of pride for many, while at the same time, a great concern for others.

Another important acknowledgment is that the use of GMOs is not only the prerogative of major economic powers, but is present in developing countries and, continues to attract new producers in all continents year by year. Brazil is a signatory to several international treaties and has developed a very broad normative base for dealing with GMOs, using the Precautionary Principle as provided for in the Constitution of the Federal Republic, in Chapter VI, article 225, which establishes that everyone has the right to an ecologically balanced environment, empowering the Public Authorities the community in their duty to defend and preserve it for present and future generations. Furthermore, paragraph 1 of the same article of the Constitution of the Federal Republic, items II and IV and V, describes the responsibilities of the Public Authorities and emphasizes that in order to meet the expectations of its constituents it is necessary to: (i) "preserve the diversity and integrity of the genetic heritage of the Country and inspect the entities dedicated to the research and manipulation of genetic material"; (ii) "require, in accordance with the law, for the installation of an activity or project potentially causing significant degradation of the environment, a prior environmental impact study, which will be publicized"; and (iii) "control the production, sale and use of techniques, methods and substances that pose a risk to life, quality of life and the environment".[3]

In addition to the Precautionary Principle, the Principles of Disclosure and Guarantee of Access to Information also serve as a guide for the Public Authorities in carrying out activities related to GMO issues, and these are some of the premises that serve as a base for the various authorities involved in these issues. Among these, we can highlight that the National Technical Biosafety Commission, CTNBio, is responsible for evaluating food safety (qualitative) related to the presence of GMOs; the Ministry of Agriculture, Livestock and Provisions; the National Health Surveillance Agency; Consumers Protection Authorities; State and Federal Public Prosecutors; and members of the National Health Surveillance System.

Although there are different normative frameworks, including laws, ordinances, and normative instructions, we consider it our responsibility, in this legal brief, to specifically assess the rules for product labeling, and the way for industry to ensure compliance with the rules in light of the position of the Brazilian Supreme Court (Federal Court of Justice, abbreviated STF) related to the obligation to inform, on product labels, the presence of GMOs only in quantities above 1 percent, as stated in Decree 4.680/2003, in a decision in Public Civil Lawsuit No. 2001.34.00.022280-6/DF, filed by the Consumer Defense Institute (IDEC) and by the Office of the Federal Prosecutor (MPF):

> It is known that among the guiding principles of consumer relations, present in the CDC, is that of proper information and of transparency, by virtue of which the supplier is obliged to provide all information about the product and/or service, its characteristics, qualities, risks, prices, in order to ensure the right to freedom of choice and equality in contracting.
>
> In the transcribed provisions (of article 6 of the CDC) there is no evident financial, environmental or security breach or concession that can be interpreted as limiting consumer access to all of the information they need, not only to ensure their safety, but

[3] Constitution of the Federal Republic, in Chapter VI, article 225.

> also their right to make the decision that they consider to be the right one, for personal reasons.
>
> Thus, even if the 1% margin proves to be technically safe, it fails to provide full knowledge of the situation, as a necessary prerequisite for the right to consciously choose among the available options, whether for cultural, religious or dietary reasons, such as the applicants claim.
>
> However, even in relation to safety, this percentage does not represent an absolute guarantee against risks (...)
>
> The preventive effort of the public authorities in the elaboration of safety laws is not denied, nor the importance of costs, but when there is incomplete knowledge of the effects of such products on human metabolism and the environment, as verified in the abundant reasoning of both parties, common sense and the law [must] prioritize health and human lives.[4]

In light of this, it is imperative for companies to redouble their efforts and attentiveness in relation to the acquisition of raw materials that are used in their products, seeking to certify the existence of or not and, if applicable, of the exact amounts of GMOs contained in these raw materials. A proper understanding of the principles set out in the previous paragraphs is vital in order to identify the risks of engaging in conduct potentially contrary to the understanding of the Brazilian Supreme Court and any consequent actions of the authorities responsible for inspecting producers of food products. At this point in time, we can affirm that Brazilian authorities remain extremely cautious about the use of GMOs in large-scale food production.

## 11.3 MEAT ALTERNATIVES AND VEGETABLE-BASED FOOD – THE "NEW" TREND

A new market is emerging in the midst of all these issues. That is, food produced from plants, or as it has become commonly known, *plant-based food*. What initially seemed to be a niche composed of idealistic producers and consumers looking for a different way to nourish the body, ended up getting the attention of large animal protein producers who identified a potential new market to be explored on an industrial scale. In fact, worldwide consumption of plant-based food grew by 27 percent in 2020, according to the Plant-Based Food Association.[5] Brazilians followed along in this trend, revealing that demand is not only from consumers who are accustomed to natural diets, but others, increasingly willing to incorporate alternative types of food into their daily routines.

As this chapter is being written, Brazilian authorities are considering the need to develop standards for regulating this novel food based on assessments from producers, consumers, and other specialists. For Brazil to be able to develop standards in this area, some important definitions will first have to be established, beginning with the nomenclature for each product developed solely from vegetable matter.

---

[4] Decree 4.680/2003, in a decision in Public Civil Lawsuit No. 2001.34.00.022280-6/DF, filed by the Consumer Defense Institute (IDEC) and by the Office of the Federal Prosecutor (MPF).

[5] https://gfi.org/press/plant-based-food-retail-sales-grow-27-percent-to-reach-7-billion-in2020/#:~:text=Corporate%20support,Plant%2Dbased%20food%20retail%20sales%20grow%2027%20percent%20to%20reach,for%20the%20third%20consecutive%20year.

For example, we already have a class of product widely referred to as "veggie burger", but can we simply adopt the classic name of a product originally developed from animal protein (burger) to make a vegetable-based correlation more commercially acceptable? Indeed, there are major parties interested and involved in this issue, and the answer to this question may determine the success or failure of many endeavors. Additionally, any potential regulation will have to define another key aspect: specifying exactly what can be classified as plant-based food. After all, how much of what is for sale on supermarket shelves today can be considered 100 percent natural or free of ingredients often added to bring texture and taste closer in line with products made from animal protein? The conclusions from the discussions that are taking place with Brazilian authorities will determine which options and dishes will be available for those who choose to include more plant-based products in their menus.

In addition, it is interesting to examine and debate different food production models, some of which have a very limited capacity to reach a large number of consumers even within Brazilian territory itself, but which are nonetheless very compelling options for catering to the demands of specific groups. Community producer gardening projects that supply high-quality products to communities in large cities are initiatives that should be highlighted, as well as organic products, which are already quite common in major restaurants throughout the country and in the households of people who want to keep things such as pesticides and other elements commonly used to treat industrial-scale crops out of their homes. At the same time, it is imperative for health authorities to warn, protect, and prevent consumers from being deceived by producers who maliciously claim to use techniques that do not depend upon these treatments when, in fact, they cannot even guarantee the origin of everything that is used in their production processes.

## 11.4 SUSTAINABLE PRODUCTION IN PRESERVATION AREAS – THE SAVE CERRADO EXPERIENCE

### 11.4.1 Habitat Protection, Environmental Stewardship, and Agroecology

Special recognition should be given to the fact that the world is increasingly focused on environmental issues. And it's not just about sustainability in the production process itself, but in looking for products whose consumption can positively impact conservation areas that are increasingly threatened with disappearance. A few years ago, these authors engaged with a project called Save Cerrado. It is an initiative from a Brazilian business leader who realized the importance of preserving this biome, which is present in a large part of the Brazilian territory, but which does not get the same kind of international coverage or attention as the Amazon Forest.

It all began when Paulo Bellonia found himself responsible, together with his family, for a property that included, in part, mandatory preservation areas in the Cerrado region of the State of Minas Gerais. The decision could not have been more opportune. Paulo and his family chose not only to abide by the stipulations of Brazilian environmental legislation, but to transform that piece of Brazil into a true symbol. And so, Save Cerrado, an organization that seeks to preserve and contribute to the sustainable development of the communities residing within it and that live off of what the Cerrado itself provides, was born.

For those unfamiliar with it, the Cerrado is one of the largest aquifers in Brazil. There are water reserves in the region that are as important as the so-called "flying rivers", formed in the Amazon region, and which bring rain to distant parts of the country. Fruits and other foods from the Cerrado are currently exported to many other countries for final consumer consumption and are also utilized in the production of medicinal products. This wealth and diversity, when managed in a responsible and sustainable manner, contributes to maintaining the biome and the livelihoods of families living in that region.

Save Cerrado functions by providing a bridge between international buyers and local producers, enabling the opening of sales channels that previously would have been absolutely impossible for local producers, usually families who plant or harvest the fruits of the Cerrado for their own sustenance, to access. In addition, Save Cerrado operates a contribution model for the "adoption" of square meters in the area it preserves. Corporations can assume responsibility for the preservation of areas in the region under the organization's responsibility, which contributes to maintaining the viability of the project.

Brazil has made progress in legislative terms and, in January 2021, Law No. 14,119, the so-called "Payment Law for Environmental Services" was enacted. This is an important tool for facilitating organizations like Save Cerrado, aiding in the effort to consolidate the dream of making it possible to preserve a private area of 180 million square meters, equivalent to more than 33 million NFL football fields, which is located in one of the most important biodiversity corridors in the Brazilian Cerrado (Figure 11.2).

### 11.4.2 Fruit from the Cerrado: Araticum, Baru, Cagaita, Piqui, and Jatobá

Being stewards of the Cerrado region entails using the resources in a responsible manner to protect the habitat, carbon sinks, and natural treasure that this region represents for the world as a whole. The biodiversity of the Cerrado is one of many global examples of how people can live in harmony with nature, protect and conserve while benefiting from the natural resources of the area. Protecting the habitat will protect animal welfare. Thus, it is worth highlighting some of the different types of fruit from the Cerrado, such as Araticum, Baru, Cagaita, Piqui, and Jatobá, which today are being exported to many different countries. It is not an exaggeration to state that at least 20 species of fruit native to the Cerrado are frequently used in regional cuisine, either through natural consumption or in the form of sweets including jellies and ice cream or in savory and spicy foods. Now they have crossed international borders these fruits are beginning to be recognized and demanded in other countries.

Araticum, a Cerrado fruit with antioxidant properties, are endemic to dry and sandy areas. It has a pulpy mass surrounding a black seed with the flavor deep inside. Then there is the case of Baru. It is important to emphasize that this is an endangered species, as its tree yields a beautiful, high-quality wood, which is in great demand. The fruit has a shell that is so hard it is practically impossible to open it without

**FIGURE 11.2** Pequi tree is more than 30 years old, inside Save Cerrado protected area.

breaking the nut inside. However, the local people of the Cerrado region have their own techniques for opening the shell and removing the nut whole for consumption. The fruit itself is high in protein and rich in omega-6 and 9, zinc, fatty acids, protein, fiber, minerals, and iron. Buriti, on the other hand, is a sweet fruit that is rich in vitamins A, B, and C, in addition to calcium, iron, and protein, and the Palm tree it grows on can reach a height of 30 meters (almost 100 feet).

The Cerrado also provides us with Cagaita, well known throughout the region, which has a somewhat tangy flavor. It is consumed after being prepared by heating it to extract the best flavor or after waiting until it is very ripe. It helps alleviate indigestion and digestive discomfort. Graviola (Soursop) is another very tasty Cerrado fruit. Rich in carbohydrates, especially fructose, it is used as a base in the production of ice cream and juices, as well as consumed fresh. Anyone adventurous enough to try it will identify a taste that resembles a mixture of strawberry and pineapple.

Finally, Pequi fruit, perhaps the best-known fruit of the Cerrado, deserves a special mention. There is no middle ground, you either love it or hate it, which does not affect its utilization as a common ingredient in many recipes, and when combined with other dishes and foods such as rice, in addition to being used for seasoning meats such as chicken. Pequi has a very hard green husk and is full of seeds in its bright yellow pulp. Special care must be taken when eating it. Biting into the fruit is risky because it contains thorns. Those who like it, however, swear that it is worth the effort, and it is known for properties that aid in the prevention of diabetes, as well as a reduction in cholesterol and obesity, according to studies carried out by Unimontes University, in Montes Claros, Minas Gerais, in partnership with the Federal University of Minas Gerais.[6] The studies attribute the benefits from consuming the fruit to the high levels of gallic acid that it contains. These are just some of the fruits the Cerrado provides and that are utilized by families that reside there in making their living from what the biome provides and what can be produced there. A visit to the Save Cerrado website, as well as other sources, can provide more information and explain more about how you can contribute to keeping this project alive.[7]

In harmony with the fruit live the animals of the Cerrado region. The best-known endemic species is the anteater. Wildlife protection goes hand-in-hand with habitat conservation. Agroecology, i.e. farming in harmony with nature, is one of the many ways to achieve these goals.

## 11.5 FOOD DONATIONS: MAKING THE MOST OF WHAT IS ALREADY PRODUCED

Reducing food waste is an obvious but often overlooked way to combat the exploitation of animals and habitat. Conversely, making the most of the food that is already produced, ensuring access for all, and its safety along every point of the supply chain (farm to fork), is a key factor in conservation efforts. Notably, food waste isn't a problem only in Brazil, although the dimension here is quite significant and fighting it will necessitate a real change in culture. Education is essential, but alone is not sufficient for ensuring that the most vulnerable groups have the food necessary to guarantee their right to life and health, as established in the Constitution of the Brazilian Republic.

In response to a worsening food insecurity situation, due to the pandemic caused by COVID-19, on June 23, 2020, Congress approved Law No. 14,016, aimed at stimulating the donation of surplus food still suitable for human consumption. Within that context, it was necessary to balance the arguments between those who believed that economic sanctions should be imposed on businesses that did not agree to donate, since this could result in a worsening of the economic crisis due to a reduction in purchased consumption, and those who believed it was more important to encourage an increase in donations, without relying on financial incentives such as tax breaks for donors, since doing so at such a critical time would be financially detrimental to tax revenues for state and local governments. The result of the debates and discussions was a law crafted to eliminate a concern that potential donors were wary

---

6 https://unimontes.br/pequi-e-vinho-podem-ajudar-no-combate-a-obesidade-revela-pesquisa-conjunta-entre-unimontes-e-ufmg/In Portuguese only.

7 To know more about the Cerrado region, and in addition, to learn now you can help in protecting the biome, visit www.savecerrado.org.

of the characterization of donations as acts of consumption, subject to the rules of commerce.

Up until that time, it had been common for decisions rendered by Brazilian courts to recognize the donor as a "supplier" and the beneficiary as a "consumer", which entailed responsibilities inherent to supplier-consumer relations, including the imposition of fines and other penalties based on the Brazilian Consumer Protection Code (Law No. 8.078/1990), which, is quite rigorous. Even if the donor took all appropriate measures to ensure the quality of the product being donated, none of this mattered in the eyes of the judges applying the previously mentioned statute. This attitude was based on the perspective regarding the objective responsibility of the supplier and the recognition of the consumer as the inferior party in the relationship. There was no exception given, even if there was some compromise in the quality of the donated product that took place subsequently to the donation itself. For example, even if only inadequate packaging of the food by the beneficiary or even by a third party in charge of the operation's logistics and, even considering that the donation had been done free of charge, without any cost to the beneficiary, no consideration would be given.

Immediately after the publication of the new law, any establishment that produced and supplied food, including unprocessed, as well as industrialized food products and ready-to-eat meals, could donate surpluses that had not been commercialized, but that could still be safely consumed by the beneficiaries. Logically, the donor still had to be concerned with aspects such as the expiration date, preservation conditions, health safety, and package integrity, even aesthetic damage to the packaging, unless this did not compromise the nutritional properties of the food.

So, in addition to restaurants and supermarkets, hospitals, hotels, and other establishments that routinely provided meals for their own workers, also became potential donors. And although no direct financial incentives were provisioned, the Law does guarantee that there will be no fiscal incidence that could encumber the donation. The act, therefore, is considered voluntary and cannot incur any additional onus other than strictly necessary for the donor. However, it is important to point out that Law No. 14.016/2020 is not a free pass for a business to dispose of leftover food without any responsibility. On the contrary, the same care that is taken in serving food on the establishment's tables must be taken when making a donation.

Finally, the intermediary in the chain of donation, whether it is an organization or other authority that gets the food delivered to those who need it, assumes all responsibility after the first act of donation by the original business provider. From this point of origin, it is the intermediary's responsibility to ensure and guarantee the quality of storage and transportation in order not to compromise the quality of what is being donated.

## 11.6 CONCLUSION

Thinking global but acting local ring true to the lessons we can learn from Brazil. There is no single solution for mitigating or putting an end to food insecurity in the world, or even, in our case, in Brazil. A country that produces food for the rest of the world cannot countenance hunger or malnutrition at home. We are living in and passing through a unique moment in history in which governments, institutions, and businesses are being put to the test. We must all dedicate ourselves to ensuring that we are not just spectators, but protagonists in the fight against food insecurity.

And how do we do that? Everything starts with information. We start by seeking to know as much as possible about the food we eat and the reality surrounding us, even though it may be distant from our daily lives. Based on the knowledge gained, critical capacity and demand for better solutions can arise.

Each aspect of production has an important contribution to make, whether locally or by providing food to the peoples abroad. And we need to free ourselves from secondary issues that often only get in the way of sustainable science and innovative technology in the search for safe production models capable of meeting the expectations we have of them. To this end, we must always seek reliable information, from appropriate sources, and resist the spreading of false ideas and messages about this or that production process. Asking questions, refining the questions, and seeking answers will help shape the system for a future we can be hopeful of.

## 11.7 EDITOR'S PROPOSED QUESTIONS FOR FUTURE DISCUSSION: SUBMERGING ENVIRONMENTAL LEGISLATION

According to a Brazilian study from 2019, various agreements are not enforced. At the end of this book, we provide Tools for Change, a comprehensive appendix of Animal Welfare Legislation from around the world, with an emphasis on major trading nations. In terms of protecting the habitat in the Amazon and Cerrado regions in Brazil, the aforementioned study reveals that the legislation is lacking in teeth. Consider the solutions proposed. How do they compare to animal welfare legislation? Do you see any parallels with the poultry protection deficits in Chapter 7?

# 12 Land as Carbon Sinks or Pollution Sources
## *International Pastoral Land Law*[1]

*Christian Mulgrew de Laire*

**CONTENTS**



---

[1] Based on the chapter by Ian Hannam, International Farm Animal, Wildlife, and Food Safety Law (Steier & Patel, Eds.) 9 Springer (2017).

DOI: 10.1201/9781003137733-20

## 12.1 INTRODUCTION

The more land is used for agriculture, including pastures, the more concerns governments and the public should have for maintaining land as carbon sinks. This textbook explores aspects of an international and national environmental law affecting pastoral land and outlines frameworks for legislative reforms to achieve sustainable use of pastoral lands. International and national legal instruments and institutional systems play a significant role in pastoral land conservation. The contribution discusses multilateral agreements that could better promote the sustainable use of pastoral land at the international level. Two national-level approaches to reform environmental law for pastoral land, in Mongolia and the People's Republic of China, may offer helpful guidelines for others to follow in environmental law reform for pastoral land management.

International and national environmental laws outline frameworks for legislative reform for sustainable use of pastoral lands. Pastoral farming, also known as livestock farming or grazing, produces livestock rather than crops. Examples include raising beef cattle and raising sheep for wool. Pastoralism is a livestock production system based on extensive land use and often some form of herd mobility. Currently, pastoralism occupies 25 percent of the earth's land (typically land that cannot grow crops), supports 200 million households, and accounts for 10 percent of the world's meat production.[2] Pastoralism is essential for the people it feeds, ecological services it provides, economies it boosts, and civilizations it sustains.[3] Unfortunately, threats and pressures associated with human population growth, economic development, land-use changes, and climate change are challenging professionals and practitioners to sustain and protect these invaluable social, cultural, economic, and ecological assets worldwide[4] (Figures 12.1 and 12.2).

Pastoral lands provide benefits to both the local and global environments that should be properly recognized in legislation.[5] Most importantly, pastoral lands provide vegetation cover protecting soil and economical production of animal fodder, firewood, and other indirect benefits.

Climate change is driving fragile pastoral systems into vulnerable conditions.[6] Socio-economic factors are fracturing large-scale pastoral ecosystems into spatially

[2] See FAO (2001), Introduction; See also FAO (2001).
[3] See Nori and Davies (2007), p. 4.
[4] See also Nori and Davies (2007), p. 6; See Abildtrup et al (2006), p. 5.
[5] See Friedel, Laycock and Bastin (2000), pp. 262–272; Neely, Bunning and Wilkes (2009), pp. 3–8.
[6] See Dong Shikui et al. (2011), pp. 10–11.

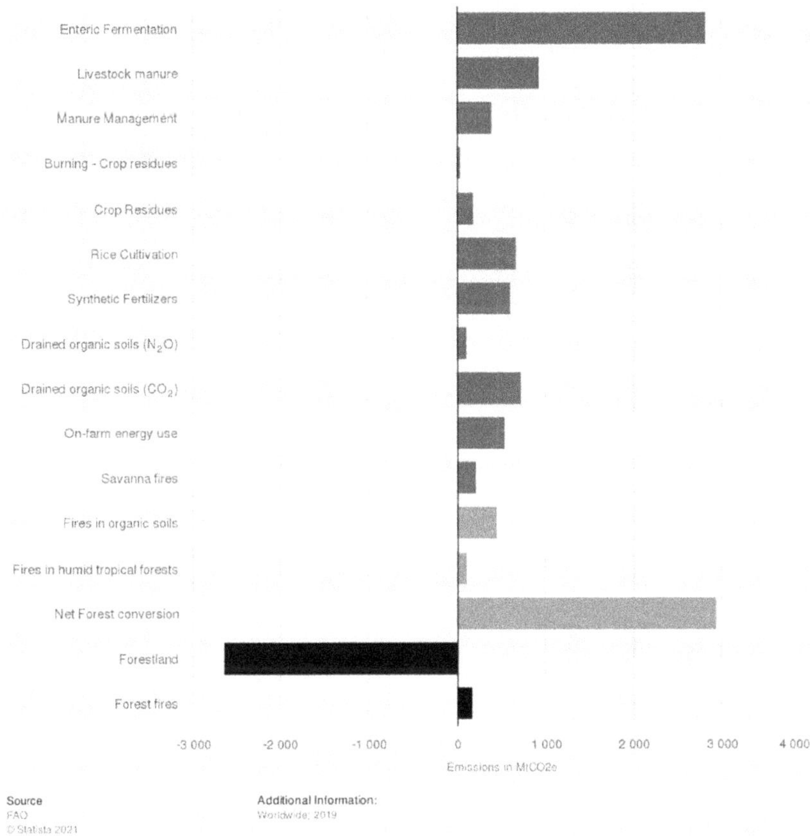


**FIGURE 12.1** FAO. (June 24, 2021). Emissions from agriculture and forest land worldwide in 2019, by component (in million metric tons of carbon dioxide equivalent) [Graph]. (In *Statista*. Retrieved December 19, 2021, from https://www-statista-com.ezproxy.neu.edu/statistics/1254324/agricultural-forest-land-emissions-worldwide-by-component/.)

isolated systems.[7] Accordingly, professionals, practitioners, and policymakers should make a system with governing and guiding institutions to enhance pastoral communities' resilience and practices. Additionally, public awareness and institutional decisions must support pastoralist efforts to adapt.

[7] See Halimova (2012), pp. 307–308.

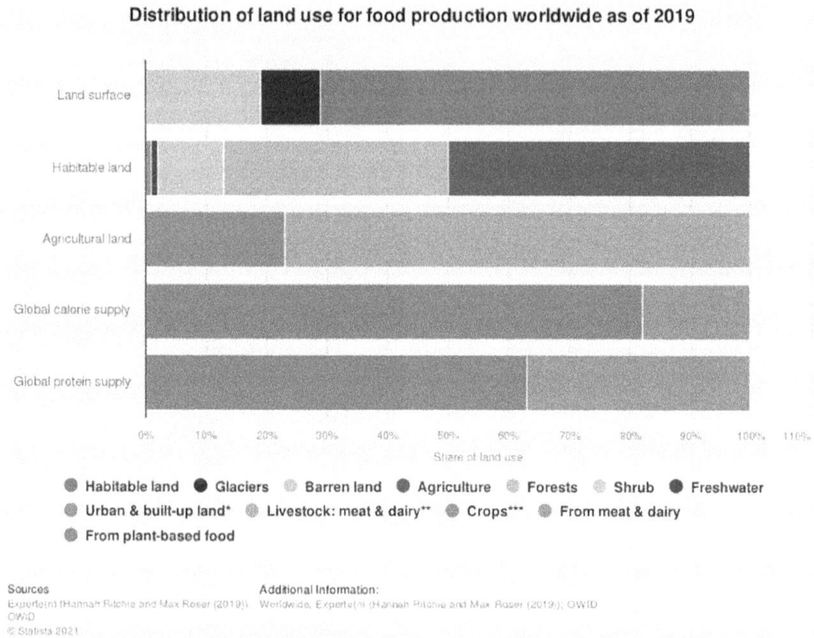


**FIGURE 12.2** OWID. (September 1, 2019). Distribution of land use for food production worldwide as of 2019 [Graph]. (In *Statista*. Retrieved December 19, 2021, from https://www-statista-com.ezproxy.neu.edu/statistics/1260751/global-land-footprint-of-food-production-by-type/.)

## 12.2 PASTORAL LAW

International law inadequately recognizes the significance of pastoral lands despite economic, human, and environmental impacts. For the purposes of this contribution, "pastoral law" is considered an area of law that includes specific laws, regulations, and legal instruments that govern the use and management of pastoral lands. Laws that regulate pastoral land conversion into non-pastoral land are also relevant. Pastoral law can regulate animal management, land consolidation, or land reallocation procedure. Additionally, other laws and regulations can indirectly but significantly regulate pastoral law. For example, land protection laws focused on soil productivity or erosion are important to keep pastoral land productive.[8]

Since the early 1900s, over 200 multilateral environmental treaties have been developed. Some contain elements that contribute to pastoral land conservation and management. Though none of these treaties is sufficient on its own, some assist by promoting activities that sustain pastoral land environments. Increased support for the least-developed nations, climate change awareness, and agricultural trade liberalism has resulted in a call to expand international environmental law. By extension,

[8] See Grossman and Brussaard (1992); Chalifour et al. (2007).

pastoral laws are being reworked and expanded.[9] Next is a brief account of the main instruments of the Millennium Assessment on dryland ecosystems and land degradation trends.[10]

### 12.2.1 The World Charter for Nature

The World Charter for Nature called on states to cooperatively establish methods to assess adverse effects on nature and implement international legal provisions to conserve and protect the environment. The Charter states land productivity shall be maintained or enhanced using measures that safeguard against forms of degradation and protect long-term fertility and organic decomposition.

### 12.2.2 The Rio Declaration and Agenda 21

The Rio Declaration on Environment and Development of 1992 established an equitable global partnership of states, key sectors of society, and individuals. The Declaration emphasized the need for an international agreement to protect the environment through effective legislation. Another key instrument, Agenda 21, discusses laws that assist with the global management of pastoral lands.[11]

The Sustainable Development Goals were part of an inclusive and transparent intergovernmental process to develop goals agreed upon by the General Assembly and provide pastoral land management.[12] Furthermore, Sustainable Development Goal 15 relates directly to pastoral land management through desertification, degradation, and drought, while other goals related to ecosystem management.[13]

### 12.2.3 Convention to Combat Desertification

Desertification and drought are problems that face most pastoral lands. The United Nations Convention to Combat Desertification (CCD) addresses land degradation and the methods to protect and manage soil and water resources of pastoral land areas.[14] The CCD defines desertification as land degradation in arid, semi-arid, and dry sub-humid areas and seeks to prevent and reduce land degradation, rehabilitate partly degraded land, and reclaim desertified land, particularly in countries that experience serious drought.

In this regard, the CCD recognizes the high concentration of developing countries experiencing severe drought and needing support from international cooperation and partnership agreements to find long-term integration strategies that focus on rehabilitation, conservation, and sustainable management of pastoral lands and water resources.[15]

---

[9] See also Chalifour et al. (2007).

[10] See White, Tunstall and Henninger (2002), p. 2.

[11] See UN (1992a), Principle 11; See UN (1992b); See Chapters 8, 38 and 39 of Agenda 21 which discuss international legal instruments and mechanisms.

[12] See UN (2014).

[13] See UN (2012) adopted at Rio+20 2012, paras. 245–251.

[14] UN (1995); See Neely, Bunning and Wilkes (2009), p. vii.

[15] See also Neely, Bunning and Wilkes (2009), p. 3.

The CCD acknowledges six program areas as providing a useful basis for combating desertification. The CCD outlines an approach for countries developing national action plans and obtaining scientific and technical cooperation and supporting measures.[16] The CCD does not specifically recognize pastoral land as an individual ecological element; however, various programs address important land degradation issues, methods to protect soil and water, and legal principles and processes for pastoral land conservation.[17] Furthermore, Article 31 provides for the development of annexes that can be used for pastoral land management and the creation of legislation.[18]

### 12.2.4 UNCCD 10-Year Strategy

The CCD 10-year strategy provides a global framework to support measures that will prevent, control, and reverse desertification, degradation, and drought. The strategic objectives to improve living conditions of affected populations, improve affected ecosystems, generate global benefits, mitigate climate change, and mobilize resources to support the implementation of the Convention are directly related to pastoral land management and guide stakeholder actions. The strategy is well-rounded and could be enshrined in national legislative frameworks for pastoral land management.[19]

### 12.2.5 Convention on Biological Diversity

The fundamental concern of the Convention on Biological Diversity (CBD) is that human activities are significantly reducing biodiversity.[20] The CBD stresses the importance of international cooperation among countries, intergovernmental organizations, and the non-governmental sector for the conservation of biological diversity and preparation to implement CBD strategies.[21]

The CBD relates to pastoral land management through conservation of biological diversity, sustainable use of its components, and equitable sharing of benefits by considering various rights over resources.[22] Additionally, similar to pastoral land conservation objectives, the CBD recognizes that countries are responsible for conserving biodiversity sustainably.

The CBD acknowledges the need for substantial investments and stresses the importance of global cooperation for biodiversity conservation, sustainable use of its components, and implementation of CBD strategies.[23] In order for the CBD to

---

16 See CCD Articles 9-21.

17 See Secretariat CCD (2000); European Commission (2000), pp. 1–5.

18 Unlike the CBD and FCCC, the CCD does not have a provision for making a protocol under the Convention. However, Article 30 provides for amendments to the CCD, and provided all parties agreed, the Convention could be amended to include a provision for the adoption of protocols. Should this happen a protocol for the management of pastoral land could be considered.

19 See Secretariat CCD (2007), *8th session, Conference of the Parties*, Madrid 3–14 September 2007, ICCD/COP (8)/16/Add.1, 23 October 2007.

20 See UNEP (1995a).

21 See CBD Article 16.

22 See CBD Article 1.

23 See CBD Article 16; See Miller and Lanou (1995); Prip et al. (2010), Part 1, pp. 10–22.

take on a more expanded and precise role in sustaining pastoral lands, specific provisions could be drafted for pastoral land and included as a protocol to the CBD. These provisions and rules should focus on the ecological effect on biodiversity and maintenance of human life, as seen with the Strategic Plan for Biodiversity's process to achieve sustainability through customized goals based on national and regional priorities and global targets.[24]

### 12.2.6 International Initiative for the Conservation and Sustainable Use of Soil Biodiversity

The International Initiative for the Conservation and Sustainable Use of Soil Biodiversity (IICSUSB) is the main international framework for action regarding soil biodiversity.[25] Raising awareness is important for pastoral land management by improving a farmer's livelihood, calling attention to their role, and building support structures.

### 12.2.7 The Strategic Plan for Biodiversity (2011–2020)

The Strategic Plan for Biodiversity (SPB) was adopted at the CBP COP-10 to provide a framework for action by CBD stakeholders. The plan is accompanied by 20 Aichi Biodiversity Targets where land health and biodiversity are cross-cutting. The Biodiversity Targets, such as reducing pollutants and sustainable agriculture, could guide the establishment of targets for sustainable use of pastoral lands since biodiversity can underpin many benefits to sustainable use of dryland areas as a cross-cutting approach.[26]

### 12.2.8 Convention on Climate Change

The key concern of the United Nations Framework Convention for Climate Change (FCCC) is that human activities are increasing the atmospheric concentration of greenhouse gases.[27] Two principal contributors are changes in land use and cover. Pastoral lands act as a carbon reservoir. Pastoral activities, such as biomass burning, using fertilizers, and livestock grazing, exacerbate land degradation and greenhouse gas emissions.[28] Agriculture and land-use change account for about 30 percent of greenhouse gas emissions blamed for global warming. Furthermore, feed efficiency for livestock in arid parts of Africa is so low that the production of one kilo of protein can contribute the equivalent of one ton of carbon dioxide.

---

[24] See Article 28, Adoption of Protocols; Decision X2, COP10.

[25] COP8 decision VI/5 2006.

[26] Decision X/2 COP10 18-29 October 2010 adopted a revised and updated Strategic Plan for Biodiversity, including the *Aichi Biodiversity Targets* for 2011–2020.

[27] See UNEP (1995b).

[28] See Squires (1998), p. 209; Squires and Glenn (1997), pp. 140–143; See Neely, Bunning, Wilkes (2009), pp. 5–13.

The FCCC does provide for changes to the terrestrial environment but is not considered the most appropriate legal vehicle for pastoral land conservation because it primarily focuses on the industrial, rather than the agricultural, sector.

The Kyoto Protocol under the FCCC includes the responsibility to promote sustainable forms of land use and recognizes the need to expand and preserve soil carbon sinks where significant emissions are related to changes in agriculture.[29] The adoption of the Paris Agreement could benefit the conservation of pastoral land by calling for international cooperation to accelerate the reduction of global greenhouse gas emissions.

### 12.2.9 UNFCC Adaptation and Mitigation

The FCCC process provides developing countries access to climate change funds to implement adaptation and mitigation activities. These activities can be based on reducing land degradation, improving livestock management, and improving human livelihood and could consider policy development, monitoring and reporting methodology, economic assessment, and capacity building. The development of an effective legal, policy, and institutional framework would be essential components of a national strategy to manage the climate change impacts on pastoral land ecosystems.[30]

### 12.2.10 Draft International Covenant on Environment and Development

The IUCN Draft Covenant ties together the environmental and developmental principles of several treaties.[31] Many articles are relevant to pastoral land management. In particular, Article 20 calls for the conservation and restoration of natural systems of biodiversity and ecology.

## 12.3 REGIONAL INSTRUMENTS

Regional instruments can guide the framing of national pastoral land legislation. For example, the African Convention on the Conservation of Nature and Natural Resource, adopted in 1968 and updated in 2003, acknowledges the principle of common responsibility for environmental management. The Convention's objectives, similar to those of pastoral land conservation, include the achievement of ecologically rational, economically sound, and socially acceptable development policies and programs.

The Convention embodies a comprehensive and integrated approach harmonizing nature with culture by providing international legal principles and best practices for pastoral land management. In this context, the African Convention provides a valuable approach applicable in other regions.

---

29 See Kyoto (1997), it is intended that a new agreement will be forged in Paris in 2015 or thereafter.
30 See Wilkes et al (2011), pp. 10–12.
31 See IUCN (2004), most of the 75 Articles have relevance to pastoral land management.

### 12.3.1 National Legal and Institutional Frameworks

Countries have used various legislative means to deal with pastoral land management, usually through a piecemeal of other laws, evident through the general nature and various mechanisms used when enforcing pastoral land management.[32] In some countries, pastoral law was used to manage rural land, providing mechanisms that achieved sustainable use of land, including enforceable conditions on land leases for grazing management, vegetation, soil, and water management.[33]

Soil degradation laws, with a focus on land utilization, can no longer adequately manage pastoral degradation problems. The conservation capabilities of the legislation became overshadowed by the objective of agricultural production, price support schemes for domestic and export needs, and land settlement and development schemes. Over time, laws developed for rangeland and pastoral land use; for example:

#### 12.3.1.1 Examples of National Laws Relating to Pastoral Management

- United States of America: Public Rangeland Improvement Law 1978, and Forest and Range Renewable Resources Planning Law 1974
- Canada: *Agricultural Land Commission Law 1979* (British Columbia)
- Australia: *South Australia Pastoral Land Management and Conservation Law 1989*
- New Zealand: *Resource Management Law 1991*
- Iceland: *Bill of Legislation on Soil Conservation 2002*
- People's Republic of China: *Grassland Law 2002*
- Mongolia: *Pastureland Law* (2015 Draft)
- Kyrgyzstan: *Law of the Kyrgyz Republic on Pastures 2009*
- Afghanistan: *Rangeland Law 2010*
- Tajikistan: *Law of the Republic of Tajikistan About Pastures2013*

In summary, observations of the characteristics of laws used for pastoral land management include the following:[34]

- Various laws have been used to rectify problems caused by poor land management and inappropriate land use.
- The physical land use overshadows many national laws associated with pastoral land.
- The primary functions of pastoral land are not well represented and fail to refer to ecological features or needs of pastoral land.
- Many laws are not clear on the purpose or objectives for pastoral land.

---

[32] See Kurucz (1993), p. 468; See Schlager and Ostrom (1992), pp. 249–251; Hannam and Boer (2002), Section 4, pp. 33–54.

[33] See e.g. *Western Land Act 1901 NSW Australia*; See Hannam (2000), pp. 168–169; Hannam (2007), pp. 8–15.

[34] See Hannam (2007), pp. 14–15.

- There is often no logical development, and laws do not contain enough elements to protect pastoral land.
- There is inconsistent use of terminology and a lack of definition and clarity.
- Some laws are reactionary rather than designed to manage pastoral land effectively.
- Some states have frameworks, but they lack the linking or coordinating mechanism to work fully.

## 12.3.2 Other Forms of Administration of Pastoral Land

The following are areas of environmental law that create a pastoral land management framework.[35]

### 12.3.2.1 Land Administration Law

There are laws in place to control land mismanagement. These include a variety of tenancy and leasehold regimes with provisions to assess land and regulate conditions of occupancy and use. There may also be provisions for forfeiture of holdings, alteration of conditions of use, and protection of land dedicated for public use.[36]

### 12.3.2.2 Biodiversity Law

Laws related to the conservation of biological diversity and ecosystems aid pastoral land through ecological sustainability and biological diversity. While these laws do not pertain to agricultural land, they protect areas in agricultural landscapes.[37]

### 12.3.2.3 Vegetation Conservation Law

This area of law is significant because it focuses on the conservation and sustainable management of vegetation and pastures. These laws promote socially, economically, and environmentally effective management of … with ecological assessments of vegetation and rules for the issue of permits.[38]

### 12.3.2.4 Forest Law

Forest laws oversee public and private land management. Some countries have applied plantation establishment, and reafforestation laws to promote the reafforestation of land in areas affected by overgrazing.[39]

### 12.3.2.5 Environmental Protection Law

These laws aim to protect the environment for human health and from human activity, preventing pollutants from affecting pastoral land.

---

[35] See also Grossman and Brussaard (1992); Flintan (2012), pp. 13–21.

[36] See Chalifour et al. (2007).

[37] See De Klemm and Shine (1993), pp. 1–24.

[38] See e.g., *Native Vegetation Act 2003* New South Wales, Australia.

[39] See Tarasofsky (1999), pp. 5–7.

#### 12.3.2.6 Environmental Planning Law

Environmental planning legislation protects human and natural environments. It also can protect pastoral land by preserving its natural character, outside features, efficient development, and maintenance and enhancement of amenity values.[40]

#### 12.3.2.7 Climate Change Law

A global review of the evidence indicates that additional support from governments, civil society, agencies, donors, and researchers for sustainable pastoral and agro-pastoral systems is needed, given the important role of pastoral lands in managing climate change.[41]

## 12.4 CASE STUDIES

The following two case studies of Mongolia and the People's Republic of China present a unique opportunity to study differences in pastoral land management response given similar institutions and ecology. These constants allow others to identify solutions toward more sustainable governance of pastoral lands.

The Mongolian example discusses a single law approach to manage a range of situations, while the PRC example discusses a comprehensive and integrated law approach to manage complex situations. The two approaches differ due to complexities surrounding pastoral land usage and political situations in the countries. Mongolia created a single law while PRC created a law within a complex of pre-existing environmental laws.

### 12.4.1 Mongolian Pastureland Law (Single Law Approach)

Grasslands and arid-grassland cover 71 percent of Mongolia and play a central role in sequestering carbon dioxide, conserving biodiversity, and supporting the livelihoods of herders and their herds.[42] In the past, the country did not value grasslands and so its regulation was avoided. Herders would compete for the best public grassland leading to unsustainable practice and degraded land.

Over time, pasturelands, or rural agricultural land with vegetation used for grazing livestock, have degraded and undermined the ecosystem services they create.[43] Mongolia passed the Pastureland law to combat this negative trend. The main aim was to create a legal framework to partition public land and land use among herders to avoid overgrazing and competition.[44] Additionally, the law created a framework to exempt herding households from income tax, discounted pasture fees, and a livestock

---

[40] See also Grossman and Brussaard (1992).

[41] See also Neely, Bunning and Wilkes (2009), pp. 25–27.

[42] Around 122 million ha of Mongolia is devoted to nomadic pastoralism: 4.6 percent of this lies in the alpine zone, 22.9 percent in the forest steppe zone, 28 percent in the steppe zone, 23.3 percent in the semi-desert zone, and 16.2 percent in the desert.

[43] See Article 3.1.6 *Land Law 2002*.

[44] See Mongolian Government (2007), pp. 1–2; See also Fernandez-Gimenez (2006), pp. 30–35.

husbandry risk fund. The policy aims to create a better environmental, social, and economic situation for herders.[45]

**Legal Concept**: The Pastureland Law has been through extensive public and governmental scrutiny to ensure a smooth legal implementation of the law's components, distinguish the functions of different levels of administration, and improve identification and management of problems associated with land degradation. In this regard, the new procedures of the Pastureland Law will support economically productive pastoral agriculture while managing pressures on the ecological environment.[46]

Land tenure, the land allocation procedure, will help overcome many problems stemming from the traditional pasture usage system.[47] The procedure allows stakeholder participation and includes identification and classification of pastureland, request for possession for livestock husbandry, and the issuance of the certificate of possession.

Experience shows that engaged stakeholders provide a more satisfactory and balanced outcome, building a stewardship ethic that will maintain the long-term stability of the law.[48] The law contains many essential elements for pastoral law. However, it could benefit from additional support systems such as operational policy, a national strategy for grassland management, the formation of local stakeholder advisory committees, providing access to finance and credit, and comprehensive education and training.[49] All programs mentioned and more would help Mongolia manage its pastoral resources despite increasingly challenging climate situation.[50]

#### 12.4.1.1 Pastoral Land Conservation and Climate Change

In line with the FCCC principle of common but differentiated responsibility and established procedures, Mongolia created a Nationally Appropriate Mitigation Action (NAMA).[51] A NAMA is for grassland and livestock management and is a measurable, reportable, and verifiable action that decreases greenhouse gas emissions.[52] By following FCCC procedures, Mongolia can receive international funding to implement NAMA.[53] One essential requirement for NAMA funding is the implementation of the Pastureland Law.[54]

---

45 See Mongolian Government (2009), pp. 1–2.

46 See also Mongolian Government (2010), pp. 51–52.

47 See also Fernandez-Gimenez (2006), p. 33.

48 See Squires (2012).

49 See Hannam ID (2007), p. 59; Mongolian Government (2009) Government policy towards herders. 4 June 2009, Resolution 39, Ulaanbaatar.

50 See Batima P (2006).

51 Mongolian Government (2010), pp. 51–52; ADB (2011) *Inception report*, ADB R-CDTA 7534, Strengthening carbon financing for regional grassland management in Northeast Asia; ADB (2013a), pp. 1–7; See also ADB (2013a), pp. 23–25.

52 Mongolian Government (2010), pp. 51–52; ADB (2011) *Inception report*, ADB R-CDTA 7534, Strengthening carbon financing for regional grassland management in Northeast Asia; ADB (2013a), pp. 1–7.

53 See KPMG (2011), p. 12.

54 See Hannam (2012), p. 58.

### 12.4.2 Management of Dry Land Ecosystems of People's Republic of China (Integrated Law Approach)

Grassland accounts for 42 percent of the country's land area and is considered multi-functional, with changing ecological-economic benefits from region to region.[55]

A 2003–2009 investigation into the legal, policy, and institutional framework of western regions (where a total of 120 million people live and a significant portion of pastoral lands are located) revealed that these regions struggle with land degradation and desertification, leading people reliant on the land to poverty.[56] The project employed an integrated ecosystem management (IEM) approach to manage land degradation with an emphasis on capacity building and technological support and to strengthen cross-sector coordination and trans-boundary management of the natural resources.[57] The investigation improved procedures for combating land degradation through (1) improving the law and policy framework to intensify institutional capacity, (2) capacity building to implement the law and policy, and (3) a supporting study program to innovate and reform environmental laws and policies. Land degradation threatens the many vulnerable communities within Western PRC that rely on pastoral land resources, but the project successfully employed the IEM approach to managing land degradation.[58]

**IEM as a Legal Concept**: IEM, as defined as "a holistic approach to address the linkages between ecosystem functions and services and human social, economic, and productive systems, has progressively developed into a set of normative principles and rules because it is a suitable framework to consider legislation in place."[59] IEM requires taking all components of an ecosystem into account, which is why the provinces each developed a framework to enact IEM properly.

Initial assessment of existing legal and policy framework requires categorizing legislative material into nine principal law areas:[60]

1. Grassland at a provincial level
2. Desertification
3. Water and soil conservation
4. Water resources
5. Forestry
6. Agriculture
7. Land administration
8. Environmental protection
9. Environmental impact assessment

---

55 See Brown, Waldron and Longworth (2008), p. 2.
56 See Ren et al. (2008); Williams et al. (2009).
57 Du and Hannam (2012), p. 2.
58 See Ren et al. (2008), p. 200; Williams et al. (2009), p. 219; See Du and Hannam (2012), pp. 3–6.
59 See UNEP (1995a); See Jiang Zehui (2007), p. 2; See Du and Hannam (2012), p. 26.
60 See Du and Hannam (2012), p. 8.

The experience of regional and provincial officials from applying the methodological procedures accommodating principles of IEM was substantial and included many lessons for other counties to follow.[61]

- IEM fulfills the commitments to various multinational conventions and establishes a framework for sustainable pastoral land development.
- The IEM approach is a cross-cutting mechanism that accommodates multiple scientific means and coordinates national and international implementation requirements relating to pastoral lands.
- It is an effective means of sustainable pastoral land use, minimization of land degradation, and a sound framework of land ownership issues.
- The flexible framework of IEM allows for implementation and reform options that change regionally.
- IEM is relevant for planning, decision-making, and evaluating ecosystem activities.

Three provinces made a commitment to adopt IEM to improve the legal framework. This provided an opportunity for PRC to introduce IEM into its legal procedures by adopting international best experiences and rules of law. As a result, the PRC has revised old laws, like the *Measures for Implementation of Grassland Law in Qinghai Province*, and introduced new measures, like *Regulations on Administration of Grassland in Tianzhu Tibet Autonomous Country*.

Local pastoral land implementation measures or management regulations specify regulations for the ownership, planning, construction, use, protection, supervision and inspection, and legal liability of grasslands. Some provinces introduced regulations governing forage-livestock balance, which continues to be a long-term task.

#### 12.4.2.1 Improving Legislation for Pastoral Land Management

The following discussions can be used as a tool to improve the capacity of both international and national environmental law frameworks concerning pastoral land.

**International Level**: Internationally, numerous treaties exist to provide for various problems that may arise, but to be more effective and protect such an important and large part of the earth in pastoral land, legally and non-legally binding instruments should be introduced.

A binding instrument could take the form of an overarching treaty, a framework treaty that links pre-existing pastoral land laws, or a protocol creating specific pastoral land rules. A binding treaty taking the form of one of the three global treaties discussed earlier would benefit all pastoral lands of the world.[62] On the other hand, this could take the form of a non-binding treaty, charter, or declaration for pastoral land.

**Process**: An international legal framework open to all interested parties, governmental, scientific, or private is essential for protecting pastoral lands. The experiences involved in the development and introduction of other existing environmental

[61] See Du and Hannam (2012), pp. 138–140.

[62] E.g., a special annex under Article 31 of the CCD, or a protocol under Article 28 of CBD.

treaties should be examined before embarking on a process for pastoral land. In general, for pastoral land, such a process could involve:

- Understanding the current pastoral land management issues to better create a future international legal framework.
- Assembling existing legal documents directly or indirectly referencing pastoral land.
- Reviewing international instruments and strategic material to see what can fit in the current pastoral land law landscape.
- Outlining a capacity-building process that includes general and technical education for people involved.

For this to work, there needs to be cooperation between international pastoral land policy, scientific organizations, and international environmental institutions. Regionally, there may be benefits to following a regional instrument that provides for the region's rangeland needs.

**National Level**: The philosophical basis to establishing or improving national pastoral land legislation is both international legal principles and preference toward the strength of regulation.[63]

**Non-regulatory Strategies**: Non-regulatory strategies for pastoral land conservation would feature elements that concentrate on: education and awareness programs; ecosystem research, assessment, and monitoring; financial support for research and extension; support and development on community pastoral land planning; development of ecologically sustainable standards and practices; and development of pastoral land management and incentive-based programs.

**Regulatory Strategies**: Regulatory strategies for pastoral land conservation would feature elements that concentrate on the development of statutory land-use plans with specific limits and targets and restraining notices when the limits are broken; issuance of licenses and permits; legally binding land-use agreements between the state and land users; and prosecution for failure to follow prescribed standards.

**Elements of National Pastoral Land Legislation**: The following are some of a set of "generic elements" that states should consider and can choose from when developing national pastoral land law frameworks:

- A comprehensive statement of the purpose of the legislation.
- Goals and objectives are drawn from global conventions, treaties, etc., with a mandate for ecologically sustainable pastoral land.
- Preparation of policy, codes, sustainability indicators, and strategy.
- Sustainable use of pastoral land can be achieved through incentives and support programs, as well as community advisory groups.
- Provisions to manage all classes of ecologically sustainable pastoral land.
- Procedures to manage natural resources generally, including provisions that protect soil, water, and biodiversity.
- A balanced distribution of management responsibilities across roles.

[63] See Hannam and Boer (2002), pp. 44–45.

- A facility for pastoral land users and occupiers to enter legal contracts for sustainable resource usage.
- A means for the community to be involved in their pastoral lands and learn best practices.
- A facility to conduct and implement pastoral land research.
- A means to act when prescribed actions and agreements are being honored.

## 12.5 CONCLUSION

Legal frameworks can effectively raise awareness and combat climate change impacts, while also improving livestock productivity and food security. Developed frameworks should include improved documentation and dissemination of information on pastoral land management through financial and non-financial incentives.[64] Using voluntary and regulatory procedures with incentives can change behavior toward adapting sustainable management and reverse land degradation, enhancing the overall livelihoods of people and livestock.

National pastoral land legislation should develop livestock policies, reduce barriers, and include procedures to build local awareness, capacity for animal husbandry, and secure tenure at community and landscape levels. Additionally, legislation should include targeted research in pastoral ecosystems and improved economic aspects.

At the administrative level, it is essential to integrate the multi-sectoral, multistakeholder, and multilevel processes to address a range of natural resources and social dimensions through a holistic approach and partnership.

International and national frameworks must appropriately address climate change and adopt parts of the United Nations FCCC relating to adaption and mitigation actions. States should pay particular attention to creating an international pastoral law instrument and a regional instrument that is more focused on the region's unique aspects of pastoral land.

## REFERENCES

Abildtrup J., Audsley E, Fekete-Farkas M, Guipponi C, Gylling M, Rosato P, Rounsevell M (2006) Socioeconomic scenario development for the assessment of climate change impacts on agricultural land use: a pairwise comparison approach. *Environ Sci Policy* 9:101–115.

ADB (Asian Development Bank) (2011) *Inception report*, ADB R-CDTA 7534, Strengthening carbon financing for regional grassland management in Northeast Asia, p. 105.

ADB (Asian Development Bank) (2013a) *Making Grasslands Sustainable in Mongolia: Adapting to Climate and Environmental Change*. ADB, Mandaluyong City, Philippines, p. 54.

ADB (Asian Development Bank) (2013b) *Making Grasslands Sustainable in Mongolia: International Experiences with Payments for Environmental Services in Grazing Lands and Other Rangelands*. ADB, Mandaluyong City, Philippines, p. 54.

64 See ADB (2013b).

African Convention (African Convention on the Conservation of Nature and Natural Resources) (2003), p. 31, see http://www.int fi sh.net/treaties/africa2003.htm; extensive analysis of the Convention is available at http://uicn.org/themes/law/pdfdocuments/EPLP56EN.pdf, accessed 5 Nov 2014.

Batima P (2006) Climate change vulnerability and adaptation in the livestock sector of Mongolia. Final report project AS06. Assessments of impact and adaptation to climate change. International START Secretariat, Washington, DC, p. 409.

Brown C, Waldron S, Longworth J (2008) *Sustainable Development in Western China: Managing People, Livestock and Grasslands in Pastoral Areas.* Edward Elgar, Cheltenham, p. 305.

Chalifour N, Kameri-Mbote J, Lin Heng Lye P, Nolon JR (eds) (2007) *Land Use Law for Sustainable Development.* IUCN Academy of Environmental Law Research Studies, Cambridge University Press, Cambridge, UK, p. 654.

De Klemm C, in collaboration with Shine C (1993) *Biological Diversity Conservation and the Law: Legal Mechanisms for Conserving Species and Ecosystems.* IUCN Environmental Law Centre and IUCN Biodiversity Programme, p. 292.

Dong S, Lu W, Liu S, Zhang XL, James P, Yi S, Li X, Li J, Li Y (2011) Vulnerability of worldwide pastoralism to global changes and interdisciplinary strategies for sustainable pastoralism. *Ecol Soc* 16(2):10.

Du Q, Hannam I (eds) (2012) *Law, Policy and Dryland Ecosystems: People's Republic of China.* IUCN, Gland, xvi + p. 140.

European Commission (2000) Analysis of national reports on the implementation of the United Nations convention to combat desertification. Turkey, Lebanon, Jordan and Syria, Drylands Program, International Institute for Environment and Development, p. 15.

Fernandez-Gimenez ME (2006) Land use and land tenure in Mongolia: a brief history and current issues. *USDA Forest Service Proceedings RMRS-P-39*, pp. 30–36.

Flintan F (2012) Making rangelands secure: Past experience and future options, Working Document Version 1.0, International Land Coalition, p. 87.

Food and Agriculture Organization of the United Nations (FAO) (2001) Pastoralism in the new millennium. Animal production and health paper No. 150, UN Food and Agriculture Organization, Rome, Italy, p. 93.

Friedel MH, Laycock WA, Bastin GN (2000) Assessing rangeland condition and trend. In: Mannetje L't, Jones RM (eds) *Field and Laboratory Methods for Grassland and Animal Production Research.* CABI, Wallingford, pp. 227–262.

Grossman M, Brussaard W (eds) (1992) *Agrarian Land Law in the Western World, Essays about Agrarian Land Policy and Regulation in Twelve Countries of the Western World.* CAB International, Wallingford, p. 280.

Halimova N (2012) Land tenure reform and implications for land stewardship. In: Squires V (ed) *Rangeland Stewardship in Central Asia.* Springer, Dordrecht, pp. 305–332.

Hannam ID (2000) Policy and law for rangeland conservation. In Arnalds, O and S Archer (eds) *Rangeland Desertification.* Kluwer Academic Publishers, London, p. 209.

Hannam ID (2007) Report to United Nations development programme Mongolia on review of draft pastureland law of Mongolia. United Nations Development Programme Sustainable Grassland Management Project, Ulaanbaatar, p. 112.

Hannam ID (2012) Legal and policy framework to support a livestock/grassland NAMA in Mongolia, Working Paper, ADB R-CDTA 7534 – Strengthening Carbon Financing for Regional Grassland Management in Northeast Asia, p. 89.

Hannam ID, Boer BW (2002) Legal and institutional frameworks for sustainable soils: a preliminary report. IUCN, Gland/Cambridge, UK, xvi + p. 88.

IUCN (The World Conservation Union) (2004) *Draft International Covenant on Environment and Development*, Third Edition: Update Text, world commission on environmental law in cooperation with international council of environmental law, IUCN Environmental Law Program, Bonn, IUCN, p. 193.

Jiang Z (2007) To implement integrated ecosystem management to accelerate combating land degradation. In: *Integrated Ecosystem Management, Proceedings of the International Workshop*, Beijing 1–2 November 2004, Global Environment Facility, Asian Development Bank and People's Republic of China Forestry Publishing House, pp. 2–6.

KPMG (2011) *Financing Low Carbon Investment in Developing Countries: Public-Private Partnership for Implementing Nationally Appropriate Mitigation Actions.* KPMG International Cooperative, Switzerland, p. 14.

Kurucz M (1993) Land protection, property rights and environmental preferences (land use control and land development). *Conn J Int Law* 8(2):467–485.

Kyoto (1997) The Kyoto protocol to the convention on climate change, the Climate Change Secretariat, UNEP, p. 20.

Miller KR, Lanou SM (1995) *National Biodiversity Planning. Guidelines Based on Early Experiences around the World.* The World Resources Institute and IUCN, Baltimore, p. 162.

Mongolian Government (2007) Draft pastureland law 20 July 2007 and brief introduction to the draft law on pastureland, p. 26.

Mongolian Government (2009) Government policy towards herders. 4 June 2009, Resolution 39, Ulaanbaatar, p. 21.

Mongolian Government (2010) Mongolian second national communication under the United Nations framework convention on climate change. Ministry of Nature, Environment and Tourism, p. 159.

Neely C, Bunning S, Wilkes A (eds) (2009) *Review for Evidence on Dryland Pastoral Systems and Climate Change, Implications and Opportunities for Mitigation and Adaptation.* Food and Agriculture Organization of the United Nations, Rome, p. 37.

Nori M, Davies J (2007) *Change of Wind or Wind of Change? Climate Change, Adaptation and Pastoralism.* The World Initiative for Sustainable Pastoralism, International Union for Conservation of Nature, Nairobi, Kenya, p. 23. http://cmsdata.iucn.org/downloads/c__documents_and_settings_hps_local_settings_application_data_mozilla_firefox_profile.pdf, accessed 10 Nov 2014.

Prip C, Gross T, Johnston S, Vierros M (2010) *Biodiversity Planning: An Assessment of National Biodiversity Strategies and Action Plans.* Institute for the Advanced Study of Sustainability, Yokohama, Japan, p 237

Ren JZ, Hu ZZ, Zhao J, Zhang DG, Hou FJ, Lin HL, Mu XD (2008) A grassland classification system and its application in China. *Rangeland J* 30(2):199–210.

Schlager E, E, Ostrom (1992) Property-rights regimes and natural resources: A conceptual analysis. *Land Econ*, 68(3):249–262.

Secretariat CCD (Secretariat of the Convention to Combat Desertification) (2000) Chart of key elements in the CCD national reports by the Countries in Asia.

Secretariat CCD (Secretariat of the Convention to Combat Desertification) (2007) Report, 8th session, *Conference of the Parties*, Madrid 3–14 September 2007, ICCD/COP (8)/16/Add.1, 23 October 2007.

Squires VR (1988) Dryland soils: their potential as a sink for carbon and as an agent in mitigating climate change. *Adv GeoEcol* 31:209–215.

Squires VR (2012) Better land stewardship: an economic and environmental imperative. In: Squires V (ed) *Rangeland Stewardship in Central Asia*. Springer, Dordrecht, pp. 31–50.

Squires VR, Glenn E (1997) Carbon sequestration in the drylands: an agenda for the twenty-first century. *World atlas of Desertification*, 2nd edn. Edward Arnold/UNEP, pp. 140–143.

Tarasofsky RG (ed) (1999) Assessing the international forest regime. IUCN Environmental Policy and Law Paper No. 37, p. 156.

UN (United Nations) (1992a) Rio Declaration, New York.

UN (United Nations) (1992b) Agenda 21, New York.
UN (United Nations) (1995) Convention to combat desertification, New York.
UN (2012) The future we want: Outcome document adopted at Rio+20, General Assembly 11 December 2012 A/RES/66/288.
UN (2014) Sustainable Development Knowledge Platform: Sustainable Development Goals, available at http://sustainabledevelopment.un.org/?menu=1300, access to 18 Aug 2014.
UNEP (United Nations Environment Program) (1995a) Convention on biological diversity, Nairobi.
UNEP (United Nations Environment Program) (1995b) United Nations framework convention on climate change.
White RP, Tunstall D, Henninger N (2002) An ecosystem approach to drylands: building support for new development policies, World Resources Institute Washington, DC Information Policy Brief No. 1, p. 14.
Wilkes A, Wang S, Tennigkeit T, Feng J (2011) Agricultural monitoring and evaluation systems: what can we learn for the MRV of agricultural NAMAs? ICRAF Working Paper No. 126. World Agroforestry Centre, Beijing, China, p. 17.
Williams A, Wang MP, Zhang M (2009) Land tenure arrangements, property rights and institutional arrangements in the cycle of rangeland degradation and recovery. In: Squires V, Lu X, Lu Q, Wang T, Yang Y (eds) *Rangeland Degradation and Recovery in China's Pastoral Lands*. CABI Publishing, Oxford, pp. 219–234.

# Part 4

## Global Legislation on Animal Welfare Overview

### Tools for Change

Morgan Boutilier

# APPENDIX 1: ZOONOTIC DISEASES AND FOOD SAFETY

| Region | Country/State | Citation | Title | Year Passed | Summary or Subsections Applicable to Regulatory Issue |
|---|---|---|---|---|---|
| Africa | Kenya | Act No. 38 of 1921 | Public Health Act | 1921 | Governs public health generally; aims to promote, safeguard, maintain, and protect the health of both humans and animals; regulates slaughterhouse conditions and other food standards for meat products<br>• Art. 54 prohibits "noxious or offensive practices that may cause damage to the lands, crops, cattle, or goods of the public or water pollution." |
| Africa | Kenya | Act No. 4 of 1965 | Animal Diseases Act | 1965 | Governs animal health and diseases; contains provisions on:<br>• notification and isolation obligations;<br>• appointment of veterinary inspectors;<br>• import and export restrictions and prohibitions; and<br>• penalties.<br>Significant new amendments and regulations issued in 2012, including:<br>• Animal Diseases Rules, regulating importation (including prescribed ports, licenses, certificates, tests, treatment, and quarantine) and restricting movement within country of animals;<br>• Animal Diseases (Birds) Rules, regulating the importation of certain birds and eggs;<br>• Diseases (Compulsory Foot and Mouth Vaccination) Rules;<br>• Animal Diseases (Control of Pig Diseases) Rules;<br>• Animal Diseases (Hatcheries) Rules;<br>• Animal Diseases (Compulsory Rinderpest Vaccination) Rules;<br>• Animal Diseases (Foot-and-Mouth Disease) Rules, ordering veterinary officers or inspectors suspecting FMD on a premises to immediately take all quarantine measures;<br>• Animal Diseases Prohibition, banning imports of live sheep and goats from countries infected with scrapie, caprine arthritis/ encephalitis and BSE as well as imports of live poultry, birds, and other products from any country reporting notifiable fowl plague or avian influenza |

| Region | Country/State | Citation | Title | Year Passed | Summary or Subsections Applicable to Regulatory Issue |
|---|---|---|---|---|---|
| Africa | Kenya | Act No. 8 of 1965 | Food, Drugs and Chemical Substances Act (Cap. 254) | 1965 | Governs food safety generally; sets out rules for marketing food and drugs; establishes Public Health (Standards) Board |
| Africa | Kenya | Act No. 7 of 1972 | Meat Control Act (Cap. 356) | 1972 | Regulates licensing, control, and regulation of slaughterhouses and meat-processing facilities; sets health, sanitary, and hygiene standards in slaughterhouses and meat processing premises; sets additional standards on other matters relative to manufacturing, processing, and marketing meat (including packing, labeling, storing, transporting, and licensing imports and exports) |
| Africa | Kenya | Act No. 17 of 2013; *Kenya Gazette* Supplement No. 29 of 25 Jan 2013 (Special Issue) | Kenya Agricultural and Livestock Research Act | 2013 | Provides administrative framework for agricultural research; provides for promoting and coordinating agricultural research activities; establishes Kenya Agricultural and Livestock Research Organization (as well as other research institutes) and a Scientific and Technical Committee |
| Africa | Nigeria | Consolidated text of Animals Diseases (Control) Decree (No. 10 of 1988) | Animal Diseases (Control) Act | 1988 | Governs animal diseases; aims to prevent the introduction and spread of infectious and contagious diseases among animals, hatcheries, and poultries; contains provisions on:<br>• exports and imports (including examination, disinfection, inoculation, and quarantine);<br>• disposal of diseased animals;<br>• compensation for owners;<br>• control of trade animals;<br>• control of hatcheries and poultry farms; and<br>• offenses and penalties |

*(Continued)*

| Region | Country/State | Citation | Title | Year Passed | Summary or Subsections Applicable to Regulatory Issue |
|---|---|---|---|---|---|
| Africa | Nigeria | Cap. N1 Laws of the Federal Republic of Nigeria, 2004 | National Agency for Food and Drugs Administration and Control (NAFDAC) Act (formerly the National Agency for Food and Drugs Administration and Control (NAFDAC) Decree No. 15 of 1993) | 1999 | Governs food safety generally |
| Africa | Nigeria | Cap. F32 Laws of the Federal Republic of Nigeria, 2004 | Food and Drugs Act (formerly the Food and Drugs Act (No. 35 of 1974)) | 2004 | Governs food safety generally |
| Africa | Nigeria | Cap. F33 Laws of the Federal Republic of Nigeria, 2004 | Food, Drugs & Related Products (Registration etc.) Act (formerly the Food, Drug and Related Products (Registration, etc.) Decree No. 19 of 1993) | 2004 | Governs food safety generally |
| Africa | Nigeria | Act No. 8 of 2014 | National Health Act | 2014 | Establishes a National Health System and the National Council on Health to facilitate and promote the provision of health services for the management, prevention and control of communicable and non-communicable diseases; requires the Federal Ministry of Health to promote the availability of hygienic food |
| Africa | South Africa | Act No. 54 of 1972 | Foodstuffs, Cosmetics and Disinfectants Act (FCD Act) | 1972 | Governs the sale, manufacture, and import (including labeling and advertising) of foodstuffs, cosmetics, and disinfectants |

(*Continued*)

| Region | Country/State | Citation | Title | Year Passed | Summary or Subsections Applicable to Regulatory Issue |
|---|---|---|---|---|---|
| Africa | South Africa | Act No. 7 of 2002 | Animal Health Act | 2002 | Governs animal health and diseases; contains provisions on:<br>• establishing animal health schemes;<br>• veterinary inspections;<br>• import and export restrictions (including quarantine stations and camps);<br>• destruction and disposal of animals known or suspected to be infected;<br>• fencing to prevent straying animals (such as near international borders or around reserves);<br>• disposal of straying animals;<br>• compensating owners of destroyed or disposed animals; and<br>• offenses and penalties<br>Implemented by various regulations, including:<br>• 2004 Control Measures Relating to Avian Influenza in Certain Areas (prohibiting export of poultry or poultry products from specified areas without permit) |
| Africa | South Africa | Act No. 61 of 2003 | National Health Act | 2003 | Governs public health generally |
| Africa | South Africa | No. R. 364 of 2012 | Publication of notice on the importation of live cloven-hoofed animals from countries or zones not recognized as free from foot and mouth disease by the World Organization for Animal Health | 2012 | Bans import of live cloven-hoofed animals from countries or zones not recognized an FMD-free by the OIE |

*(Continued)*

| Region | Country/State | Citation | Title | Year Passed | Summary or Subsections Applicable to Regulatory Issue |
|---|---|---|---|---|---|
| Americas | Brazil | Law No. 24645 of 1934/Decreto No. 24.645, de 10 de julho de 1934 | Law No. 24645/1934 on Animal welfare/ *Decreto 24.645/1934 estabelece medidas de proteça-o aos animais* | 1934 | Governs animal welfare generally |
| Americas | Brazil | Act No. 5027 of 1966/Lei No. 5.027 de 1996; *Coletâ nea da Legislaça-o Federal de Meio Ambiente*, Bras´ılia 1992, pp. 132–144 | Act No. 5027 establishing the Sanitary Code/*Lei No. 5.027 estabelece o Código Sanitário* | 1966 | Establishes Sanitary Code to regulate environmental and food quality standards; covers all matters related to human health protection, including food sanitation and quality requirements; lays out veterinary procedures for detecting animal diseases |
| Americas | Brazil | Law No. 8171 of 1991/Decreto No. 8.171 de 1991; *Legislaça-o Federal de Meio Ambiente*, Vol. I, Bras´ılia 1996, pp. 81–91 | Law No. 8171 on Agricultural policy/ *Decreto No. 8.171 sobre política agrícola* | 1991 | Establishes Brazil's basic framework for animal health protection; specifies sanitary controls to protect human health from animal diseases as well as sanitary measures for animal feed<br>Implemented by various acts and regulations, including:<br>• Decree No. 5.741 establishing a Unified System for Farming and Cattle Health (30/03/2006);<br>• Law No. 13.496 creating the Plant and Livestock Protection System and related Agency (ADAGRI) (02/07/2004); and<br>• Law No. 12.097 on the Marking Procedure for Bovine and Buffalo Meat Certification (24/11/2009) |
| Americas | Brazil | Instruction No. 3/2000; Instruç-ao Normativa 3/2000 | Instruction No. 3/2000—Animal welfare at slaughtering/*Instruça-o Normativa* 3/2000 *Abate humanitário e o bem-estar animal em bovinos* | 2000 | Regulates slaughterhouse conditions |

(*Continued*)

| Region | Country/State | Citation | Title | Year Passed | Summary or Subsections Applicable to Regulatory Issue |
|---|---|---|---|---|---|
| Americas | Brazil | Instruction No. 1/2002/Instruç-ao Normativa 1/2002 | Instruction No. 1/2002: establishing the Brazilian System of Identification and Certification of Bovine and Bubaline Products (SISBOV)/*Instruça-o Normativa 1/2002 estabelece (Sistema Brasileiro de Identificaça-o e Certificaça-o de Origem Bovina e Bubalina (SISBOV)* | 2002 | Establishes SISBOV as Brazil's national animal identification system |
| Americas | Brazil | Instruction No. 6/2004/Instruç-ao Normativa 6/2004 de 08 de Janeiro de 2004 | Instruction No. 6/2004 – National Program for the control and eradication of animal brucellosis and tuberculosis/*Instruça-o Normativa 6/2004 Programa Nacional de Controle e Erradicaça-o da Brucelose e Tuberculose Anima* | 2004 | Approves technical regulations for the National Program for the Control and Eradication of Animal Brucellosis and Tuberculosis |

(*Continued*)

| Region | Country/State | Citation | Title | Year Passed | Summary or Subsections Applicable to Regulatory Issue |
|---|---|---|---|---|---|
| Americas | Brazil | Circular No. 175/2005/CGPE/DIPOA, Circular No. 176/2005 CGPE/DIPOA | Circulars Nos. 175 and 176/2005 on Instructions for fiscalization and auditing in self controlled industries/*Circular 175 e 176/2005 sobre procedimentos de Verificaça-o dos Programas de Autocontrole* | 2005 | Serve as operative guides for HACCP inspectors in all sectors of the food industry; incorporate and reference certain European and American USDA regulations |
| Americas | Brazil | D.O.U. de 31/03/2006, p. 82 | Law No. 5741/2006 on Unified System of Attention to Agricultural and Livestock Sanity (SUASA): HACCP and Risk Analysis/*Decreto 5.741/2006 organiza o Sistema Unificado de Atença-o á Sanidade Agropecuária, e dá outras providências (SUASA)* | 2006 | Lays out measures for Brazil's approach to HACCP plans in food industries; calls for developing risk analysis programs for food production systems |
| Americas | Brazil | Instruction No. 44/2007; Instruç-ao Normativa 44/2007; Publicado no *Dia'rio Oficial da Unia-*o No. 191, – quarta-feira, 3 de outubro de 2007, seç-ao 1, pa´g 2 a 10 | Instruction No. 44/2007—Foot and Mouth Disease plan/*Instruça-o Normativa* 44/2007 *diretrizes gerais para a Erradicaça-o e a Prevença-o da Febre Aftosa* | 2007 | Sets out measures for controlling FMD |

(*Continued*)

| Region | Country/State | Citation | Title | Year Passed | Summary or Subsections Applicable to Regulatory Issue |
|---|---|---|---|---|---|
| Americas | Brazil | Instruction No. 49/2008/Instruç-ao Normativa 49/2008 | Instruction No. 49/2008—BSE risk categories/ *Instruça-o Normativa 49/2008 Regulamenta o procedimento para identificaça-o, reconhecimento, delimitaça-o, demarcaça-o, desintrusa-o, titulaça-o e registro das terras ocupadas por remanescentes das comunidades dos quilombos de que tratam o Art. 68 do Ato das Disposiço˜es Constitucionais Transitórias da Constituiça-o Federal de 1988 e o Decreto n$^0$ 4.887, de 20 de novembro de 2003* | 2008 | Sets out measures for controlling FMD; defines BSE risk categories |

(*Continued*)

| Region | Country/State | Citation | Title | Year Passed | Summary or Subsections Applicable to Regulatory Issue |
|---|---|---|---|---|---|
| Americas | Brazil | Complementary Law No. 197/Lei Complementar No. 197 | Complementary Law No. 197 on the Code of Sanitary Surveillance, Surveillance on Zoonoses, Agricultural and Livestock Inspection of the Municipality of Rio de Janeiro and adding provisions to Title V of Book 1 of Law No. 691 of 24 December 1984 – Municipal Tax Code./ *Dispõe sobre o Código de Vigilância Sanitária, Vigilância de Zoonoses e de Inspeção Agropecuária do Município do Rio de Janeiro e acrescenta dispositivos ao Título V do Livro Primeiro da Lei nº 691, de 24 de dezembro de 1984 – Código Tributário Municipal.* | 2018 | Establishes the Code of Sanitary Surveillance, Surveillance on Zoonoses, Agricultural and Livestock Inspection of the Municipality of Rio de Janeiro; requires the Municipality to implement the code and adopt measures that prevent risks and diseases arising from the production and circulation of goods and services. |
| Americas | Canada | R.S.C., 1985, c. F-12 | Fish Inspection Act | 1985 | Governs inspections of fish and marine plants; addresses export or import of fish and containers<br>Implemented by Fish Inspection Regulations (C.R.C., c. 802) |
| Americas | Canada | R.S.C., 1985, c. F-27 | Food and Drugs Act | 1985 | Governs food safety generally Implemented by Food and Drugs Regulations (C.R.C., c. 870) |

(*Continued*)

| Region | Country/State | Citation | Title | Year Passed | Summary or Subsections Applicable to Regulatory Issue |
|---|---|---|---|---|---|
| Americas | Canada | R.S.C., 1985, c. 25 (1st Supp.) | Meat Inspection Act | 1985 | Contains provisions on:<br>• import, export, and interprovincial trade in meat products;<br>• registration of establishments;<br>• inspection of animals and meat products in registered establishments; and<br>• standards for those establishments and for animals slaughtered and meat products prepared in those establishments<br>Implemented by Meat Inspection Regulations (SOR/90-288) |
| Americas | Canada | S.C. 1990, c. 21 | Health of Animals Act | 1990 | Governs animal diseases and health; contains provisions on:<br>• notification requirements;<br>• prohibitions on keeping diseased animals and bringing diseased animals to market;<br>• import restrictions and prohibitions (although the import, export, and interprovincial trade in meat and marine products are covered by the Meat Inspection Act and Fish Inspection Act, respectively);<br>• exportation requirements;<br>• infected place and control zone declarations;<br>• inspectors and officers, including analysts, inspectors, and veterinary inspectors;<br>• inspections;<br>• search and seizure powers, including warrant requirements;<br>• disposition, disposal, and treatment of animals and seized materials;<br>• compensation; and<br>• offenses and punishments |
| Americas | Mexico | DOF 04-06-2014, publicada en el *Diario Oficial de la Federación* el 7 de febrero de 1984 | General Health Law/*Ley General de Salud* | 1984 | Governs public health generally |

(*Continued*)

| Region | Country/State | Citation | Title | Year Passed | Summary or Subsections Applicable to Regulatory Issue |
|---|---|---|---|---|---|
| Americas | Mexico | DOF 09-04-2012, publicada en el *Diario Oficial de la Federación* el 1$^{0}$ de julio de 1992 | Federal Law of Metrology and Normalization/*Ley Federal sobre Metrología y Normalización* | 1992 | Provides basic legal framework for Mexico's regulatory process and calls for rules to protect human, animal and plant, and environmental health; provides for two types of food safety standards:<br>1. Official Mexican Standards (NOMs), which are mandatory regulations; and<br>2. Mexican Standards (NMXs), which are voluntary standards |
| Americas | Mexico | NOM-008-ZOO-1994 | NOM-008-ZOO-1994: slaughterhouse construction/ *Especificaciones zoosanitarias para la construcción y equipamiento de establecimientos para el sacrificio de animales y los dedicados a la industrialización de productos cárnicos* | 1994 | Regulates slaughterhouse construction |
| Americas | Mexico | NOM-009-ZOO-1994 | NOM-009-ZOO-1994: hygienic handling of meat/ *Proceso sanitario de la carne* | 1994 | Regulates hygienic handling of meat |

(*Continued*)

| Region | Country/State | Citation | Title | Year Passed | Summary or Subsections Applicable to Regulatory Issue |
|---|---|---|---|---|---|
| Americas | Mexico | NOM-128-SSA1-1994 | NOM-128-SSA1-1994: hazard analysis and HACCP in seafood processing/*Que establece la aplicación de un sistema de análisis de riesgos y control de puntos críticos en la planta industrial procesadora de productos de la pesca* | 1994 | Regulates hazard analysis and HACCP in seafood processing |
| Americas | Mexico | NMX-F-605-NORMEX-2004 | NMX-F-605-2004: "H-seal program" | 2004 | Establishes voluntary "H-seal program," which aims to reduce and prevent foodborne illness in all food-service operations |
| Americas | Mexico | NOM-194-SSA1-2004 | NOM-194-SSA1-2004: hygiene standards for slaughterhouses and preparation, warehousing, transportation, and sale of meats/*Especificaciones sanitarias en los establecimientos dedicados al sacrificio y faenado de animales para abasto, almacenamiento, transporte y expendio* | 2004 | Addresses hygiene standards for municipal slaughterhouses as well as the preparation, warehousing, transport, and sale of meat products (including vertical standards for food associated with international trade) |

(*Continued*)

| Region | Country/State | Citation | Title | Year Passed | Summary or Subsections Applicable to Regulatory Issue |
|---|---|---|---|---|---|
| Americas | Mexico | DOF 07-06-2012, publicada en el *Diario Oficial de la Federación* el 25 de julio de 2007 | Federal Law for Animal Health/*Ley Federal de Sanidad Animal* | 2007 | Governs animal health; contains provisions on:<br>• animal welfare;<br>• animal feed and pharmaceuticals;<br>• best practices in livestock production, slaughter, and processing;<br>• harmonization with international standards;<br>• import requirements (including inspection and quarantine requirements based on animal and plant health risks)<br>Reformed in 2012 to implement HACCP in federal slaughterhouses and meat-processing facilities |
| Americas | Mexico | NOM-251-SSA1-2009 | NOM-251-SSA1-2009: food re-call procedures, traceability, and food processing HACCP/*Prácticas de higiene para el proceso de alimentos, bebidas o suplementos alimenticios* | 2009 | Sets out:<br>• food recall and traceability procedures; and<br>• recommendations for implementing HACCP in all food processing operations<br>Based on *Codex Alimentarius*; does not include microbiological specifications<br>Set to be revised in 2016 to enhance traceability control and possibly make HACCP mandatory |
| Americas | United States | 21 U.S.C. §§ 601-695 | Federal Meat Inspection Act (FMIA) | 1906 | Established inspection standards for all meat processing plants conducting business across state lines |
| Americas | United States | 21 U.S.C. § 301 et seq. | Federal Food, Drug, and Cosmetic Act (FFDCA) | 1938 | Governs food safety generally; authorizes federal Food and Drug Administration (FDA) to inspect all livestock and poultry species not listed in Federal Meat Inspection Act or Poultry Products Inspection Act |
| Americas | United States | 21 U.S.C. § 451 et seq., P.L. 85-172 | Poultry Products Inspection Act (PPIA) | 1957 | Requires USDA FSIS to inspect slaughtering and processing facilities for domesticated birds (defined to include chickens, turkeys, ducks, geese, guinea fowl, and ratites) |

(*Continued*)

| Region | Country/State | Citation | Title | Year Passed | Summary or Subsections Applicable to Regulatory Issue |
|---|---|---|---|---|---|
| Americas | United States | 21 U.S.C. § 601 et seq.;, P.L. 90-201 | Wholesome Meat Act ("Equal to" Law) | 1967 | Amends the 1906 Federal Meat Inspection Act; regulates federal meat inspection; requires states to have programs "equal to" the federal program |
| Americas | United States | P.L. 90-492 | Wholesome Poultry Products Act | 1968 | Amends the 1957 Poultry Products Inspection Act; requires states to have programs "equal to" federal program |
| Americas | United States | FDA Food Code 2013, Report number PB2013-110462 [Online] | FDA Food Code | 1993 | Model code providing scientifically-based technical and legal basis for regulating retail/food service sectors |
| Americas | United States | 7 U.S.C. §§ 8301-8322 | Animal Health Protection Act (AHPA) | 2002 | Governs animal health and disease prevention; creates Pest and Disease Response Fund; contains provisions on:<br>• import and export restrictions;<br>• interstate movement;<br>• seizure, quarantine, and disposal;<br>• inspections, seizures, and warrants;<br>• detection, control, and eradication;<br>• veterinary accreditation and training;<br>• cooperation with other federal, state, local, and international bodies; and<br>• criminal and civil penalties |
| Americas | United States | 6 U.S.C. § 101 et seq., P.L. 107-296, 116 Stat. 2135 | Homeland Security Act | 2002 | Created the Department of Homeland Security (DHS); Sec. 310 transfers Plum Island Animal Disease Center (federal research facility on animal diseases) from Department of Agriculture (USDA) to DHS |
| Americas | United States | 42 U.S.C. 262(a);, P.L. 107-188 | Public Health Security and Bioterrorism Preparedness and Response Act (Bioterror ism Act) | 2002 | Empowers Centers for Disease Control and Prevention (CDC) and Department of Health and Human Services (DHHS) to assist Department of Agriculture (USDA) with coordinating zoonotic disease surveillance |

*(Continued)*

| Region | Country/State | Citation | Title | Year Passed | Summary or Subsections Applicable to Regulatory Issue |
|---|---|---|---|---|---|
| Americas | United States | 9 C.F.R. §§ 49-56 | Cooperative Control and Eradication of Livestock or Poultry Diseases Regulations | 2006 | Set up cooperative program between states and the USDA to control and eradicate listed diseases; require prompt destruction (usually by burial or burning) of animals and materials affected by or exposed to disease, as well as the disinfection of all affected premises, conveyances, and materials |
| Americas | United States | 9 C.F.R. §§ 60, 65 | Country of Origin Labelling (COOL) Regulations | 2009 | Require retailers to notify customers with information about the source of certain foods |
| Americas | United States | 21 U.S.C. § 2201 et seq., P.L. 111-353 | Food Safety Modernization Act (FSMA) | 2011 | Covers food safety generally; strengthens the FDA's focus on disease prevention; contains provisions on vaccines and animal trade measures; Sect. 109 calls on FDA to work with the USDA, DHHS, and DHS to come up with proposals on how to best prevent animal or plant disease outbreaks and food contamination |
| Americas | United States | 9 C.F.R. §§ 91-99 | Exportation and Importation of Animals (Including Poultry) and Animal Products Regulations | 2012 | Set forth a number of import and export restrictions, as well as inspection and handling protocols, for animal products |
| Americas | United States | 9 C.F.R. § 86 | Animal Disease Traceability Regulations | 2013 | Set up a national identification and traceability system for livestock transported across state borders; allow states and tribes to develop their own preferred tracking systems for administering the program within their borders |
| Asia | China | Adopted at the 26th Meeting of the Standing Committee of the Eighth National People's Congress on July 3, 1997 | Law of Animal Disease of the People's Republic of China (Animal Disease Law) | 1997 | Governs animal diseases; aims to:<br>• prevent, control, and eliminate animal diseases;<br>• developing a breeding industry; and<br>• safeguarding human health<br>Contains provisions on:<br>• Production;<br>• trade restrictions;<br>• control and ceasing of disease outbreaks;<br>• quarantine measures; and<br>• inspection procedures |

(Continued)

| Region | Country/State | Citation | Title | Year Passed | Summary or Subsections Applicable to Regulatory Issue |
|---|---|---|---|---|---|
| Asia | China | Adopted at the 26th Meeting of the Standing Committee of the Eighth National People's Congress of Japan August 30, 2007 | Animal Epidemic Prevention Law of the People's Republic of China (2015). | 2008 | Aims to:<br>• Strengthen animal epidemic prevention;<br>• promote the development of the breeding industry.<br>• and protect human health.<br>The Veterinary Administrative Department is responsible for conducting risk assessments and carrying out the laws provisions, which cover:<br>• quarantine of animals and animal products;<br>• animal diagnosis and treatment;<br>• animal epidemic prevention supervision and administration;<br>• safeguard measures;<br>• legal liabilities;<br>Amended in 2015 to add new measures including compulsory vaccinations; and stipulations for veterinarians to pass qualification tests. |
| Asia | China | Adopted at the Seventh Session of the Standing Committee of the 11th National People's Congress of the People's Republic of China on February 28, 2009 | Law of Food Safety of the People's Republic of China (Food Safety Law) | 2009 | Governs food safety generally; contains provisions covering entire food chain (from production to advertising); establishes Food Safety Commission and a food risk monitoring system; lays out rules and standards on:<br>• pesticides;<br>• fertilizers;<br>• feeding;<br>• breeding programs;<br>• hygienic requirements;<br>• HACCP requirements;<br>• food inspections;<br>• import and export requirements;<br>• food safety accident response; and<br>• legal liabilities |

*(Continued)*

| Region | Country/State | Citation | Title | Year Passed | Summary or Subsections Applicable to Regulatory Issue |
|---|---|---|---|---|---|
| Asia | China | Adopted at the executive meeting of the State Administration of Quality Supervision, Inspection and Quarantine on March 10, 2010 (Order No. 136) | Measures for the Supervision and Administration of the Inspection and Quarantine of Imported and Exported Meat Products | 2010 | Seek to strengthen China's meat product inspection and quarantine procedures; establish State Administration of Quality Supervision, Inspection, and Quarantine, which is charged with:<br>• preventing epidemic animal diseases from entering or exiting the country; and<br>• ensuring the safety of agricultural production and animal husbandry practices with respect to both animal and human health |
| Asia | India | Act No. 23 of 1940 | Drugs and Cosmetics Act | 1940 | Does not govern food safety |
| Asia | India | Act No. 37 of 1954 | Prevention of Food Adulteration Act | 1954 | Aims to prevent food adulteration and protect consumer health; provides for a Central Committee for Food Standards and a Central Food Laboratory for testing food samples |
| Asia | India | Act No. 34 of 2006 | Food Safety and Standards Act | 2006 | Governs food safety generally; establishes Food Safety and Standards Authority<br>Implemented by 2011 Food Safety and Standards Regulations |
| Asia | India | Act No. 59 of 1960 | The Prevention of Cruelty to Animals Act | 1960 | Prevents unnecessary pain and suffering, and prohibits cruel transport and confinement of animals. |

*(Continued)*

| Region | Country/State | Citation | Title | Year Passed | Summary or Subsections Applicable to Regulatory Issue |
|---|---|---|---|---|---|
| Asia | India | Act No. 27 of 2009 | Prevention and Control of Infectious and Contagious Diseases in Animals Act | 2009 | Governs animal diseases; contains provisions on:<br>• veterinary officer appointments;<br>• disease reporting;<br>• segregating infected animals;<br>• notifying on controlled and disease-free areas;<br>• vaccinations;<br>• prohibitions on animal movement;<br>• precautionary measures for controlled areas;<br>• check posts and quarantine camps;<br>• declarations of infected areas;<br>• euthanasia;<br>• postmortem examinations;<br>• seizures and removals;<br>• escape prevention; and<br>• enforcement<br>Implemented by 2010 Prevention & Control of Infectious & Contagious Diseases in Animal Rules |
| Asia | Israel | 1940, P.O. 1065, Suppl. I, 239 | Public Health Ordinance | 1940 | Governs public health generally |
| Asia | Israel | Codex 92, 5712, p. 128;<br>FAOLEX No: LEX-FAOC028210 | Improvement of Agricultural Production (Livestock) Law | 1952 | Aims to improve livestock offspring yield and increase disease resistance; Sect. 4 addresses inspections<br>Implemented by 2001 Improvement of Agricultural Production (Livestock) (Dairy Cows) Regulations |

(*Continued*)

| Region | Country/State | Citation | Title | Year Passed | Summary or Subsections Applicable to Regulatory Issue |
|---|---|---|---|---|---|
| Asia | Israel | *Dinim* Vol. 15 p. 8924 | Animal Diseases Ordinance | 1985 | Governs animal diseases; contains provisions on:<br>• disease prevention measures;<br>• notification requirements;<br>• destruction of infected animals;<br>• spread prevention;<br>• sanitation;<br>• compensation;<br>• declaration of infected areas;<br>• quarantines;<br>• imports and exports;<br>• vaccines;<br>• seizures; and<br>• penalties<br>Implemented by various regulations, including:<br>• 1964 Animal Disease Regulations (Elimination of Beef Tuberculosis);<br>• 1976 Animal Disease Regulation (Brucellosis in Cattle);<br>• 1976 Animal Disease Regulation (Registration, Marking and Transportation of Cattle);<br>• 1982 Animal Disease Regulations (Animal Transportation in Israel);<br>• 1989 Animal Disease Regulations (Elimination of Brucellosis Disease in Sheep);<br>• 1988 Animal Disease Regulations (Import and Export of Animal Products);<br>• 1994 Animal Diseases Regulations (Testing of Biological Residues); and<br>• 2001 Animal Disease Regulations (Feeding of Animals) |

*(Continued)*

| Region | Country/State | Citation | Title | Year Passed | Summary or Subsections Applicable to Regulatory Issue |
|---|---|---|---|---|---|
| Asia | Israel | See Amnon Carmi and Mohammed Saif-Alden Wattad, *Medical Law in Israel* (Kluwer Law International, 2010) | Veterinarians Law | 1991 | • Regulates work of veterinarians; contains provisions on: examinations;<br>• Diagnoses; and<br>• certifying carcasses as fit for human consumption |
| Asia | Japan | Act No. 233 of Dec 14, 1947 | Food Sanitation Act | 1947 | Sets out measures to prevent human health hazards from food consumption |
| Asia | Japan | Act No. 166 of May 31, 1951 | Act on Domestic Animal Infectious Dis eases Control | 1951 | • Governs animal disease prevention and control; primarily concerns domestic livestock industry; contains provisions on: disease prevention;<br>• notification obligations;<br>• inspections;<br>• slaughter practices;<br>• exports and imports (including quarantines);<br>• and penalties |
| Asia | Japan | Act No. 114 of 1953 | Slaughterhouse Act | 1953 | Regulates hygienic and sanitary practices in slaughterhouses |
| Asia | Japan | Act No. 70 of 1990 | Poultry Slaughtering Business Control and Poultry Meat Inspection Act | 1990 | Regulates poultry slaughtering and inspections |
| Asia | Japan | Act No. 70 of 2002 | Law on Special Measures Against Bovine Spongiform Encephalopathy | 2002 | Requires mandatory tracking and traceback for cattle: every cow must be identified with an ear tag, and producers must submit data on each animal to be stored in central database of all domestic herds |

*(Continued)*

| Region | Country/State | Citation | Title | Year Passed | Summary or Subsections Applicable to Regulatory Issue |
|---|---|---|---|---|---|
| Asia | Japan | Act No. 72 of 2003 | Law for Special Measures. Concerning the Management and Relay of Information for the Individual Identification of Cattle (Beef Traceability Law) | 2003 | Mandates tracking of all domestically raised cattle |
| Asia | Japan | Act No. 48 of 2003 | Food Safety Basic Law | 2003 | Establishes basic food safety principles |
| Asia | Russia | Law No. 4979-I of 1993 | Law No. 4979-I on veterinary medicine | 1993 | Sets out general requirements for preventing and eliminating animal disease as well as veterinary provisions for the safety of animal products; provides for rules and standards on veterinary inspections of meat-producing and dairy enterprises; calls for Federal Special Programs to prevent and control (including via quarantines) animal disease; Sect. 5 addresses protecting the population against common animal and human diseases |
| Asia | Russia | Law No. 52-FZ of 1999 | Federal Law No. 52-FZ on sanitary and epidemiological well-being of the population | 1999 | Governs public health generally |
| Asia | Russia | No. 13-8-01/2-8 of 1999 | Veterinary and sanitary requirements<br>No. 13-8-01/2-8 of 1999 regarding import to the Russian Federation of fish products, aquaculture and thermally treated aquatic products | 1999 | Sets stringent requirements for imported fish and aquaculture products: such products must come from factories within administrative territories free from African pig-plague for the past 3 years and free from FMD for the past 12 months |

*(Continued)*

| Region | Country/State | Citation | Title | Year Passed | Summary or Subsections Applicable to Regulatory Issue |
|---|---|---|---|---|---|
| Asia | Russia | No. 13-8-01/1-1 of 2001 | Veterinary and sanitary requirements No. 13-8-01/1-1 regarding import to the Russian Federation of pedigree cattle and common cattle | 2001 | Sets stringent requirements for imports of pedigree and common cattle: cattle must not be vaccinated from brucellosis, FMD, and leptospirosis and must come from infectious animal disease free farms and administrative territories |
| Asia | Russia | No. 13-8-01/1-4 of 2001 | Veterinary and sanitary requirements No. 13-8-01/1-4 regarding import to the Russian Federation of cattle sheep and goats destined for slaughtering | 2001 | Sets stringent requirements for imports of cattle sheep and goats destined for slaughtering: cattle must not be vaccinated from brucellosis, FMD, and leptospirosis and must come from infectious animal dis ease free farms and administrative territories |
| Australia/ New Zealand | Australia | Act No. 47 of 1982 | Export Control Act | 1982 | Controls export of prescribed goods, including animals, plants, fish, and food; provides for veterinary accreditations under approved export programs relating to eligible live animals and other goods |
| Australia/ New Zealand | Australia | Act No. 71 of 1983 | Meat Inspection Act | 1983 | Regulates meat inspections |
| Australia/ New Zealand | Australia | Act No. 130 of 1989 | Exotic Animal Disease Control Act | 1989 | Governs control and eradication of exotic animal diseases; establishes Exotic Animal Disease Preparedness Consultative Council and Exotic Animal Disease Preparedness Trust Account |
| Australia/ New Zealand | Australia | Act No. 23 of 1991 | National Cattle Disease Eradication Reserve Act | 1991 | Establishes National Cattle Disease Eradication Reserve to make payments to States for the purpose of eradicating any endemic cattle disease |
| Australia/ New Zealand | Australia | Act No. 206 of 1997 | Meat and Livestock Industry Act | 1997 | Regulates meat and livestock industry |

*(Continued)*

| Region | Country/State | Citation | Title | Year Passed | Summary or Subsections Applicable to Regulatory Issue |
|---|---|---|---|---|---|
| Australia/ New Zealand | Australia | Act No. 61 of 2015 | Biosecurity Act | 2015 | Governs diseases and pests that may cause harm to human, animal or plant health, or to the environment in Australia.<br>Regulates goods, aircraft, and vessels entering Australian territory and enables the governor-general of Australia to declare biosecurity emergencies and for biosecurity officers to ensure compliance with the act and carry out activities to manage biosecurity risks. |
| Australia/ New Zealand | Australia (Australian Capital Territory) | A2005-18 | Animal Diseases Act | 2005 | Governs animal diseases at a regional level (and is more comprehensive than the national act); aims to prevent and control outbreaks of both endemic and exotic animal diseases; contains provisions on:<br>• declaring infected places;<br>• endemic diseases;<br>• quarantine areas;<br>• restricted areas; and<br>• notification and quarantine requirements<br>Implemented by Animal Diseases Regulations (which include rules on identifying infected stock and measures against New-castle disease) |
| Australia/ New Zealand | New Zealand | 1993 No. 95 | Biosecurity Act | 1993 | Governs animal diseases; aims to exclude, eradicate, and effectively managing pests and unwanted organisms; contains provisions on:<br>• imports;<br>• surveillance;<br>• inspections;<br>• quarantines; and<br>• declarations of restricted or control areas<br>Part VII covers "exigency actions" aimed at "the effective prevention, management, or eradication of unwanted organisms if emergencies or other exigencies occur"<br>Implemented by:<br>• 1998 Biosecurity (National Bovine Tuberculosis Pest Management Strategy) Order<br>• 1999 Biosecurity (Animal Identification Systems) Regulations;<br>• 2005 Biosecurity (Meat and Food Waste for Pigs) Regulations |

| Region | Country/State | Citation | Title | Year Passed | Summary or Subsections Applicable to Regulatory Issue |
|---|---|---|---|---|---|
| Australia/ New Zealand | New Zealand | 1997 No. 87 | Agricultural Compounds and Veterinary Medicines Act | 1997 | Governs use of agricultural compounds; aims to prevent and manage risks to public health, animal welfare, primary produce trade, and agricultural security |
| Australia/ New Zealand | New Zealand | 1999 No. 93 | Animal Products Act | 1999 | Covers exports of potentially dangerous animals and animal products; contains provisions on:<br>• risk management programs;<br>• animal product standards and specifications; and<br>• exports of animal materials and products<br>Implemented by 2000 Animal Products Regulations, which lay out examining, sampling, testing, packaging, storage, handling, and record-keeping requirements |
| Australia/ New Zealand | New Zealand | 2012 No. 2 | National Animal Identification and Tracing Act | 2012 | Establishes national animal identification and tracing system to help manage risks to human health from food-borne zoonoses (and other diseases) |
| Australia/ New Zealand | New Zealand | 2014 No. 32 | The Food Act | 2014 | Governs food safety generally; supersedes the 1981 Food Act and implements a sliding scale where higher-risk producers operate under more stringent food safety requirements. |
| Europe | European Union | O.J. L. 147, 31.5.2001, p. 1; consolidated version at 02001R0999-20210907 | Commission Regulation (EC) No 999/2001 of 22 May 2001 laying down rules for the prevention, control and eradication of certain transmissible spongiform encephalopathies | 2001 | Governs management of prevention, control, and response measures to transmissible spongiform encephalopathies (TSEs); contains provisions on:<br>• eradication and destruction;<br>• diagnostic testing;<br>• animal nutrition and feed bans;<br>• BSE (and other TSE) monitoring and surveillance programs; import controls; and<br>• breeding<br>Has been updated and amended numerous times since 2001 |

*(Continued)*

| Region | Country/State | Citation | Title | Year Passed | Summary or Subsections Applicable to Regulatory Issue |
|---|---|---|---|---|---|
| Europe | European Union | O.J. L. 31, 1.2.2002, pp. 1–24; consolidated version at 02002R0178-20210526 | Commission Regulation (EC) No 178/2002 of the European Parliament and of the Council of 28 January 2002 laying down the general principles and requirements of food law, establishing the European Food Safety Authority and laying down procedures in matters of food safety | 2002 | Establishes the European Food Safety Authority (EFSA) as an independent source of information on all areas affecting food safety, including emerging risks |
| Europe | European Union | O.J. L. 139, 30.4.2004, pp. 1–54 | Commission Regulation (EC) No 852/2004 of the European Parliament and of the Council of 29 April 2004 on the hygiene of foodstuffs ["General Food Hygiene Rules"] | 2004 | Lay down common principles and definitions for national and EU-level food law; calls for an integrated approach to food safety; sets out role of HACCP principles |
| Europe | European Union | O.J. L. 139, 30.4.2004, pp. 55–206 | Commission Regulation (EC) No 853/2004 laying down specific hygiene rules for food of animal origin in order to guarantee a high level of food safety and public health ["Animal-Origin Food Hygiene Rules"] | 2004 | Lay out sanitary requirements for producers handling animal or animal products intended for human consumption; require food business operators to ensure that imported animal-origin products meet several conditions; takes a sectoral approach to standard-setting |

*(Continued)*

| Region | Country/State | Citation | Title | Year Passed | Summary or Subsections Applicable to Regulatory Issue |
|---|---|---|---|---|---|
| Europe | European Union | O.J. L. 338, 22.12.2005, pp. 60–83 | Commission Implementing Regulation (EU) 2015/1375 of 10 August 2015 laying down specific rules on official controls for *Trichinella* in meat | 2005 | Implements rules for the sampling of carcasses of species susceptible to *Trichinella* infection, for the determination of the status of holdings and compartments and conditions for the import of meat into the Union; provides for reference methods for the detection of *Trichinella* in samples of carcasses. |
| Europe | European Union | O.J. L. 322, 7.12.2007, pp. 12–30 | Commission Regulation (EC) No 1441/2007 of 5 December 2007 amending regulation (EC) no 2073/2005 on microbiological criteria for foodstuffs | 2007 | Lays down microbiological criteria for certain microorganisms (like *Salmonella*) in foodstuffs; implements Regulation (EC) No 852/2004 |
| Europe | European Union | O.J. L 010 15.1.2009, pp. 7 | Council Directive 2008/119/EC of 18 December 2008 laying down minimum standards for the protection of calves | 2008 | Lays down minimum standards (including requirements for holding facilities) for protecting calves confined for rearing and fattening |
| Europe | European Union | O.J. L. 300, 14.11.2000, pp. 1–33 | Commission Regulation (EC) No 1069/2009 of the European Parliament and of the Council of 21 October 2009 laying down health rules as regards animal by-products and derived products not intended for human consumption and repealing Regulation (EC) No 1774/2002 (Animal by-products Regulation) | 2009 | Regulates the movement, processing, and disposal of animal by-products; sets out principles for categorizing animal by-products according to risk |

*(Continued)*

| Region | Country/State | Citation | Title | Year Passed | Summary or Subsections Applicable to Regulatory Issue |
|---|---|---|---|---|---|
| Europe | European Union | O.J. L. 54, 26.2.2011, pp. 1; consolidated version at 2011R0142 — EN — 19.08.2011 — 001.001 | Commission Regulation (EC) No 142/2011 of the European Parliament and of the Council of 25 February 2011 implementing Regulation (EC) No 1069/2009 of the European Parliament and of the Council laying down health rules as regards animal by-products and derived products not intended for human consumption and implementing Council Directive 97/78/EC as regards certain samples and items exempt from veterinary checks at the border under that Directive | 2011 | Implements Regulation (EC) No 1069/2009 and lays down further health rules |

(*Continued*)

| Region | Country/State | Citation | Title | Year Passed | Summary or Subsections Applicable to Regulatory Issue |
|---|---|---|---|---|---|
| Europe | European Union | O.J. L 084 31.3.2016, pp. 1 | Regulation (EU) 2016/429 on transmissible animal diseases and amending and repealing certain acts in the area of animal health ("Animal Health Law") | 2016 | Streamlines animal health acts and lays down rules for the prevention and control of transmissible animal diseases with provisions on:<br>• The prioritization and categorization of diseases of Union concern and for the establishment of responsibilities for animal health<br>• Detection, notification and reporting of diseases, surveillance, eradication programs and disease-free status<br>• disease awareness, preparedness and control<br>• Approval of establishments and transporters<br>• Traceability of animals and products of animal origin<br>• Imports and exports of animals and products of animal origin<br>• non-commercial movements of pet animals into a Member State from another Member State or from a third country or territory<br>• the emergency measures to be taken in the event of a disease emergency situation |
| Europe | European Union | O.J. L 95, 7.4.2017, pp. 1–142 | Regulation (EU) 2017/625 of the European Parliament and of the Council of 15 March 2017 on official controls and other official activities performed to ensure the application of food and feed law, rules on animal health and welfare, plant health and plant protection products | 2017 | Helps ensure enforcement of EU rules for animals and goods at any stage of production and distribution by providing for audits of good hygiene practices and HACCP-based procedures along the entire food production chain; requires Member States to integrate their control plans into the Information Management System for Official Controls (IMSOC); requires official certification when appropriate to ensure compliance with EU rules on animals and goods. |

*(Continued)*

| Region | Country/State | Citation | Title | Year Passed | Summary or Subsections Applicable to Regulatory Issue |
|---|---|---|---|---|---|
| Europe | European Union | O.J. L 261, 14.10.2019, pp. 37–96; consolidated version at 02019R1715–20211201 | Commission Implementing Regulation (EU) 2019/1715 of 30 September 2019 laying down rules for the functioning of the information management system for official controls and its system components (the IMSOC Regulation) | 2019 | Lays down specific conditions and procedures applicable to the transmission of notifications and supplementary information for the Rapid alert system for food and feed (RASFF), including procedures for the establishment and use of the computerized system for Union notification and reporting of diseases set up and managed by the commission. |
| Europe | European Union—Germany | *Bundesgesetzblat*, Part II, No. 57, 14 August 2002, pp. 3082–3104 | Act Reorganizing the Regime of Consumer Protection and Food Safety (Federal Office for Consumer Protection and Food Safety Act)/*Gesetz zur Neuorganisation des gesundheitlichen Verbraucherschutzes und der Lebensmittelsicherheit* | 2002 | Establishes a Federal Office for Consumer Protection and Food Safety within the Federal Ministry of Consumer Protection, Food and Agriculture |
| Europe | European Union—Germany | *BfR-Gesetz* vom 6. August 2002 (BGBl. I S. 3082), das zuletzt durch § 44 Absatz 1 des Gesetzes vom 22. Mai 2013 (BGBl. I S. 1324) gea¨ndert worden ist | Federal Agency Risk Assessment Act/*Gesetz u€ber die Errichtung eines Bundesinstitutes fu€r Risikobewertung (BfR-Gesetz—BfRG)* | 2002 | Establishes the Federal Institute for Risk Assessment within the Federal Ministry of Consumer Protection, Food and Agriculture; lays out agency's duties, which include preparing scientific reports on food safety issues (including animal feed and pharmaceuticals for animal use) and cooperating with the European Food Safety Authority (EFSA) |

*(Continued)*

| Region | Country/State | Citation | Title | Year Passed | Summary or Subsections Applicable to Regulatory Issue |
|---|---|---|---|---|---|
| Europe | European Union—Germany | Federal Ministry of Food and Agriculture (BMEL) [Online] | German *Codex Alimentarius* (German Food Code)/*Der Codex Alimentarius* | 2003 | Lays out set of guidelines describing the production and marketing of foodstuffs; based on international standards; published by Federal Minister for Consumer Protection, Food and Agriculture |
| Europe | European Union—Germany | *Bundesgesetzblat*, Part I, No. 20, 27 April 2006, pp. 945–980 | German Foodstuff and Feedstuff Codex/ *Lebensmittel-, Bedarfsgegensta€nde- und Futtermittelgesetzbuch (Lebensmittel- und Futtermittelgesetzbuch—LFGB)* | 2005 | Amends and re-organizes various foodstuff and feedstuffs regulations; implements various EU directives; aims to:<br>• protect consumer heath;<br>• ensure the safe use of animal feed;<br>• protect the environment; and<br>• ensure that foodstuff of animal origin is of a high quality<br>• Part 15 references the German Food Code<br>• Part 7 covers inspection; Part 8 covers monitoring and<br>• Part 9 covers imports and exports |
| Europe | European Union—Germany | *Bundesgesetzblat*, Part I, No. 49, 30 October 2006, pp. 2355–2375 | Animal Vaccine Ordinance/ *Verordnung u€ber Sera, Impfstoffe und Antigene nach dem Tierseuchengesetz (Tierimpfstoff- Verordnung)* | 2006 | Regulates production and use of serum, inoculum, and antigen according to the (now repealed) Animal Diseases Act<br>• Part 2 covers vaccine production;<br>• Part 3 covers vaccine authorizations; and<br>• Part 7 covers vaccine delivery and use |

(*Continued*)

| Region | Country/State | Citation | Title | Year Passed | Summary or Subsections Applicable to Regulatory Issue |
|---|---|---|---|---|---|
| Europe | European Union—Germany | *Bundesgesetzblatt Jahrgang* 2013 Teil Nr. 25, ausgegeben zu Bonn am 27. Vom 22. Mai 2013., pp. 1324–1347 | Animal Health Act/*Gesetz zur Vorbeugung vor und Beka€mpfung von Tierseuchen (Tiergesundheitsgesetz—TierGesG)* | 2013 | Governs animal health; repeals Animal Diseases Act; contains provisions on:<br>• measures to prevent and control animal disease;<br>• special protective measures;<br>• immunological veterinary drugs;<br>• intra-Community movement, import, export, and transit of animals;<br>• compensation for loss of animals;<br>• data collection; and<br>• penalties and fines;<br>Implemented by various regulations and ordinances, including:<br>• 2013 Lower Saxony Herpes Virus Ordinance;<br>• 2014 Saxony Animal Disease Act Implementation Law; and<br>• 2014 Nordrhein-Westfalen Animal Products Disposal Act Implementation Law |
| Europe | United Kingdom (England) | S.I. No. 183 of 2006 | Foot-and-Mouth Disease (Control of Vaccination) (England) Regulations | 2006 | Carry out in England parts of Council Directive 2003/85/EC (on measures for controlling FMD) relating to vaccinations against FMD; provide for vaccination programs; declare vaccination zones and vaccination surveillance zones; amended for continued operation by the Zoonotic Disease Eradication and Control (Amendment) (EU Exit) Regulations 2018 |
| Europe | United Kingdom (England) | S.I. No. 3255 of 2009 | Official Feed and Food Controls (England) Regulations | 2009 | Regulate compliance with feed and food law and official controls on imports of certain feed and food of non-animal origin; amended for continued operation by the Official Controls (Animals, Feed and Food) (England) (Amendment) (EU Exit) Regulations 2018 |

*(Continued)*

| Region | Country/State | Citation | Title | Year Passed | Summary or Subsections Applicable to Regulatory Issue |
|---|---|---|---|---|---|
| Europe | United Kingdom (England) | S.I. No. 801 of 2010 | Transmissible Spongiform Encephalopathies (England) Regulations | 2010 | Carry out in England Regulation (EC) No 999/2001 of the European Parliament and of the Council laying down rules for the prevention, control and eradication of certain transmissible spongiform encephalopathies; amended for continued operation by the Animal By-Products and Transmissible Spongiform Encephalopathies (England) (Amendment) (EU Exit) Regulations 2018 |
| Europe | United Kingdom (England) | S.I. No. 2503 of 2010 | Animal Feed (England) Regulations | 2010 | Implement EU-level legislation on the marketing and use of feed and feed additives; amended for continued operation by the Official Controls (Animals, Feed and Food) (England) (Amendment) (EU Exit) Regulations 2018 |
| Europe | United Kingdom (England) | S.I. No. 1197 of 2011 | Trade in Animals and Related Products (England) Regulations | 2011 | Establish system for intra-Community and international trade in live animals and genetic material; empower Secretary of State to prohibit import into England of any animal or product in the event of a disease outbreak outside the UK<br>Enforced by the Secretary of State, port health authorities, local authorities, and the<br>United Kingdom Border Agency<br>Amended for continued operation by the Trade in Animals and Related Products (Amendment) (EU Exit) Regulations 2018 |
| Europe | United Kingdom (England) | S.I. No. 2952 of 2013 | Animal By-Products (Enforcement) (England) Regulations | 2013 | Carry out in England Regulation (EC) No 1069/2009 and Regulation (EU) No 142/2011 (laying down health rules for animal by-products and derived products); amended for continued operation by The Animal By-Products and Transmissible Spongiform Encephalopathies (England) (Amendment) (EU Exit) Regulations 2018 |

(Continued)

| Region | Country/State | Citation | Title | Year Passed | Summary or Subsections Applicable to Regulatory Issue |
|---|---|---|---|---|---|
| Europe | United Kingdom (England) | S.I. No. 2996 of 2013 | Food Safety and Hygiene (England) Regulations | 2013 | Enforce EU Hygiene Regulations on controlling food business operators with respect to hygienic conditions in producing and placing of food on the market; place authority in the Food Standards Agency; allow Secretary of State to issue codes of recommended practice to food authorities<br>• Sch. 5 covers meat from poultry or lagomorphs;<br>• Sch. 6 covers raw milk; and<br>• Sch. 8 covers emergency slaughters<br>Amended for continued operation by the Official Feed and Food Controls (England) (Miscellaneous Amendments) Regulations 2019 |
| Europe | United Kingdom (England) | S.I. No. 1855 of 2014 | Food Information (England) Regulations | 2014 | Regulates information provided by food business operators to the public<br>Implemented by various regulations, including the Eggs and Chicks (England) Regulations 2009 (S.I. No. 2163 of 2009)<br>Amended for continued operation by The Nutrition (Amendment etc.) (EU Exit) Regulations 2019 |

*(Continued)*

| Region | Country/State | Citation | Title | Year Passed | Summary or Subsections Applicable to Regulatory Issue |
|---|---|---|---|---|---|
| Europe | United Kingdom (England, Wales) | 1981, c. 22 | Animal Health Act | 1981 | Governs animal disease prevention and eradication; contains provisions on:<br>• veterinary services;<br>• eradication areas;<br>• biosecurity;<br>• animal movement and transport;<br>• imports and exports;<br>• national contingency plans in event of disease outbreaks, vaccinations;<br>• treatment;<br>• slaughter and destruction;<br>• inspections and powers of entry;<br>• seizures and disposals; and<br>• animal welfare<br>Contains specific provisions on certain diseases (like FMD and rabies)<br>• Sect. 29 addresses control of zoonoses<br>• Sects. 31 and 32 address slaughter; amended in 2002 in direct response to BSE outbreak to extend government's power to slaughter infected and potentially infected animals<br>Amended for continued operation by the Farriers and Animal Health (Amendment) (EU Exit) Regulations 2019 |
| Europe | United Kingdom (England, Wales, Scotland) | 1990, c. 16 | Food Safety Act | 1990 | Govern food safety generally |

*(Continued)*

| Region | Country/State | Citation | Title | Year Passed | Summary or Subsections Applicable to Regulatory Issue |
|---|---|---|---|---|---|
| Europe | United Kingdom (England, Wales, Scotland) | S.I. No. 539 of 1995 | Fresh Meat (Hygiene and Inspection) Regulations | 1995 | Govern slaughterhouses and conditions for marketing fresh meat |
| Europe | United Kingdom (England, Wales, Scotland) | S.I. No. 540 of 1995 | Poultry Meat, Farmed Game Bird Meat and Rabbit Meat (Hygiene and Inspection) Regulations | 1995 | Govern slaughterhouses and conditions for marketing fresh meat from poultry, farmed game birds, and rabbits |
| Europe | United Kingdom (England, Wales, Scotland) | S.I. No. 3279 of 2004 | General Food Regulations | 2004 | Implement and amend the Food Safety Act; Art. 18 governs traceability (including obligations on food operators); Art. 19 governs food safety requirements for food operators |
| Europe | United Kingdom (Scotland) | 2006 asp 11 | Animal Health and Welfare (Scotland) Act | 2006 | Governs animal and welfare in Scotland; amends Animal Health Act of 1981 provisions on disease outbreaks of livestock Transmissible Spongiform Encephalopathies (TSEs); establishes breeding program to breed TSE resistance; provides for measures to prevent the spread of animal diseases, including slaughtering and issuing "Biosecurity codes"; contains further provisions on animal welfare aimed at preventing cruelty and animal distress; amended for continued operation by the Animal Welfare (EU Exit) (Scotland) (Amendment) Regulations 2019 |

*(Continued)*

| Region | Country/State | Citation | Title | Year Passed | Summary or Subsections Applicable to Regulatory Issue |
|---|---|---|---|---|---|
| Europe | United Kingdom (Scotland) | S.S.I. No. 177 of 2012 | Trade in Animals and Related Products (Scotland) Regulations | 2012 | Establish system for intra-Community and international trade in live animals and genetic material; empower Scottish Ministers to prohibit importation into England of any animal or product in the event of a disease outbreak outside the EU; amended for continued operation by the Animal Health (EU Exit) (Scotland) (Amendment) Regulations 2019 |
| Europe | United Kingdom (Scotland) | S.S.I. No. 307 of 2013 | Animal By-Products (Enforcement) (Scot land) Regulations | 2013 | Carry out in Scotland Regulation (EC) No 1069/2009 and Regulation (EU) No 142/2011 (laying down health rules for animal by-products and derived products); amended for continued operation by the Animal Health (EU Exit) (Scotland) (Amendment) Regulations 2019 |
| International | Members of *Codex Alimentarius* Commission | Codex Standards, updated Nov. 21, 2022 [Online] | FAO/WHO *Codex Alimentarius* (Food Code) | 1963 | Lays out a series of general and specific food safety standards that aim to protect consumer health and ensure fair practices in the global food trade; relevant standards include the:<br>• General Principles of Food Hygiene;<br>• Code of Hygienic Practice for Meat;<br>• Code of Hygienic Practice for Milk and Milk Products; and<br>• Code of Practice on Good Animal Feedings<br>All standards are voluntary, although they are widely adopted and followed in countries around the world |
| International | Members of OIE | 2021 OIE—*Aquatic Animal Health Code*, 23rd Edition [Online] | OIE Aquatic Animal Health Code | 1995 | Lays out a series of science-based standards, guidelines, and recommendations for international, national, and regional bodies to adopt<br>All standards are voluntary, although they are widely adopted and followed in countries around the world |

(*Continued*)

| Region | Country/State | Citation | Title | Year Passed | Summary or Subsections Applicable to Regulatory Issue |
|---|---|---|---|---|---|
| International | Members of OIE | 2021 OIE—*Terrestrial Animal Health Code*, 29th Edition [Online] | OIE Terrestrial Animal Health Code | 1968 | Lays out a series of science-based standards, guidelines, and recommendations for international, national, and regional bodies to adopt; contains provisions on:<br>• animal welfare;<br>• food safety;<br>• disease detection and notification;<br>• veterinary services (including the use of anti-microbial agents); and safe trade (including import risk analysis)<br>All standards are voluntary, although they are widely adopted and followed in countries around the world |
| International | Parties to Convention | S. Treaty Doc. No. 103-20 (1993), 31 I.L.M. 818 | Convention on Biological Diversity (CBD) | 1992 | Addresses all aspects of biodiversity; increasingly concerned with linkages between biodiversity, zoonotic diseases, and human health; has set up a series of “One Health” networks in collaboration with the WHO to promote research on the human-animal-ecosystem interface |
| International | Parties to Protocol | 37 I.L.M. 22 (1998) | Kyoto Protocol to the United Nations Framework Convention on Climate Change | 1997 | Sets binding GHG reduction targets; can have indirect impacts on animal diseases: countries that fulfill their Kyoto obligations by launching Clean Development Mechanism (CDM) projects (like forest plantations) in developing countries could create new human-animal-ecosystem interfaces that impact disease transmission |

*(Continued)*

| Region | Country/State | Citation | Title | Year Passed | Summary or Subsections Applicable to Regulatory Issue |
|---|---|---|---|---|---|
| International | Member States of the World Trade Organization (WTO) | 31 U.S.T. 405, 1186 U.N.T.S. 276 | Agreement on Technical Barriers to Trade (TBT Agreement) | 1995 | Aims to facilitate global trading markets; prohibits nations from setting technical regulations, standards, testing, and certification procedure requirements that create unnecessary obstacles to trade OIE, FAO, and WHO standards serve as reference organizations for animal health and food safety measures |
| International | Member States of the World Trade Organization (WTO) | 1867 U.N.T.S. 483 | Agreement on the Application of Sanitary and Phytosanitary Measures (SPS Agreement) | 1995 | Sets similar constraints as the TBT Agreement on Member States' food safety, animal, and plant health measures OIE, FAO, and WHO standards serve as reference organizations for animal health and food safety measures |

# APPENDIX 2: ENVIRONMENTAL PROTECTION AND CLEAN ENERGY OVERLAPS

| Region | Country/State | Citation | Title | Year Passed | Summary or Subsections Applicable to Regulatory Issue |
|---|---|---|---|---|---|
| Africa | Kenya | Act No. 8 of 1999, as amended by Act No. 6 of 2006, Act No. 17 of 2006, Act No. 5 of 2007, Act No. 6 of 2009 | Environmental Management and Co-ordination Act (EMCA) (Cap. 387) | 1999 | Governs environmental management generally<br>• Sect. 87 prohibits disposing of any wastes "in a manner that would cause pollution to the environment or ill health to any person."<br>• Section 87(4) obligates "every person whose activities generate waste to employ measures essential to minimize wastes through treatment, reclamation and recycling." |
| Africa | Kenya | L.N. No. 121 of 2006; printed in *Kenya Gazette Supplement* No. 69 of 29 Sept 2006, pp. 1–63 | Environmental Management and Coordination (Waste Management) Regulations (Cap. 387) | 2006 | Implement the EMCA; define general rules for waste management and specific rules for certain waste streams (including solid waste, industrial waste, hazardous waste, and pesticides and toxic substances); embrace "cleaner production" principles |
| Africa | Nigeria | S.I.15 of 1991 | National Environmental Protection (Management of Solid and Hazardous Wastes) Regulations | 1991 | Address the collection, treatment, and disposal of solid and hazardous wastes from industrial and municipal sources; contain highly detailed provisions for certain waste streams<br>Part II envisions a multi-pronged approach with information-sharing and monitoring among private operators, industries, and public agencies |
| Africa | Nigeria | Act No. 25 of 2007 | National Environmental Standards and Regulations Enforcement Agency (Establishment) Act (NESREA) | 2007 | Establishes the National Environmental Standards and Regulations Enforcement Agency as the major federal body in charge of environmental protection; repeals the now-defunct Federal Environmental Protection Agency (FEPA) |

*(Continued)*

| Region | Country/State | Citation | Title | Year Passed | Summary or Subsections Applicable to Regulatory Issue |
|---|---|---|---|---|---|
| Africa | South Africa | Act No. 73 of 1989, as amended by the National Environmental Laws Amendment Act of 2009 | Environment Conservation Act | 1989 | Governs environmental protection generally; sets out national environmental conservation policy<br>• Part IV contains provisions on waste management<br>• Part V covers agriculture and MSW disposal |
| Africa | South Africa | No. R. 543 of 2002 | Plastic Bags Regulations | 2002 | Prohibit the manufacture, trade and commercial distribution of specified plastic bags<br>Enacted under section 24 (D) of the Environment Conservation Act |
| Africa | South Africa | No. R. 625 of 2003 | Plastic Carrier Bags and Plastic Flat Bags Regulations | 2003 | Prohibit the manufacture, trade and commercial distribution of domestically produced and imported plastic carrier bags and plastic flat bags Enacted under section 24 (D) of the Environment<br>Conservation Act |
| Africa | South Africa | Act No. 68 of 2008 | Consumer Protection Act | 2008 | Puts packaging disposal and product labelling burdens on industry<br>• Sect. 22 requires that packaging labels identify the type of packaging and its recyclability.<br>• Sect. 59 requires suppliers to accept return of goods at no charge to consumer if any national legislation prohibits the disposal or deposit of the goods or their components into a common waste collection system. |
| Africa | South Africa | Act No. 59 of 2008, as amended by Act No. 26 of 2014 | National Environmental Management: Waste Act (NEM: WA) | 2008 | Governs waste management; sets up waste management hierarchy; embraces "polluter pays"/EPR principle sets goals of:<br>• 25% diversion of recyclables from landfills by 2016; and<br>• ultimate phase-out of biodegradable waste from landfills<br>Primarily implemented by the 2011 National Waste Management Strategy (NWMS). |

(Continued)

| Region | Country/State | Citation | Title | Year Passed | Summary or Subsections Applicable to Regulatory Issue |
|---|---|---|---|---|---|
| Americas | Brazil | Lei n. 12.305/10 | National Policy of Solid Waste/*Política Nacional de Resíduos So'lidos* (PNRS) | 2010 | Governs solid waste management; focuses on sustainable measures (including packaging requirements); calls for integrated, multi-level management plans |
| Americas | Brazil (Sergipe) | Lei n. 7.465 de 20 de Julho de 2012; Publicado no *Diário Oficial* No 26529, do dia 24/07/2012 | Law No. 7.645 on the Obligation of Using Biodegradable Packaging/Lei No. 7.645 de Dispõe sobre a utilizaç-ao obrigatória de embalagens biodegrad*á*veis, e d*á* providências correlatas | 2012 | Requires use of biodegradable packaging for materials used for commercial purposes within the State of Sergipe |
| Americas | Canada | S.C. 1999, c. 33 | Canadian Environmental Protection Act (CEPA) | 1999 | Aims to protect environmental and human health through pollution prevention; embraces "polluter pays"/EPR principle<br>Part 7 contains general waste management and prevention provisions (although most of the relevant authority is delegated to provincial governments) |
| Americas | Canada (Manitoba) | C.C.S.M., c. W40 | Waste Reduction and Prevention Act | 1990 | Aims to reduce and prevent waste production and disposal consistent with sustainable development principles |
| Americas | Canada (Manitoba) | Man. Reg. 195/2008 | Packaging and Printed Paper Stewardship Regulation | 2008 | Requires affected companies to register as product stewards and to remit fees to cover up to 80% of the cost of municipal recycling programs |

(*Continued*)

| Region | Country/State | Citation | Title | Year Passed | Summary or Subsections Applicable to Regulatory Issue |
|---|---|---|---|---|---|
| Americas | Canada (Ontario) | R.S.O. 1990, c. E.18 | Environmental Assessment Act | 1990 | Regulates environmental assessments<br>Part II addresses municipal waste disposal |
| Americas | Canada (Ontario) | R.S.O. 1990, c. E.19 | Environmental Protection Act | 1990 | Aims to protect and conserve the natural environment<br>• Part V covers waste management<br>• Part IX covers packaging and container waste Implemented by Regulation on General Waste Management (R.R.O. 1990; O. Reg. 362) |
| Americas | Canada (Ontario) | O. Reg 104/94 | Regulation on Packaging Audits and Packaging Reduction Work Plans | 1994 | Requires audits of certain manufacturers and importers to examine impacts of packaging on waste management; requires packaging reduction plans under some circumstances to reduce the amount of waste resulting from packaging |
| Americas | Canada (Ontario) | O. Reg. 101/94 | Regulation on Recycling and Composting of Municipal Waste | 1994 | Requires certain municipalities to implement "blue box" waste management systems<br>• Part III addresses municipal waste recycling depots<br>• Part IV addresses waste recycling sites |
| Americas | Canada (Ontario) | O. Reg. 102/94 | Regulation on Waste Audits and Waste Reduction Work Plans | 1994 | Requires certain businesses (retailers, schools, hotels, manufacturers, etc.) and construction projects to undergo waste audits, and, under some circumstances, implement plans to reduce, reuse and recycle waste |
| Americas | Canada (Ontario) | O. Reg. 103/94 | Regulation on Industrial, Commercial and Institutional Source Separation Programs | 1994 | Requires certain businesses (retailers, schools, hotels, manufacturers, etc.) to implement programs to facilitate source separation of waste for reuse or recycling |
| Americas | Canada (Ontario) | S.O. 2002, c. 6 | Waste Diversion Act | 2002 | Promotes reducing, reusing and recycling waste; establishes "Waste Diversion Ontario" corporation to develop, implement, and operate waste diversion programs for designated wastes |

(*Continued*)

| Region | Country/State | Citation | Title | Year Passed | Summary or Subsections Applicable to Regulatory Issue |
|---|---|---|---|---|---|
| Americas | Canada (Ontario) | O. Reg. 273/02 | Blue Box Waste Regulation | 2002 | Designates "blue box" waste as waste consisting of any of the following materials, or any combination of them: (1) glass; (2) metal; (3) paper; (4) plastic; (5) textiles; establishes "Stewardship Ontario" corporation |
| Americas | Canada (Ontario) | O. Reg. 101/07 | Waste Management Projects Regulation | 2007 | Implements parts of Environmental Assessment Act (Ontario)<br>Part II concerns landfill siting, waste disposal, and energy/fuel use |
| Americas | Canada (Prince Edward Island) | R.S.P.E.I. 1988, c. E-9 | Environmental Protection Act | 1988 | Aims to manage, protect, and enhance the environment<br>Sects. 13 to 19 cover waste treatment |
| Americas | Canada (Prince Edward Island) | P.E.I. Reg. EC691/00 | Waste Resource Management Regulations | 2000 | Carry out Sect. 25 of Environmental Protection Act (Prince Edward Island); contain provisions on waste disposal and waste management, including requirements for composting facilities and recycling plants |
| Americas | Mexico | DOF 19-06-2007 | General Law for the Prevention & Integral Management of Wastes/*Ley General Para la Prevención y Gestión Integral de los Residuos* (LGPGIR) (Federal Waste Law) | 2003 | Governs waste management; provides the country's legal framework on packaging and recycling activities; embraces "polluter pays"/EPR principle<br>Implemented by 2006 Regulation for the General Law for the Prevention & Integral Management of Wastes/*Reglamento de la Ley General Para la Prevencio'n y Gestio'n Integral de los Residuos* (DOF 30-11-2006) |
| Americas | Mexico (Federal District) | *Gaceta Oficial del Distrito Federal* 22-04-2003 | Solid Waste Law of the Federal District/*Ley de Residuos Sólidos del Distrito Federal* | 2003 | Regulates solid waste management in Mexico City<br>Implemented by the Regulation for the Solid Waste Law of the Federal District/*Reglamento de la Ley de Residuos So'lidos del Distrito Federal* (DOF 07-10-2008) |

(*Continued*)

| Region | Country/State | Citation | Title | Year Passed | Summary or Subsections Applicable to Regulatory Issue |
|---|---|---|---|---|---|
| Americas | Mexico (Federal District) | *Gaceta Oficial del Distrito Federal* 10-02-2004 | Amendment of the Solid Waste Law of the Federal District/ *Decreto por el que se Reforman los Artículos Tercero, Cuarto, Séptimo y Octavo Transitorios de la Ley de Residuos Sólidos del Distrito Federal* | 2004 | Requires households to separate organic and inorganic wastes<br>Amends the Solid Waste Law of the Federal District |
| Americas | Peru | Law No. 27314 | General Law on Solid Waste (The Waste Act) | 2000 | Establishes rights, duties, powers, and responsibilities of society as a whole to ensure the sanitary and environmentally sound management and handling of solid waste; embraces principles of minimization, environmental risk prevention, and human health and welfare protection; sets norms for recycling operations<br>Implemented by the Rules of the Waste Act (Supreme Decree No 057-2004-PCM) |
| Americas | United States | 21 U.S.C. § 301 et seq. | Federal Food, Drug, and Cosmetic Act (FFDCA) | 1938 | Regulates food safety, including approval of food packaging materials |
| Americas | United States | 42 U.S.C. § 4321 et seq. | National Environmental Policy Act (NEPA) | 1969 | Requires environmental reviews of major federal actions (including food packaging material approvals) |
| Americas | United States | 42 U.S.C. § 6901 et seq. | Resource Conservation and Recovery Act (RCRA) | 1976 | Regulates waste management and recycling; provides general mandates, with most authority delegated to state and local governments |

(*Continued*)

| Region | Country/State | Citation | Title | Year Passed | Summary or Subsections Applicable to Regulatory Issue |
|---|---|---|---|---|---|
| Americas | United States | 21 C.F.R. § 170 et seq., 42 FR 14483 | Federal Food, Drug, and Cosmetics Act (FFDCA), Food Additive Regulations | 1977 | Regulates food contact substances in packaging |
| Americas | United States | 42 U.S.C. § 1791 | Bill Emerson Good Samaritan Food Donation Act | 1996 | Encourages food donations by shielding donors from liability |
| Americas | United States (California – City of San Francisco) | Ordinance No. 100-09 (June 9, 2009) | Mandatory Recycling and Composting Ordinance | 2009 | Sets goal of "virtually zero" landfill waste by 2020 in City of San Francisco |
| Americas | United States (Vermont) | Act No. 148, 10 V.S.A. § 6602 | An act relating to establishing universal recycling of solid waste. (Vermont's Universal Recycling (UR) Law) | 2012 | Bans landfilling all food scraps; bans disposal of recyclable materials (metal, glass, certain plastics, and paper/cardboard); amended in 2019 to ban retailers and food establishments from providing single-use plastic items like bags and straws |
| Asia | China | Order of the President No. 72, adopted in the 28th Meeting of the Standing Committee of the 9th National People's Congress of the People's Republic of China on June 29, 2002 | Law of the People's Republic of China on Promotion of Cleaner Production (Cleaner Production Promotion Law) | 2002 | Aims to:<br>• promote cleaner production;<br>• increase the efficiency of the utilization rate of resources;<br>• reduce and avoid the generation of pollutants;<br>• protect and improve environments;<br>• ensure the health of human beings; and<br>• promote the sustainable development of the economy and society<br>Requires "cleaner production audits" of certain manufacturers |

*(Continued)*

| Region | Country/State | Citation | Title | Year Passed | Summary or Subsections Applicable to Regulatory Issue |
|---|---|---|---|---|---|
| Asia | China | Announcement No. 63 of 2008 | Announcement of Ministry of Environmental Protection promulgating the cleaner production standard for wine industry | 2008 | Implements the Cleaner Production Promotion Law for the country's wine producers; provides requirements on:<br>• production techniques;<br>• resource and energy use;<br>• waste recycling; and<br>• environmental management |
| Asia | China | Order of the President No. 4, passed in the 4th meeting of the Standing Committee of the 11th National People's Congress of the People's Republic of China on August 29, 2008 | Circular Economy Promotion Law of the People's Republic of China | 2008 | Aims to promote sustainable development in the nation; introduces policies and instruments for reducing resource consumption and encouraging recycling; embraces a top-down, command-and-control approach |
| Asia | India | Act No. 29 of 1986 | An Act to provide for the protection and improvement of environment and for matters connected therewith (Environment (Protection) Act) | 1986 | Establishes the powers of the Central Government to protect and improve the environment |
| Asia | India | S.O. 908(E) | Municipality Solid Waste (Management and Handling) Rules | 2000 | Form the backbone of India's solid waste management policy; charge each municipal authority with developing infrastructure for collecting, storing, segregating, transporting, processing, and disposing of MSW; encourage composting and banning of biodegradable wastes from landfills |

*(Continued)*

| Region | Country/State | Citation | Title | Year Passed | Summary or Subsections Applicable to Regulatory Issue |
|---|---|---|---|---|---|
| Asia | India | S.O. 249(E) | Plastic Waste (Management and Handling) Rules | 2011 | Prohibit use of recycled or compostable plastic in carry bags used for storing, carrying, dispensing, or packaging food products<br>Implement Sects. 3, 6, and 25 of the Environment (Protection) Act |
| Asia | India (Maharashtra) | See Maharashtra Pollution Control Board, Circular No. MPCB/RO(HQ)/B: 1508 of 14/03/2011 | Maharashtra Plastic Carry Bags (Manufacture and Use) Rules | 2006 | Supplement the Central Plastic Waste (Management and Handling) Rules within Maharashtra |
| Asia | India (Tamil Nadu) | FAOLEX No: LEX-FAOC052632 | A Bill to provide for prohibition of sale, storage, transport and use of certain plastic articles for the protection of environment and public health and for matters connected there with or incidental thereto (Tamil Nadu Plastic Articles (Prohibition of Sale, Storage, Transport and Use) Act) | 2002 | Prohibits the sale, storage, transport and use of certain plastic articles (including carry bags, cups, plates, and tumblers); forbids restaurant owners from using plastic articles in their establishments |
| Asia | Israel | Law No. 5753-1993 | Collection and Disposal of Waste for Recycling Law | 1993 | Sets out principles and basic legal framework for recycling in Israel; authorizes municipalities to pass bylaws specifying procedures for waste collection, disposal, and recycling |

(*Continued*)

| Region | Country/State | Citation | Title | Year Passed | Summary or Subsections Applicable to Regulatory Issue |
|---|---|---|---|---|---|
| Asia | Israel | Law No. 5759-1999 | Deposit on Beverage Containers Law | 1999 | Requires a fully refundable deposit on most single-serve beverage containers Implemented by 2001 Regulations<br>Carried out by ELA Recycling Corporation (a private non-profit organization) |
| Asia | Israel | Law No. 5771-2011 | Regulation of Processing of Packaging Law (Packaging Law) | 2011 | Regulates both production and processing of packaging material waste; embraces "polluter pays"/EPR principle; aims to:<br>• minimize packaging volume;<br>• prevent waste dumping; and<br>• encourage recycling<br>Increases local taxes for waste storage in landfills (with revenues to go toward recycling facilities and infrastructure); sets concrete recycling targets, including zero landfilling waste by 2020 |
| Asia | Japan | Act No. 112 of June 16, 1995 | Act on the Promotion of Sorted Collection and Recycling of Containers and Packaging (Container and Packaging Recycling Law) | 1995 | Forms basis for all official recycling efforts in Japan; aims to reduce garbage levels and use recycled containers and packages efficiently |
| Asia | Japan | Law No. 176 of 2001 | Act on Promotion of Recycling of Food Circulation Resources (Food Recycling Law) | 2001 | Calls for reducing food waste through recycling food resources; requires municipalities to create "recycling loops" for feed and fertilizers |

(*Continued*)

| Region | Country/State | Citation | Title | Year Passed | Summary or Subsections Applicable to Regulatory Issue |
|---|---|---|---|---|---|
| Asia | Russia | Law No. 89-FZ | Federal Law on industrial and consumer waste (Law on Waste) | 1998 | Russia's primary federal waste management regulation A proposed 2014 bill would amend the law to allow for landfill diversion targets and to embrace "polluter pays"/EPR principles |
| Asia | Russia (Astrakhan) | Regional Law No. 40/2004-OZ | Law on Domestic and Non-domestic Waste | 2004 | Implements federal Law on Waste |
| Asia | Russia (Buryatia) | Regional Law No. 1254-IV | Law on Domestic and Non-domestic Waste | 2010 | Implements federal Law on Waste |
| Asia | South Korea | As amended by Act No. 13 038, Jan. 1, 2015 | Wastes Control Act | 1986 | Provides basic framework on waste management; requires Minister of Environment to prepare master waste management plan every 10 years |
| Asia | South Korea | As amended by Act No. 8957, Mar. 21, 2008 | Act on Promotion of Saving and Recycling of Resources | 1992 | Provides basic framework for waste recycling; requires Minister of Environment to prepare recycling plan every 5 years |
| Australia/New Zealand | Australia | Law No. 21 of 2012 | Heavy Vehicle National Law (HVNL) | 2012 | Aims to achieve greater efficiencies in the country's commercial transportation sector, with provisions on vehicle specifications, curfews, load limitations, and speed and travel time restrictions |
| Australia/New Zealand | Australia (Australian Capital Territory) | Act No. a2001-31 | Waste Minimisation Act | 2001 | Governs waste management at a provincial level; embraces "polluter pays"/EPR principle; work to:<br>• establish a waste management hierarchy;<br>• promote environmentally responsible transporting, reprocessing and handling of waste;<br>• seek integrated waste planning and services |
| Australia/New Zealand | Australia (New South Wales) | Act No. 58 of 2001 | Waste Avoidance and Resource Recovery Act | 2001 | Promotes waste avoidance and resource recovery; creates a Waste Fund |

*(Continued)*

| Region | Country/State | Citation | Title | Year Passed | Summary or Subsections Applicable to Regulatory Issue |
|---|---|---|---|---|---|
| Australia/ New Zealand | Australia (New South Wales) | LW 17 October 2014 (2014 No. 666) | Protection of the Environment Operations (Waste) Regulation | 2014 | Provides specifications on waste management, transport, and disposal; enacted under the Protection of the Environment Operations Act 1997 |
| Australia/ New Zealand | Australia (Northern Territory) | Act No. 71 of 1998 | Waste Management and Pollution Control Act | 1998 | Aims to reduce waste generation, increase re-use and recycling, and effectively manage waste disposal<br>Part 6 provides for environmental audits Implemented by Waste Management and Pollution Control (Administration) Regulations. |
| Australia/ New Zealand | Australia (Queensland) | FAOLEX No: LEX-FAOC107497 | Waste Reduction and Recycling Act | 2011 | Aims at promoting waste avoidance and reduction as well as resource recovery and efficiency actions; establishes shared responsibility between government, business and industry and the community for waste management; creates Waste and Environment Fund; levies taxes at certain waste disposal sites<br>Implemented by 2011 Waste Reduction and Recycling Regulation |
| Australia/ New Zealand | Australia (South Australia) | *Gazette* 6.5.2004, p. 1222 | An Act to establish a statutory corporation, Zero Waste SA, with the function of reforming waste management in the State; and for other purposes (Zero Waste SA Act) | 2004 | Establishes a statutory corporation, Zero Waste SA, to reform waste management in the State; aims to eliminate waste generation as well as the landfilling of waste; promotes resource recovery and recycling<br>Implemented by Zero Waste SA Regulations 2006 |

(*Continued*)

| Region | Country/State | Citation | Title | Year Passed | Summary or Subsections Applicable to Regulatory Issue |
|---|---|---|---|---|---|
| Australia/ New Zealand | Australia (Tasmania)b | Act No. 83 of 1994 | Environmental Management and Pollution Control Act (EMPCA) | 1994 | Governs environment protection and pollution control generally<br>Relevant implementing regulations include:<br>• Environmental Management and Pollution Control (Controlled Waste Tracking) Regulations (2010); and<br>• Environmental Management and Pollution Control (Waste Management) Regulations (2010) |
| Australia/ New Zealand | Australia (Western Australia) | Act No. 36 of 2007 | Waste Avoidance and Resource Recovery Act | 2007 | Establishes Waste Authority charged with setting out a long-term strategy for continuous improvement of waste services, waste avoidance, and resource recovery and for waste reduction, resource recovery, and landfill diversion; calls for yearly business plans, waste plans and of product stewardship plans |
| Australia/ New Zealand | New Zealand | Act No. 89 of 2008 | Waste Minimisation Act | 2008 | Governs waste management; calls for levy on landfill waste and mandatory waste reporting; establishes Waste Advisory Board<br>• Part 1, Sect. 3 encourages "waste minimisation and a decrease in waste disposal in order to protect the environment from harm and provide environmental, social, economic, and cultural benefits"<br>• Part II encourages (or sometimes requires) product stewardship schemes (including bottle deposit systems) that hold all parties involved in a product's lifecycle responsible for managing end-of-life environmental impacts |

*(Continued)*

| Region | Country/State | Citation | Title | Year Passed | Summary or Subsections Applicable to Regulatory Issue |
|---|---|---|---|---|---|
| Europe | European Union | O.J. L. 365, 31.12.1994, pp. 10–23 | European Parliament and Council Directive 94/62/EC of 20 December 1994 on packaging and packaging waste (Packaging and Packaging Waste Directive) | 1994 | Aims to protect the environment and maintain robust internal EU trading market; sets recycling targets for packaging waste; mandates environmental "essential requirements" for all packaging; embraces "polluter pays"/EPR principle |
| Europe | European Union | O.J. L. 182, 16.7.1999, pp. 1–19 | Council Directive 1999/31/EC of 26 April 1999 on the landfill of waste (Landfill Directive) | 1999 | Sets short- and long-term landfill diversion targets for Member States, including goals for reducing amount of biodegradable municipal waste going to landfills |
| Europe | European Union | O.J. L. 338, 13.11.2004, pp. 4–17 | Framework Regulation EC 1935/2004 on Food Contact Materials | 2004 | Sets requirements for food contact materials; requires good manufacturing practices; requires materials to be traceable throughout the production chain |
| Europe | European Union | O.J. L. 312, 22.11.2008, pp. 3–30 | Directive 2008/98/EC of the European Parliament and of the Council of 19 November 2008 on waste (Waste Framework Directive) | 2008 | Establishes basic waste management principles for EU Member States; provides general legislative framework for collecting, transporting, recovering, and disposing of waste<br>• Art. 4 establishes waste management hierarchy<br>• Art. 22 addresses composting |

*(Continued)*

| Region | Country/State | Citation | Title | Year Passed | Summary or Subsections Applicable to Regulatory Issue |
|---|---|---|---|---|---|
| Europe | European Union | O.J. L. 54, 26.2.2011, pp. 1–254 | Commission Regulation (EU) No 142/2011 implementing Regulation (EC) No 1069/2009 of the European Parliament and of the Council laying down health rules as regards animal by-products and derived products not intended for human consumption and implementing Council Directive 97/78/EC as regards certain samples and items exempt from veterinary checks at the border under that Directive (Animal By-Products (ABP) Rules) | 2011 | Regulates materials of animal-origin not consumed by people (including certain by-products of the food production chain, such as fallen stock) |
| Europe | European Union | O.J. L. 150, 14.6.2018, p. 100–108 | Directive (EU) 2018/850 of the European Parliament and of the Council of 30 May 2018 amending Directive 1999/31/EC on the landfill of waste | 2018 | Requires member States to take the necessary measures so that by 2035, the amount of municipal waste disposed of in landfills is reduced to 10% or less of the total amount of municipal waste generated. |

(Continued)

| Region | Country/State | Citation | Title | Year Passed | Summary or Subsections Applicable to Regulatory Issue |
| --- | --- | --- | --- | --- | --- |
| Europe | European Union | O.J. L. 150, 14.6.2018, pp. 109–140 | Directive (EU) 2018/851 of the European Parliament and of the Council of 30 May 2018 amending Directive 2008/98/EC on waste | 2018 | Requires member states to improve their waste management systems and efficiency of resource use. Its provisions cover:<br>• Measures to prevent waste generation<br>• Handling of MSW, including the Member States' obligation to separately collect paper, metal, plastic, and glass waste.<br>• Incentives for the application of the waste hierarchy, such as landfill and incineration charges or pay-as-you-throw schemes.<br>• Measures to promote the re-use of products and to encourage the development of recycled product suitable for re-use and recycling.<br>• Requirements for producer responsibility schemes. |
| Europe | European Union | O.J. L. 150, 14.6.2018, pp. 141–154 | Directive (EU) 2018/852 of the European Parliament and of the Council of 30 May 2018 amending Directive 94/62/EC on packaging and packaging waste | 2018 | Requires member states to meet targets, including by the end of 2030, at least 70% by weight of all packaging waste must be recycled. Member states may meet these targets through measures that encourage an increase in the share of reusable packaging placed on the market, including the use of deposit-return schemes, setting qualitative or quantitative targets, using economic incentives, and setting up a minimum percentage of reusable packaging placed on the market every year for each packaging stream |

(*Continued*)

| Region | Country/State | Citation | Title | Year Passed | Summary or Subsections Applicable to Regulatory Issue |
|---|---|---|---|---|---|
| Europe | European Union—Germany | *Federal Law Gazette I*, p. 2705, as amended by the Act of 3 May 2000 (*Federal Law Gazette I*, p. 632) | An act for Promoting Closed Substance Cycle Waste Management and Ensuring Environmentally Compatible Waste Disposal/Gesetz zur F€orderung der Kreislaufwirtschaft und Sicherung der umnweltverträglichen Beseitigung von Abfällen (Closed Substance Cycle and Waste Management Act) | 1994 | Establishes Germany's national waste management policy framework; advocates waste avoidance, recovery, and environmentally compatible waste disposal; embraces "polluter pays"/EPR principle |
| Europe | European Union—Germany | *Federal Law Gazette I*, p. 2379, as amended by the Fifth Amending Ordinance of 2 April 2008 (*Federal Law Gazette I*, p. 531) | Ordinance on the Avoidance and Recovery of Packaging Waste of 21 August 1998/Verpackungs-verordnung—VerpackV (Packaging Waste Ordinance or Packaging and Waste Avoidance Law) | 1998 | Implements the EU Packaging and Packaging Waste Directive; requires retailers to install recycling bins for primary and secondary pack aging; imposes mandatory beverage deposits; prohibits incineration as waste disposal option; embraces "polluter pays"/EPR principle |

(*Continued*)

| Region | Country/State | Citation | Title | Year Passed | Summary or Subsections Applicable to Regulatory Issue |
|---|---|---|---|---|---|
| Europe | United Kingdom | No. 1640 of 2015 | The Packaging (Essential Requirements) Regulations 2015 | 2015 | Requires producers to ensure packaging satisfies certain "essential requirements" related to minimum volume and weight, reusability, and minimization of heavy metals |
| Europe | United Kingdom | No. 188 of 2019 | The Waste (Miscellaneous Amendments) (EU Exit) (No. 2) Regulations | 2019 | Ensures the continued operation of the waste regime after the UK left the EU |
| Europe | United Kingdom (England and Wales) | No. 1889 of 2012 | Waste (England and Wales) (Amendment) Regulations | 2012 | Implemented the EU Waste Framework Directive; requires waste collection authorities to collect waste paper, metal, plastic, and glass separately (from both household and commercial/industrial sources) |
| Europe | United Kingdom (England, Scotland, and Wales) | No. 871 of 2007 | Producer Responsibility Obligations (Packaging Waste) Regulations (England, Scotland, and Wales) | 2007 | Implemented the EU Packaging and Packaging Waste Directive; address industry's role in meeting the UK's recovery and recycling targets; establish a statutory producer responsibility regime by placing legal obligation on businesses that make or use packaging to ensure that a proportion of the packaging is recovered and recycled; embraces "polluter pays" and "collective producer responsibility" principles |
| Europe | United Kingdom (Northern Ireland) | No. 198 of 2007 | Producer Responsibility Obligations (Packaging Waste) Regulations (Northern Ireland) | 2007 | Implemented the EU Packaging and Packaging Waste Directive in Northern Ireland |
| Europe | United Kingdom (Scotland) | S.S.I. No. 148 of 2012 | The Pollution Prevention and Control (Scotland) Regulation | 2012 | Implemented the EU Waste Framework Directive; provide for the separate collection and treatment of "wet" food waste from other "dry" recyclable waste streams |

*(Continued)*

| Region | Country/State | Citation | Title | Year Passed | Summary or Subsections Applicable to Regulatory Issue |
|---|---|---|---|---|---|
| International | Parties to Protocol | 37 I.L.M. 22 (1998) | Kyoto Protocol to the United Nations Framework Convention on Climate Change | 1997 | Sets binding GHG reduction targets, potentially achievable through agriculture and food sector reforms |
| International | Member States of the World Trade Organization (WTO) | 31 U.S.T. 405, 1186 U.N.T.S. 276 | Agreement on Technical Barriers to Trade (TBT Agreement) | 1995 | Aims to facilitate global trading markets; prohibits setting technical regulations, standards, testing, and certification procedure requirements that create unnecessary obstacles to trade |
| International | Member States of the World Trade Organization (WTO) | 1867 U.N.T.S. 483 | Agreement on the Application of Sanitary and Phytosanitary Measures (SPS Agreement) | 1995 | Sets similar constraints as the TBT Agreement on Member States' food safety, animal, and plant health measures |

# APPENDIX 3: HABITAT LOSS, AGROBIODIVERSITY, AND INCIDENTAL WILDLIFE LOSS

| Region | Country/State | Citation | Title | Year Passed | Summary or Subsections Applicable to Regulatory Issue |
|---|---|---|---|---|---|
| Africa | Kenya | Act No. 8 of 1955, as amended by Act No. 6 of 2012 | Agriculture Act (Cap. 318) | 1955 | Provides rules on good agricultural practices, including rules on:<br>• agricultural land use and preservation;<br>• sound agricultural development; and<br>• marketing of agricultural products |
| Africa | Kenya | Act No. 7 of 2005 | Forest Act | 2005 | Provides for the establishment, development and sustainable management (including conservation) and rational utilization of forest resources for the socioeconomic development of the country; establishes Forest Conservation Committee (FCC); requires public consultation for all major forest decisions |
| Africa | Kenya | L.N. No. 166 of 2009 | Agriculture (Farm Forestry) Rules (Cap. 318) | 2009 | Implements the Agriculture Act; aims to help conserve water, soil, and biodiversity and to protect riverbanks, shorelines, and wetlands; requires farmers to:<br>• establish and maintain farm forestry on at least 10% of each of their agricultural land holdings;<br>• ensure that tree species planted do not have adverse effects on water sources, crops, livestock, soil fertility, or the neighborhood; and<br>• ensure that tree species planted are not invasive<br>Orders District Agricultural Committee to:<br>• identify land at risk of degradation;<br>• establish necessary conservation measures (including planting trees) and<br>• prepare and oversee annual seedling production plans |

*(Continued)*

| Region | Country/State | Citation | Title | Year Passed | Summary or Subsections Applicable to Regulatory Issue |
|---|---|---|---|---|---|
| Africa | Kenya | Act No. 47 of 2013 | Wildlife (Conservation and Management) Act | 2013 | Devolves wildlife governance and decision-making to county-level; calls for the creation of County Wildlife Conservation and Compensation Committees<br>Devolves wildlife conservation and management to landowners and managers; recognizes wildlife conservation as a form of landuse with equal recognition as other land uses (such as agriculture); provides for establishing Wildlife Conservancies<br>Seeks to provide better access to benefits arising from wildlife conservation; |
| Africa | Nigeria | Law No. 41 of 1955 (Chapter 57 L.R.S.N. 1999) | Forest Law | 1956 | Along with the Land Use Decree and National Parks Decree, helps establish a network of protected areas by providing for forest reserve and protected forest declarations |
| Africa | Nigeria | Decree No. 6 of 1978 | Land Use Act | 1978 | Along with the Forest Law and National Parks Decree, helps establish a network of protected areas; vests all non-Federal agency held land in military governors of each State to hold in trust for the people |
| Africa | Nigeria | Act No. 11 of 1985 | Endangered Species (Control of International Trade and Traffic) Act | 1985 | Implements CITES; prohibits the hunting, capture, or trade in endangered animal species listed within the act's First Schedule and restricts the taking and trade of threatened animal species listed in the act's Second Schedule<br>Has a broader scope than CITES (also covers domestic taking of listed species) |
| Africa | Nigeria | Decree No. 101 of 1991 | National Parks Decree | 1991 | Along with the Forest Law and Land Use Decree, helps establish a network of protected areas by setting up 6 national parks in the country; establishes Board for National Parks Service to carry out in-situ flora and fauna conservation |

*(Continued)*

| Region | Country/State | Citation | Title | Year Passed | Summary or Subsections Applicable to Regulatory Issue |
|---|---|---|---|---|---|
| Africa | South Africa | Act No. 73 of 1989 | Environment Conservation Act | 1989 | Governs natural resources protection; provides for declaring protected natural areas or limited development areas<br>Part V covers agricultural and land use and transformation issues |
| Africa | South Africa | Act No. 84 of 1998 | National Forests Act | 1998 | Provides for managing and conserving public forests Chapter 3 contains specific measures to combat deforestation |
| Africa | South Africa | Act No. 10 of 2004 | National Environmental Management Biodiversity Act | 2004 | Governs biodiversity preservation; aims to protect ecosystems and species that are threatened or in need of protection; establishes South African National Biodiversity Institute<br>Aims to support:<br>• the use of indigenous biological resources in a sustainable manner; and<br>• the fair and equitable sharing of benefits arising from bioprospecting involving indigenous resources |
| Africa | South Africa | No. R. 152 of 2007 | Threatened or Protected Species Regulations | 2007 | Carry out the National Environmental Management Biodiversity Act; prohibit certain restricted activities involving listed threatened or protected species |
| Americas | Brazil | Decreto n. 76.623/75 de novembro de 1975; Legislaçao *Federal de Meio Ambiente*, Vol. I, Brasília 1996. pp. 372–382 | Decree No. 76.623 laying down a list of flora and fauna endangered species, according to the convention on the international trade/*Decreto No. 76.623 Promulga a convencao sobre comercio internacional das especies da flora e fauna selvagem em perigo de extincao* | 1975 | Implements CITES; lists all Brazilian flora and fauna endangered species Most listed species found in the Atlantic forest and the cerrado grasslands (which have largely been converted to cattle ranches and industrial soy farms) Number of listed species tripled in 2008 |

(*Continued*)

| Region | Country/State | Citation | Title | Year Passed | Summary or Subsections Applicable to Regulatory Issue |
|---|---|---|---|---|---|
| Americas | Brazil | Decreto n. 4.703, de 21 de Maio de 2003 | Decree No. 4.703 on the National Programme on Biological Diversity (PRONABIO) and on the National Commission on Biodiversity/*Decreto No. 4.703, Dispõe sobre o Programa Nacional da Diversidade Biolo, gica—PRONABIO e a Comissa~o Nacional da Biodiversidade, e da, outras providências* | 2003 | Establishes the Program on Biological Diversity, which promotes a partnership between civil society and the government for conserving organic diversity and sustainably using its components |
| Americas | Brazil | Lei n. 12.651, de 25 de Maio de 2012 | Law No. 12.651 on the Protection of Native Forests/*Lei No. 12.651, dispõe sobre a proteça-o da vegetaça-o nativa* | 2012 | Governs the protection and sustainable use and exploitation of native forests and other indigenous plants; recognizes the importance of preserving national forests, habitats, biodiversity, soil, and water resources for future generations; establishes permanent protected forest areas as the primary tool for protecting biodiversity in Brazil<br>Promotes rural economic development<br>Amended by Law No. 12.727 of 2012 to establish general norms on:<br>• plant protection;<br>• permanent preservation areas and Legal Reserve Areas;<br>• forestry exploitation; and<br>• control over forestry origin products |

*(Continued)*

| Region | Country/State | Citation | Title | Year Passed | Summary or Subsections Applicable to Regulatory Issue |
|---|---|---|---|---|---|
| Americas | Brazil | Lei n. 12.805, de 29 de Abril de 2013 | Law No. 12.805 creating the National Policy for Livestock, Agro-forestry and Sylvo-pastoralism (ILPF)/*Lei No. 12.805 nstitui a Política Nacional de Integração Lavoura- Pecua, ria-Floresta* | 2013 | Establishes National Policy for Livestock, Agro-forestry, and Sylvo-pastoralism; aims to improve the productivity of agro-forestry income-generating activities; sets a number of goals, including:<br>• seeking alternatives to traditional monoculture in deforested areas;<br>• mitigating deforestation caused by native forest conversion into pasture or agricultural areas;<br>• contributing to the maintenance of permanent preservation areas and of legal reserves;<br>• promoting environmental education;<br>• promoting rehabilitation of degraded areas of pasture through sustainable production systems; and<br>• promoting diversification of production systems through incorporating forestry resources |
| Americas | Canada | R.S.C. 1985, c W-9 | Canada Wildlife Act | 1985 | Governs wildlife protection; covers both species and habitat protection; provides for declaring "national wildlife areas"<br>Implemented by Wildlife Area Regulations |
| Americas | Canada | S.C. 1992, c. 52 | Wild Animal and Plant Protection and Regulation of International and Interprovincial Trade Act (WAPPRIITA) | 1992 | Implements CITES<br>Has a broader scope than CITES; applies not only to species on the CITES control list, but also to:<br>• foreign species whose capture, possession, and export are prohibited or regulated by laws in their country of origin;<br>• Canadian species whose capture, possession, and transportation are regulated by provincial or territorial laws; and<br>• species whose introduction into Canadian ecosystems could endanger Canadian species |

(*Continued*)

| Region | Country/State | Citation | Title | Year Passed | Summary or Subsections Applicable to Regulatory Issue |
|---|---|---|---|---|---|
| Americas | Canada | S.C. 1994, c. 22 | Migratory Birds Convention Act | 1994 | Aims to conserve migratory bird populations by regulating potentially harmful human activities; requires permits for all activities affecting wild birds or their eggs or nests<br>Implemented by Migratory Bird Sanctuary Regulations, which grant sanctuary status to habitat areas important to migratory birds |
| Americas | Canada | S.C. 2002, c 29 | Species at Risk Act | 2002 | Governs endangered species protection; aims to support the recovery of wildlife species that are extirpated, endangered, or threatened as a result of human activity; provides for critical habitat designations; establishes a Committee on the Status of Endangered Wildlife in Canada |
| Americas | Canada (British Columbia) | S.B.C 2003, c. 58 | Integrated Pest Management Act | 2003 | Part 2 prohibits or restricts the use of pesticides that cause or may cause "unreasonable adverse effects" |
| Americas | Mexico | *Diario Oficial de la Federacio, n*, 28 de enero de 1988, últimas reformas publicadas DOF 05-11-2013 | General Law on Ecological Balance and Environmental Protection/ *Ley General del Equilibrio Ecolo, gico y Proteccio, n al Ambiente* (LGEEPA) | 1988 | Establishes:<br>• Natural Protected Areas as the country's primary means of protecting endangered species<br>• Restoration Areas to addresses problems of desertification and land degradation; calls for eco-logical restoration programs in these areas<br>Title III, Chap. 2 calls for sustainable agriculture practices |
| Americas | Mexico | *Diario Oficial de la Federacio, n*, 3 de julio de 2000, última reforma publicada DOF 26-01-2015 | General Wildlife Law/*Ley General de Vida Silvestre* (LGVS) | 2000 | Governs wildlife and biodiversity preservation (including endangered species and habitats); contains provisions on:<br>• sustainable use of wild-life;<br>• wildlife diseases;<br>• species at risk and critical habitats; and<br>• controlling nuisance species |

*(Continued)*

| Region | Country/State | Citation | Title | Year Passed | Summary or Subsections Applicable to Regulatory Issue |
|---|---|---|---|---|---|
| Americas | United States | 16 U.S.C. §§ 3371-3378 | Lacey Act | 1900 | Aims to combat trafficking in illegal plant and animal species; forbids anyone to import, export, sell, acquire, or purchase fish, wildlife or plants that are taken, possessed, transported, or sold:<br>1. in violation of U.S. or Indian law, or<br>2. in interstate or foreign commerce involving any fish, wildlife, or plants taken possessed or sold in violation of State or foreign law |
| Americas | United States | 16 U.S.C. §§ 703-712, Ch. 128 | Migratory Bird Treaty Act | 1918 | Prohibits taking of Native American bird species |
| Americas | United States | 16 U.S.C. § 1131 | National Wilderness Preservation System (Wilderness Act) | 1964 | Creates formal mechanism for designating federal wilderness areas; prohibits disruptive human activities (including intensive agriculture) in such areas |
| Americas | United States | 42 U.S.C. § 4321 et seq | National Environmental Policy Act (NEPA) | 1969 | Requires environmental reviews of proposed major federal actions; requires consideration of impacts of these actions on species listed under the Endangered Species Act (ESA) |
| Americas | United States | 16 U.S.C. §§ 1361-1407 | Marine Mammal Protection Act | 1972 | Extends ESA-like protections to non-listed marine mammal species |
| Americas | United States | 16 U.S.C. § 1531 et seq. | Endangered Species Act (ESA) | 1973 | Governs endangered species protection; provides for:<br>• "listing" endangered and threatened plant and animal species<br>• prohibiting the "taking" of listed animal species<br>• designating critical habitat areas |
| Americas | United States | 16 U.S.C. §§ 2901-2911 | Fish and Wildlife Conservation Act (Non-game Act) | 1980 | Grants assistance to states for implementing conservation plans and programs for non-game fish and wildlife |
| Americas | United States | 16 U.S.C. §§ 4901-4916 et seq. | Wild Bird Conservation Act (WBCA) | 1992 | Prohibits import of CITES-listed bird species; encourages wild bird conservation programs in countries of origin |

(*Continued*)

| Region | Country/State | Citation | Title | Year Passed | Summary or Subsections Applicable to Regulatory Issue |
|---|---|---|---|---|---|
| Americas | United States | 7 U.S.C. § 136 et seq. | Federal Insecticide, Fungicide, and Rodenticide Act (FIFRA) | 1996 | Regulates pesticide distribution, sale, and use |
| Americas | United States | H.R. 2, P.L. 115-334 | Agriculture Improvement Act of 2018 (2018 Farm Bill) | 2018 | Governs agricultural and food policy at the federal level; aims to:<br>• expanding markets for agricultural products at home and abroad;<br>• strengthening conservation efforts;<br>• creating new opportunities for local and regional food systems; and<br>• grow the bio-based economy<br>Updates previous 2014 Farm Bill ("Agricultural Act of 2014") |
| Asia | China | Adopted at the Seventh Meeting of the Standing Committee of the Sixth National People's Congress on September 20, 1984 | Forestry Law of the People's Republic of China | 1984 | Aims to protect natural forest ecosystems<br>Article 1 establishes various goals, including:<br>• protecting, cultivating, and rationally exploiting forest resources;<br>• accelerating territorial afforestation;<br>• using forests for water storage, soil conservation, climate regulation, and environmental improvement; and<br>• supplying forest products to the people |

(*Continued*)

| Region | Country/State | Citation | Title | Year Passed | Summary or Subsections Applicable to Regulatory Issue |
|---|---|---|---|---|---|
| Asia | China | Adopted at the 24th Meeting of the Standing Committee of the Fifth National People's Congress and promulgated by Order No. 9 of the Standing Committee of the National People's Congress on August 23,1982 | Marine Environmental Law of the People's Republic of China | 1982 | Governs marine protection Article 1 establishes various goals, including:<br>• protecting the marine environment and resources;<br>• preventing pollution damage;<br>• maintaining ecological balances;<br>• safeguarding human health; and<br>• promoting the development of marine programs |
| Asia | China | Adopted at the 11th Meeting of the Standing Committee of the Sixth National People's Congress on June 18, 1985 | Grasslands Law of the People's Republic of China | 1985 | Article 1 establishes various goals, including:<br>• protecting, developing, and making rational use of grasslands;<br>• improving ecological environments;<br>• maintaining the diversity of living things;<br>• modernizing animal husbandry; and<br>• promoting the sustainable development of the economy and society |

*(Continued)*

| Region | Country/State | Citation | Title | Year Passed | Summary or Subsections Applicable to Regulatory Issue |
|---|---|---|---|---|---|
| Asia | China | Adopted at the Fourth Meeting of the Standing Committee of the Seventh National People's Congress and promulgated by Order No. 9 of the President of the People's Republic of China on November 8, 1988 | Protection of Wildlife Law of the People's Republic of China | 1988 | Governs biodiversity preservation; provides general principles for wildlife protection; applies to terrestrial and aquatic species that are rare or endangered, as well as terrestrial species which are beneficial or of important economic or scientific value |
| Asia | China | Adopted at the 24th Meeting of the Standing Committee of the Sixth National People's Congress on January 21, 1988 | Water Law of the People's Republic of China | 1988 | Article 1 establishes various goals, including:<br>• rational development, utilization, preservation, and protection of water to prevent and control water disasters; and<br>• sustainable utilization of water resources to meet the needs of national economic and social development |

(*Continued*)

| Region | Country/State | Citation | Title | Year Passed | Summary or Subsections Applicable to Regulatory Issue |
|---|---|---|---|---|---|
| Asia | China | Adopted at the 11th Meeting of the Standing Committee of the Seventh National People's Congress on December 26, 1989, promulgated by Order No. 22 of the President of the People's Republic of China on December 26, 1989 | Environmental Protection Law of the People's Republic of China | 1989 | Article 1 establishes various goals, including:<br>• protecting and improving the people's environment and the ecological<br>• environment;<br>• preventing and controlling pollution and other public hazards;<br>• safeguarding human health; and<br>• facilitating the development of socialist modernization |
| Asia | China | Promulgated by the State Council on December 18, 1989 | Regulations on the Prevention of Forest Plant Diseases and Insect Pests | 1989 | Aims to protect natural forest ecosystems from plant diseases and insect pests |
| Asia | China | Adopted at the 20th Meeting of the Standing Committee of the Seventh National People's Congress of the People's Republic of China on June 29, 1991 | Water and Soil Conservation Law of the People's Republic of China | 1991 | Aims to protect natural forest ecosystems by conserving water and soil resources<br>Article 1 establishes various goals, including:<br>• preventing and controlling soil erosion;<br>• protecting and rationally utilizing water and soil resources;<br>• mitigating disasters from floods, droughts, and sandstorms;<br>• improving ecological environments; and<br>• developing production |

(*Continued*)

| Region | Country/State | Citation | Title | Year Passed | Summary or Subsections Applicable to Regulatory Issue |
|---|---|---|---|---|---|
| Asia | China | Adopted at the 2nd Meeting of the Standing Committee of the Eighth National People's Congress on July 2, 1993 | Agriculture Law of the People's Republic of China | 1993 | Article 49 provides that "the country protects Intellectual Property such as Plant Variety and Geographical Indication" |
| Asia | China | Promulgated by Decree No. 204 of the State Council of the People's Republic of China on September 30, 1996 | Regulation on Protection of Wild Plants | 1996 | Article 1 establishes various goals, including:<br>• protecting, developing, and rationally utilizing wild plant resources; and<br>• preserving biodiversity and maintaining ecological balances |
| Asia | China | Adopted at the 2nd Meeting of the Standing Committee of the Ninth National People's Congress on April 29, 1998 and promulgated by Order No. 4 of the President of the People's Republic of China on April 29, 1998 | Fire Protection Law of the People's Republic of China – Fire Prevention Regulations | 1998 | Aims to protect natural forest ecosystems from fires |

*(Continued)*

| Region | Country/State | Citation | Title | Year Passed | Summary or Subsections Applicable to Regulatory Issue |
|---|---|---|---|---|---|
| Asia | China | Adopted at the 24th Executive Meeting of the State Council on September 2, 1994, promulgated by Decree No. 167 of the State Council of the People's Republic of China on October 9, 1994 | Regulations of the People's Republic of China on Nature Reserves (Nature Reserves Regulations) | 2000<br>1994 | Provide for constructing and managing nature reserves, including protected areas |
| Asia | China | Adopted at the 23rd meeting of the Standing Committee of the Ninth National People's Congress of the People's Republic of China on Aug. 31, 2001 | Desert Prevention and Transformation Law of the People's Republic of China | 2001 | Article 1 establishes various goals, including:<br>• preventing land desertification;<br>• transforming desertified land;<br>• protecting the safety of the environment; and<br>• promoting the sustainable development of both the economy and society |

(Continued)

| Region | Country/State | Citation | Title | Year Passed | Summary or Subsections Applicable to Regulatory Issue |
|---|---|---|---|---|---|
| Asia | China | Adopted at the 66th executive meeting of the State Council on December 6, 2002; Order of the State Council of the People's Republic of China (No. 367) | Regulations on Restoring Farmland to Forest | 2002 | Incentivize farmers to convert their agricultural lands back to forest through subsidies (including food, seed and sampling, and general living subsidies) |
| Asia | China (Hong Kong) | L.N. 206 of 2006 | Protection of Endangered Species of Animals and Plants Ordinance (Chapter 586) | 2006 | Gives effect to CITES in Hong Kong |
| Asia | India | Act No. 53 of 2001 | Protection of Plant Varieties and Farmers' Rights Act | | Sets out basic framework for protecting rights relating to breeding new plant varieties (including the conservation, selection, and improvement of farmers' varieties) |
| Asia | India | Act No. 53 of 1972 | Wildlife (Protection) Act | 1972 | Governs the protection of wild animals, birds, and plants; contains provisions on:<br>• restricting the hunting of listed wild animal species;<br>• restricting the picking or uprooting of listed plant species; and<br>• establishing wildlife sanctuaries, National Parks, and Closed Areas<br>Chapter V regulates the trade or commerce in wild animals, animal articles, and trophies<br>Applies to all Indian States except for Jammu and Kashmir |

*(Continued)*

| Region | Country/State | Citation | Title | Year Passed | Summary or Subsections Applicable to Regulatory Issue |
|---|---|---|---|---|---|
| Asia | India | Act No. 18 of 2003 | Biological Diversity Act | 2003 | Implements CBD; governs biodiversity conservation; promotes sustainable use of biological resources and the equitable sharing of benefits arising from such use |
| Asia | India (Andhra Pradesh) | Act No. 11 of 1997 | Andhra Pradesh Farmers' Management of Irrigation Systems Act | 1997 | Lays out the powers and responsibilities of farmers' organizations in managing irrigation systems<br>Sect. 16 provides that farmers' organizations' primary aim is to promote the secure distribution of water and to maintain irrigation systems, while achieving efficient water utilization and environmental protection goals |
| Asia | India (Kerala) | Act No. 28 of 2008 | Kerala Conservation of Paddy Land and Wetland Act | 2008 | Aims to:<br>• regulate the conversion and development of paddy fields into agricultural lands;<br>• protect wetland areas;<br>• ensure food security; and<br>• sustain ecological systems Amended in 2011 by Act No. 14 of 2011 |
| Asia | India (Rajasthan) | Act No. 21 of 2000 | Rajasthan Farmers' Participation in Management of Irrigation Systems Act | 2000 | Lays out the powers and responsibilities of farmers' organizations in managing irrigation systems; contains provisions on:<br>• promoting and securing the distribution of water;<br>• adequately maintaining the irrigation system;<br>• efficiently and economically utilizing water to optimize agricultural production; and<br>• protecting the environment and ensuring ecological balances |

*(Continued)*

| Region | Country/State | Citation | Title | Year Passed | Summary or Subsections Applicable to Regulatory Issue |
|---|---|---|---|---|---|
| Asia | India (Tamil Nadu) | Act No. 7 of 2000 | Tamil Nadu Farmers' Management of Irrigation Systems Act | 2000 | Lays out the powers and responsibilities of farmers' organizations in managing irrigation systems |
| Asia | Israel | P.O. Suppl. I, 1 | Forest Ordinance | 1926 | Governs forest protection; allows Ministry of Agriculture to declare any non-privately owned forests as protected State forests (where grazing and other harmful activities are prohibited); embraces "polluter pays" principle with penalties (including fines or imprisonment) for environmental harm |
| Asia | Israel | Law No. 5715-1955 | Israel Wildlife Protection Law | 1955 | Implements CITES; requires hunting licenses for game hunting or pest extermination |
| Asia | Israel | Law No. 5719-1959 | Water Law | 1959 | Creates framework for controlling and protecting water resources; governs water use for agricultural purposes |
| Asia | Israel | Law No. 5758-1998 | National Parks, Nature Reserves, National Sites and Memorial Sites Law | 1963 | Provides basic legal structure for protecting Israel's natural habitats, natural assets, wildlife, and sites of scientific, historic, architectural, and educational interest; establishes united Nature and National Parks Protection Authority, whose powers include declaring and maintaining nature reserves and national parks |
| Asia | Israel | *Laws of the State of Israel*, Vol 21, 1966/67, from 24.10.1966 to 25.9.1967, pp. 102–105-110. *Dinim* Vol. 10, pp. 5337 | Agricultural Settlement (Restriction on Use of Agricultural Land and of Water) Law | 1967 | Establishes personal water use quotas for agricultural land owners or operators |

(*Continued*)

| Region | Country/State | Citation | Title | Year Passed | Summary or Subsections Applicable to Regulatory Issue |
|---|---|---|---|---|---|
| Asia | Japan | Law No. 32 of 1918 | Wildlife Protection and Hunting Law | 1918 | Aims to:<br>• protect birds and mammals;<br>• increase populations of birds and mammals; and<br>• control pests through implementing wildlife protection projects and hunting controls |
| Asia | Japan | Law No. 85 of 1972 | Nature Conservation Law | 1972 | Provides basic framework for nature conservation in Japan; aims to protect and manage natural resources and natural ecosystems; provides for declaring:<br>• wilderness areas (where human activities are forbidden);<br>• nature conservation areas; and<br>• prefectural nature conservation areas |
| Asia | Japan | Law No. 75 of 1992 | Law for the Conservation of Endangered Species of Wild Fauna and Flora | 1992 | Governs endangered species protection; prohibits taking or transporting (as well as importing and exporting) endangered species; contains provisions on:<br>• designating natural habitat conservation areas;<br>• registering international endangered species; and<br>• establishing natural habitat rehabilitation programs to maintain viable populations of endangered species |
| Asia | Japan | Act No. 58 of 2008 | Basic Act on Biodiversity | 2008 | Implements CBD and UNFCCC; attempts to synthesize multiple legal instruments on biodiversity conservation and to integrate biodiversity values into all tiers (from national to local) of decision-making; encourages sustainable land and resource use practices that avoid or minimize impacts on biodiversity<br>Adopted in accordance with the Basic Environment Law (Act No. 91 of 1993) |

(*Continued*)

| Region | Country/State | Citation | Title | Year Passed | Summary or Subsections Applicable to Regulatory Issue |
|---|---|---|---|---|---|
| Asia | Russia | Law No. 52-FZ of 1995 | Federal Law of the Russian Federation on Wildlife (Law on Wildlife) | 1995 | Governs wildlife protection; declares all wildlife to be the national property of Russian people and to be under protection; encourages sustainable use and management of natural wildlife resources |
| Asia | Russia | Ministerial Decree No. 158 of 19 Feb 1996 | Ministerial Decree No. 158 of 19 February 1996 on the catalogue of endangered wildlife species ("Red Book") of the Russian Federation | 1996 | Provides for compiling the Red Data Book of the Russian Federation (RDBRF), a basic state document listing rare and endangered species of wild animals, plants, and fungi<br>Complements the Federal Law on Environmental Protection (No. 7-FZ of Jan 10, 2002) and the Law on Animal World (No. 52-FZ of Apr 24, 1995), which provide further a legislative basis for the RDBRF |
| Asia | Russia | Ministerial Decree No. 1010 of Aug 13, 1997 | Ministerial Decree No. 1010 on strengthening the protection of wildlife species and their natural habitat on the territory of forestry | 1997 | Implements the Federal Law on Wildlife; authorizes the Federal Forest Service and its territorial branches to carry out wildlife species and natural habitat protection activities in forest territories |
| Asia | Russia | Ministerial Decree No. 13 of Jan 6, 1997 | Ministerial Decree No. 13 Enforcing the Regulations on Hunting of Endangered Wildlife Species Under Protection of the Russian Federation | 1997 | Aims to restrict hunting of endangered wildlife species by imposing more severe rules (such as stricter hunting permit requirements) |
| Asia | Russia | Law No. 136-FZ of Oct 25, 2001 | Land Code | 2001 | Defines agricultural lands by law; divides all of Russia's lands into 7 categories by use; delegates most land management authority to small, local elected bodies<br>Provides that designated agricultural lands (~15% of all land) can only be used for agricultural production (all other uses are prohibited) |

*(Continued)*

| Region | Country/State | Citation | Title | Year Passed | Summary or Subsections Applicable to Regulatory Issue |
|---|---|---|---|---|---|
| Australia/ New Zealand | Australia | No. 91 of 1999 | Environmental Protection and Biodiversity Conservation Act (EPBC Act) | 1999 | Establishes a national environmental protection framework, with a focus on conserving biodiversity; calls for ecologically sustainable development and use of natural resources; covers:<br>• nationally endangered, threatened, and migratory species (including international trade in wildlife and wildlife products);<br>• nationally threatened ecological communities;<br>• wetlands of international importance; and<br>• world and national heritage properties (including the Great Barrier Reef) Provides for a streamlined national environmental assessment and approvals process;<br>Promotes use of Indigenous peoples' biodiversity knowledge<br>Delegates matters of state and local significance to states and territories |
| Australia/ New Zealand | Australia | Act No. 92 of 1999 | Environmental Reform (Consequential Provisions) Act | 1999 | Repeals and amends a number of acts relevant to the enactment of the Environmental Protection and Biodiversity Conservation Act (EPBC) Act<br>Drafted in harmony with EPBC Act |
| Australia/ New Zealand | New Zealand | 1949 No. 19 | Forests Act | 1949 | Governs forest management and timber activities; defines sustainable forest management<br>Does not apply to planted indigenous forests |
| Australia/ New Zealand | New Zealand | 1953 No. 31 | Wildlife Act | 1953 | Sets up "default" protected status for majority of New Zealand's vertebrate species: instead of listing species that are specifically protected, it provides that *all* vertebrate species that are *not* listed are protected |
| Australia/ New Zealand | New Zealand | 1977 No. 66 | Reserves Act | 1977 | Provides for declaring and managing national reserves<br>Part 3 concerns farming, grazing, and afforestation |

(*Continued*)

| Region | Country/State | Citation | Title | Year Passed | Summary or Subsections Applicable to Regulatory Issue |
|---|---|---|---|---|---|
| Australia/ New Zealand | New Zealand | 1980 No. 66 | National Parks Act | 1980 | Provides for establishing national parks<br>Part I(4) concerns preserving indigenous plants and animals |
| Australia/ New Zealand | New Zealand | 1987 No. 65 | Conservation Act | 1987 | Governs biodiversity conservation; sets forth conservation management strategies (CMSs) and conservation management plans (CMPs), with an emphasis on "protection" rather than "sustainable management"; sets up a hierarchy of activities occurring on public conservation land:<br>1. intrinsic value;<br>2. non-commercial recreation; and<br>3. tourism Complements the National Parks Act and the Reserves Act by providing for specially protected areas, including: conservation parks, wilderness areas, ecological areas, sanctuary areas, watercourse areas, amenity areas, and wildlife management areas Establishes the Department of Conservation and the Department of Fish and Game |
| Australia/ New Zealand | New Zealand | 1989 No. 18 | Trade in Endangered Species Act | 1989 | Provides for managing, conserving, and protecting endangered, threatened, and exploited species |
| Australia/ New Zealand | New Zealand | S.R. 2007/354 | Forests (Permanent Forest Sink) Regulations | 2007 | Set out various requirements for establishing and maintaining permanent forest sinks (defined as any eligible forest actively established on Kyoto-compliant land)<br>Enacted pursuant to sections 67Y and 67ZL of the Forests Act |
| Europe | European Union | O.J. L. 20, 26.1.2010, pp. 7–25 | Directive 2009/147/EC of the European Parliament and of the Council of 30 November 2009 on the conservation of wild birds (Birds Directive) | 1979 | Implements CBD; establishes comprehensive protection scheme for all wild bird species naturally occurring in the EU Complements the Habitats Directive |

*(Continued)*

| Region | Country/State | Citation | Title | Year Passed | Summary or Subsections Applicable to Regulatory Issue |
|---|---|---|---|---|---|
| Europe | European Union | O.J. L. 206, 22.7.1992, pp. 7–50 | Council Directive 92/43/EEC of 21 May 1992 on the conservation of natural habitats and of wild fauna and flora (Habitats Directive) | 1992 | Implements CBD in EU; cornerstone of EU nature conservation policy; aims at to preserve, protect, and improve the quality of the environment, including conserving of natural habitats and of wild fauna and flora; establishes Natura 2000 habitat preservation network |
| Europe | European Union | O.J. L. 327, 22.12.2000, pp. 1–73 | Directive 2000/60/EC of the European Parliament and of the Council establishing a framework for Community action in the field of water policy (Water Framework Directive or WFD) | 2000 | Article 1 aims to regulate water as a "heritage" Article 16 calls for integrating sustainable water management into policy areas, such as agriculture and fisheries |
| Europe | European Union | O.J. L. 162, 30.4.2004, pp. 18–28 | Council Regulation (EC) No 870/2004 of 24 April 2004 establishing a Community program on the conservation, characterization, collection and utilization of genetic resources in agriculture and repealing Regulation (EC) No. 1467/94 | 2004 | Article 1 declares that "[b] iological and genetic diversity in agriculture is essential to the sustainable development of agricultural production and of rural areas" and calls for measures "to conserve, characterize, collect and utilize the potential of that diversity in a sustainable way to promote the aims of the common agricultural policy (CAP)" |

(*Continued*)

| Region | Country/State | Citation | Title | Year Passed | Summary or Subsections Applicable to Regulatory Issue |
|---|---|---|---|---|---|
| Europe | European Union | O.J. L. 347, 20.12.2013, pp. 549–607 | Regulation (EU) No. 1306/2013 of the European Parliament and of the Council on the financing, management and monitoring of the common agricultural policy and repealing Council Regulations (EEC) No. 352/78, (EC) No. 165/94, (EC) No. 2799/98, (EC) No. 814/2000, (EC) No. 1290/2005 and (EC) No. 485/2008 | 2013 | Sets out financing rules under the Common Agricultural Policy (CAP), including expenditure on rural development programs |
| Europe | European Union | O.J. L. 347, 20.12.2013, pp. 487–548 | Regulation (EU) No. 1305/2013 of the European Parliament and of the Council on support for rural development by the European Agricultural Fund for Rural Development (EAFRD) and repealing Council Regulation (EC) No. 1698/2005 (Rural Development Regulations) | 2013 | Calls for reforms to the Common Agricultural Policy (CAP) to take place starting in January 2014 (through 2020)<br>Article 2 calls for establishing a rural development policy "to accompany and complement direct payments and market measures of the CAP" |

(*Continued*)

| Region | Country/State | Citation | Title | Year Passed | Summary or Subsections Applicable to Regulatory Issue |
|---|---|---|---|---|---|
| Europe | United Kingdom | 1981, c. 69 | Wildlife and Countryside Act | 1981 | Implemented EU Birds Directive in the UK; aims to:<br>• protect native plant and animal species (especially those that are threatened or endangered);<br>• control the release of non-native species;<br>• regulate the import and export of endangered species; and<br>• establish national nature reserves (including marine nature reserves and national parks)<br>Calls for a compulsory five-year review of listed species, which were greatly expanded via the Wildlife and Countryside Act 1981 (Variation of Schedule 9) (England and Wales) Order 2010;<br>Amended for continued operation by the Animal Health, Invasive Alien Species, Plant Breeders' Rights and Seeds (Amendment etc.) (EU Exit) Regulations 2019 |
| Europe | United Kingdom | 1996, c. 3 | Wild Mammals (Protection) Act | 1996 | Provides for the protection of wild mammals from certain cruel acts |
| Europe | United Kingdom | 2006, c. 16 | Natural Environment and Rural Communities Act | 2006 | Establishes bodies (e.g., Natural England, Country-side Council of Wales, and Scottish Natural Heritage) to manage affairs of natural environment and rural communities; covers:<br>• wildlife;<br>• sites of special scientific interest;<br>• National Parks; and<br>• the Broads |
| Europe | United Kingdom (England and Wales) | 2000, c. 37 | Countryside and Rights of Way Act (CRoW Act) | 2000 | Regulates public access to the countryside; lays out rules on nature conservation and wildlife protection, including restrictions on driving mechanically propelled vehicles off-roads (i.e., on fields) |

*(Continued)*

| Region | Country/State | Citation | Title | Year Passed | Summary or Subsections Applicable to Regulatory Issue |
|---|---|---|---|---|---|
| Europe | United Kingdom (England and Wales) | 2007, c. 23 | Sustainable Communities Act (Chapter 23) | 2007 | Promotes sustainable local communities |
| Europe | United Kingdom (England) | S.I. 2003/1235 (replacing S.I. 2001/3139) | Organic Farming (England Rural Development Programme) Regulations | 2003 | Implements EU Rural Development Regulations; provides for aid payments to farmers who introduce organic farming methods on their lands and comply with certain environmental conditions |
| Europe | United Kingdom (Northern Ireland) | S.R. No. 172 of 2008 | Organic Farming Regulations (Northern Ireland) | 2008 | Provide for the payment of grants to farmers who agree to introduce organic farming methods and comply with the management requirements and standards of good agricultural and environmental conditions (similar to the Organic Farming (England Rural Development Programme) Regulations) |
| Europe | United Kingdom (Scotland) | 2003 asp 2 | Land Reform (Scotland) Act | 2003 | Establishes statutory public rights of access to land for recreational and other purposes |
| Europe | United Kingdom (Scotland) | 2011, asp 6 | Wildlife and Natural Environment (Scotland) Act | 2011 | Applies Wildlife and Countryside Act in Scotland |
| International | Parties to Convention | UNEP/CBD/COP/DEC/X/1, 29 Oct. 2010 | Nagoya Protocol on Access and Benefit-Sharing | 2010 | Aims to ensure fair and equitable sharing of benefits arising out of the utilization of genetic resources; creates incentives to use genetic resources sustainably and equitably<br>Adopted by CBD COP 10 |

*(Continued)*

| Region | Country/State | Citation | Title | Year Passed | Summary or Subsections Applicable to Regulatory Issue |
|---|---|---|---|---|---|
| International | Parties to Convention | 996 U.N.T.S. 245, T.I.A.S. No. 11,084 | Convention on Wetlands of International Importance especially as Waterfowl Habitat (Ramsar Convention) | 1971 | Seeks to conserve and wisely use all wetlands through local and national actions and international cooperation; incorporates sustainable development goals; instructs nations to limit destructive agricultural practices (like using fertilizers and pesticides in reclaimed wetlands) and encourage sustainable ones (like using flood-resistant crops in naturally functioning floodplains) |
| International | Parties to Convention | 27 U.S.T. 37, 1037 U.N.T.S. 151 | Convention for the Protection of the World Cultural and Natural Heritage (World Heritage Convention) | 1972 | Aims to conserve natural and cultural sites of interest to the international community; provides for listing designated World Heritage Sites, which can include forests and agricultural landscapes |
| International | Parties to Convention | 27 U.S.T. 1087, 12 I.L.M 1085 | Convention on International Trade in Endangered Species of Wild Fauna and Flora (CITES) | 1973 | Aims to ensure that international wildlife trade does not threaten the survival of listed plant and animal species; sets controls on trade in listed species (and products derived from such species) |
| International | Parties to Convention | S. Treaty Doc. No. 103–20 (1993), 31 I.L.M. 818 | Convention on Biological Diversity (CBD) | 1992 | Addresses all aspects of biodiversity<br>COP 3's Decision III/11 established a multi-year work program on agrobiodiversity<br>COP 5's Decision V/5 made agrobiodiversity a thematic work program |

(*Continued*)

| Region | Country/State | Citation | Title | Year Passed | Summary or Subsections Applicable to Regulatory Issue |
|---|---|---|---|---|---|
| International | Parties to Convention | 33 I.L.M. 1328 | United Nations Convention to Combat Desertification in Those Countries Experiencing Serious Drought and/or Desertification, Particularly in Africa (UNCCD) | 1994 | Sets out a universal legal policy and advocacy framework for combatting desertification and promoting sustainable development; aims to reverse and prevent desertification/land degradation and mitigate the effects of drought in affected areas |
| International | Parties to Convention (and some non-Party range states) | 29 U.S.T. 4647, 1134 U.N.T.S. 97 | Convention on the Conservation of Migratory Species of Wild Animals (CMS or Bonn Convention) | 1979 | Seeks to provide a global platform for conserving and sustainably utilizing migratory animals and their habitats; instructs all States falling within an animal's migratory range to coordinate conservation measures |
| International | Parties to Protocol | 37 I.L.M. 22 (1998) | Kyoto Protocol to the United Nations Framework Convention on Climate Change | 1997 | Sets binding GHG reduction targets, which are potentially achievable through measures aimed at preventing habitat loss (e.g., forest protection policies to preserve "carbon sinks") and restricting intensive agricultural practices |
| International | Parties to Treaty | 2400 U.N.T.S. 303 | International Treaty on Plant Genetic Resources for Food and Agriculture (IT PGRFA) | 2001 | Aims to:<br>• recognize farmers' contribution to crop diversity;<br>• establish a global system to provide farmers, plant breeders, and scientists with access to plant genetic materials; and<br>• ensure that benefits are shared with the countries where the resources originated Drafted in harmony with CBD |

## APPENDIX 4: MARINE AND (OVER-)FISHING

| Region | Country/State | Citation | Title | Year Passed | Summary or Subsections Applicable to Regulatory Issue |
|---|---|---|---|---|---|
| Africa | Kenya | Act No. 5 of 1989 | Fisheries Act (Cap. 378) | 1989 | Sets out basic principles for fisheries development, management, exploitation, utilization, and conservation; calls for:<br>• closed seasons;<br>• prohibited fishing areas;<br>• gear and use restrictions;<br>• fish size and age limits;<br>• fisheries access limits; and<br>• prohibiting fishing for marine mammals<br>Part IV calls for imprisonment of foreign vessel owners caught illegally fishing in Kenya's waters<br>Implemented by:<br>• 1991 Fisheries (General) Regulations (Cap. 378), providing specific measures on fishing gear, landing, protection of breeding areas, and protection of marine mammals and turtles;<br>• 2003 Fisheries (Prohibitions) Regulations (Cap. 378), covering lobster fishing and trade; and<br>• 2010 Prawn Fishery Management Plan (Cap. 378), calling for zoning restrictions, mesh size regulations, vessel use limitations, and mandatory use of Turtle Excluder Devices |
| Africa | Nigeria | Act No. 71 of 1992 | Sea Fisheries Act | 1992 | Governs control, regulation, and protection of sea fisheries in Nigeria's territorial waters; prohibits certain fishing methods Implemented by Sea Fisheries (Fishing) Regulations, which prohibit fishing within first 5 nautical miles of continental shelf and place further restrictions on trawling, crustacean size, landing requirements and dumping, and mesh net requirements |

*(Continued)*

| Region | Country/State | Citation | Title | Year Passed | Summary or Subsections Applicable to Regulatory Issue |
|---|---|---|---|---|---|
| Africa | South Africa | Act No. 74 of 1995 | Sea Fishery Act | 1988 | Outlines basic regulatory framework for fishing in the country's fishing zones; creates Sea Fishery Advisory Committee, a Quota Board, and a Sea Fishery Fund |
| Africa | South Africa | Act No. 15 of 1994 | Maritime Zones Act | 1994 | Establishes South Africa's maritime zones |
| Africa | South Africa | Act No. 18 of 1998 | Marine Living Resources Act | 1998 | Amends Sea Fishery Act; provides for:<br>• fisheries management plans;<br>• fishing priority areas;<br>• marine protected areas;<br>• gear and use restrictions;<br>• closed seasons;<br>• and other prohibited activities Provides further rules on:<br>• trawl fishing, pursine fishing (and other types net and line fishing);<br>• bag limits;<br>• fish size and mass limits;<br>• landing, disposal, and dumping requirements;<br>• minimizing environmental impacts; and<br>• disease control<br>Implemented by various regulations and orders, including Prince Edward Islands Marine protected area regulation (No. R. 252 of 2012) |

*(Continued)*

| Region | Country/State | Citation | Title | Year Passed | Summary or Subsections Applicable to Regulatory Issue |
|---|---|---|---|---|---|
| Americas | Brazil | Law No. 11.959/ Lei no. 11.959, de 29 de Junho de 2009 | Law No. 11.959 on Sustainable Development Policy on Fishery and Aquaculture/*Lei No. 11.959 de Dispõe sobre a Política Nacional de Desenvolvimento Sustentável da Aquicultura e da Pesca, regula as atividades pesqueiras, revoga a Lei no 7.679, de 23 de novembro de 1988, e dispositivos do Decreto- Lei no 221, de 28 de fevereiro de 1967, e dá outras providências.* | 2009 | Sets out guidelines for developing fishery and aquaculture sectors in Brazil Implemented by various regulations and orders, including:<br>• Order No. 5, creating the Permanent Committee for Fishing and Sustainable Use of Shrimps (CPG);<br>• Order No. 36-N, establishing protective measures (TED) for marine turtles during pink shrimp fishing;<br>• Order No. 56, establishing minimum size for swordfish fishing;<br>• Order No. 97, establishing a closed season period for motor-powered trawl fishing for certain shrimps' species;<br>• Order No. 98, establishing a periodical closed season for sardine fishing; and<br>• Order No. 120-N, regulating sardine fishing via prescriptions on authorized fishing methods, fish size, season, and protected zones |
| Americas | Canada | R.S.C. 1985, c. C-33 | Coastal Fisheries Protection Act | 1985 | Prohibits foreign fishing vessels in Canadian fisheries waters; prohibits vessels from introducing fish from outside, non-territorial waters, unless authorized |

*(Continued)*

| Region | Country/State | Citation | Title | Year Passed | Summary or Subsections Applicable to Regulatory Issue |
|---|---|---|---|---|---|
| Americas | Canada | R.S.C. 1985, c. F-14 | Fisheries Act | 1985 | Governs fisheries management; sets out general fish habitat protection and pollution prevention measures, as well as general prohibitions on certain fishing activities, with special provisions on seal hunting, salmon fishing, and lobster fisheries<br>Implemented by:<br>• 1985 Atlantic Fishery Regulations (S.O. R./86-21), which mandate restrictions on gear, seasons, areas, and tagging;<br>• 1993 Fishery (General) Regulations (S.O. R./93-53), which advocate periods and variation of close times, fishing quotas, size and weight limits, and orders to remove obstructions to the passage of fish; and<br>• 2007 Pacific Fishery Management Area Regulations (SOR/2007–77) |
| Americas | Canada | S.C. 1996, c. 31 | Oceans Act | 1996 | Establishes Canada's maritime zones<br>• Part II lays out an oceans management strategy<br>• Part III requires the Minister of Fisheries and Oceans to prepare a national strategy for the management of estuarine, coastal, and marine ecosystems<br>• Section 35 calls for establishing Marine Protected Areas |
| Americas | Canada | S.C. 1998, c. 10 | Canada Marine Act | 1998 | Regulates Canada's ports and other matters related to maritime trade and transport |
| Americas | Canada | S.C. 2001, c. 26 | Canada Shipping Act | 2001 | Regulates shipping and navigation in Canada |
| Americas | Canada | S.C. 2002, c. 18 | National Marine Conservation Areas Act | 2002 | Aims to protect and conserve representative marine areas for the benefit, education, and enjoyment of the people of Canada |

(*Continued*)

| Region | Country/State | Citation | Title | Year Passed | Summary or Subsections Applicable to Regulatory Issue |
|---|---|---|---|---|---|
| Americas | Mexico | DOF 08-01-1986, publicada en el *Diario Oficial de la Federación* el 8 de enero de 1986 | Federal Law of the Sea/ *Ley Federal del Mar* | 1985 | Sets out basic rules governing maritime areas under national jurisdiction<br>Part 4 contains measures on marine environment protection and preservation |
| Americas | Mexico | DOF 04-01-1994, publicada en el *Diario Oficial de la Federación* el 4 de enero de 1994 | Law on Navegation/*Ley de Navegación* | 1994 | Establishes Mexico's EEZ; grants the Ministry of the Navy sovereignty over territorial waters |
| Americas | Mexico | DOF 24–07–2007, publicada en el *Diario Oficial de la Federación* el 24 de julio de 2007 | General Law for Sustainable Fisheries and Aquaculture/*Ley General de Pesca y Acuacultura Sustentables* | 2007 | Promotes integrated and sustainable development of the fisheries and aquaculture sectors; provides a basic framework for the management, conservation, protection, reforestation, protection, and rehabilitation, and sustainable use of fisheries and aquaculture resources; creates a National Council of Fisheries and Aquaculture<br>Article 37 addresses fisheries management programs |
| Americas | United States | 16 U.S.C. §§ 3371-3378 | Lacey Act | 1900 | Prohibits fish or wildlife transactions and activities that violate state, federal, Native American tribal, or foreign law |
| Americas | United States | 16 U.S.C. §§ 661-667e | Fish and Wildlife Coordination Act | 1934 | Directs the Bureau of Fisheries to use impounded waters for fish-culture stations and migratory-bird resting and nesting areas; requires consultation with the Bureau of Fisheries prior to construction of any new dams in order to allow the dams to provide for fish migration |

*(Continued)*

| Region | Country/State | Citation | Title | Year Passed | Summary or Subsections Applicable to Regulatory Issue |
|---|---|---|---|---|---|
| Americas | United States | 22 U.S.C. §§ 1971-1979 | Fisherman's Protective Act—Pelly Amendment | 1967 | Authorizes the President to embargo wildlife products (including fish) if "nationals of a foreign country are engaging in trade or taking that diminishes the effectiveness of an international program in force with respect to the United States for the conservation of endangered or threatened species" |
| Americas | United States | 42 U.S.C. § 4321 et seq. | National Environmental Policy Act (NEPA) | 1969 | Requires environmental reviews of proposed major federal agency actions, including some fisheries management decisions (such as significant changes to a fishery management plan) |
| Americas | United States | 16 U.S.C. §§ 1361-1407 | Marine Mammal Protection Act (MMPA) | 1972 | Contains provisions on protecting marine mammals and sea turtles from habitat destruction, vessel strikes, and incidental capture as bycatch |
| Americas | United States | 16 U.S.C. §§ 1431-1445 | National Marine Sanctuaries Act (NMSA) | 1972 | Designates certain marine environments as sanctuaries because of their environmental, historical, or cultural importance |
| Americas | United States | 33 U.S.C. §§ 1401-1445 and 16 U.S.C. §§ 1431-1445 | Marine Protection, Research and Sanctuaries Act of 1972 (MPRSA or Ocean Dumping Act) | 1972 | Regulates ocean dumping of industrial wastes, sewage sludge, and other wastes through a permit program |
| Americas | United States | 16 U.S.C. § 1531 et seq. | Endangered Species Act (ESA) | 1973 | Lists over 100 endangered and threatened marine species that fall under NOAA's jurisdiction |
| Americas | United States | 16 U.S.C. §§ 1801-1891d | Magnuson-Stevens Fishery Conservation and Management Act (Magnuson-Stevens Act or MSA) | 1976 | Governs marine fisheries management; establishes the US EEZ and regulates fish populations within this zone; modeled on principles of scientific management and best-stewardship practices; establishes a regional public-private management system Reauthorized by the 1996 Sustainable Fisheries Act and the Magnuson-Stevens Act Reauthorization Act (MSRA) of 2006; has been up for another reauthorization since 2014 |

(Continued)

| Region | Country/State | Citation | Title | Year Passed | Summary or Subsections Applicable to Regulatory Issue |
|---|---|---|---|---|---|
| Americas | United States | 7 U.S.C. § 136 et seq. | Federal Insecticide, Fungicide, and Rodenticide Act (FIFRA) | 1996 | Regulates pesticide use; has been invoked to instill "no-spray buffer zones" on the use of pesticides in or near waters that host threatened or endangered species |
| Americas | United States | 16 U.S.C. § 6401 et seq | Coral Reef Conservation Act (CRCA) | 2000 | Aims to preserve, sustain, and restore the condition of coral reef ecosystems; calls for scientific research and conservation programs on coral reef ecosystems |
| Americas | United States | 33 U.S.C. §§ 857-19 | Oceans Act | 2000 | Establishes the United States Commission on Ocean Policy to promote responsible stewardship of fisheries and other marine resources |
| Americas | United States (Northern Marianas Islands) | P.L. 17-27 | An Act to Prohibit any Person from Possessing, Selling, Offering for Sale, Trading, or Distributing Shark Fins in the CNMI | 2011 | Provides measures to protect sharks from commercial overfishing |

(*Continued*)

| Region | Country/State | Citation | Title | Year Passed | Summary or Subsections Applicable to Regulatory Issue |
|---|---|---|---|---|---|
| Asia | China | Adopted at the 14th Meeting of the Standing Committee of the Sixth National People's Congress on January 20, 1986 and promulgated by Order No. 34 of the President of the People's Republic of China on January 20, 1986 | Fisheries Law of the People's Republic of China | 1986 | Governs fisheries resources management in China; aims to:<br>• protect and "reasonably" use fishery resources; and<br>• protect fishery workers' rights<br>Call for a national fishing quota system Part IV concerns enhancing and protecting fishery resources<br>Implemented by 1987 Regulations for the Implementation of the Fisheries Law of the People's Republic of China, which call for additional conservation measures |
| Asia | China | Adopted at the third session of the Standing Committee of the Ninth National People's Congress, 26 June 1998 | China's Exclusive Economic Zone and Continental Shelf Act | 1998 | Implements UNCLOS; establishes China's rights over its EEZ and continental shelf |

*(Continued)*

| Region | Country/State | Citation | Title | Year Passed | Summary or Subsections Applicable to Regulatory Issue |
|---|---|---|---|---|---|
| Asia | China (Hong Kong) | Cap. 171 (Oct 12, 1962), as amended by E.R. 2 of 2012 | To promote the conservation of fish and other forms of aquatic life within the waters of Hong Kong and to regulate fishing practices and to prevent activities detrimental to the fishing industry (Fisheries Protection Ordinance (Cap. 171)) | 1962 | Provides rules for protecting fish and other forms of aquatic life in Hong Kong waters |
| Asia | China (Hong Kong) | Cap. 476 (June 1, 1995) | An Ordinance to provide for the designation, control and management of marine parks and marine reserves, and for purposes connected therewith (Marine Parks Ordinance (Cap. 476)) | 1995 | Provides the legal framework for designating, controlling, and managing marine parks and marine reserves in Hong Kong waters |
| Asia | China (Taiwan) | As amended by Presidential Order Hua-Tsung (1) No. 10200156101 on August 21, 2013 | Fisheries Act | 1929 | Provides rules for conserving and "rationally" using aquatic resources Taiwanese waters |

(*Continued*)

| Region | Country/State | Citation | Title | Year Passed | Summary or Subsections Applicable to Regulatory Issue |
|---|---|---|---|---|---|
| Asia | India | Act No. 4 of 1897 | Indian Fisheries Act | 1897 | Establishes basis for Union and State territories to create their own fishery laws and rules: as a result, every Indian State has its own Fisheries Act suited to local conditions (with no uniform national approach) [examples of such State laws are included in this Appendix] |
| Asia | India | Act No. 13 of 1972 | Marine Products Export Development Authority Act (MPEDA) | 1972 | Establishes the Marine Products Export Development Authority to address export standards, marketing, promotion of prawn fisheries, and training in fisheries sector |
| Asia | India | Act No. 80 of 1976 | Territorial Waters, Continental Shelf, Exclusive Economic Zone and other Maritime Zones Act | 1976 | Establishes India's EEZ and declares India's territorial waters |
| Asia | India | Act No. 42 of 1981 | Maritime Zones of India (Regulation of Fishing by Foreign Vessels) Act | 1981 | Does not address conservation or management of fisheries resources |
| Asia | India (Arunachal Pradesh) | Act No. 5 of 2006 | Arunachal Pradesh Fisheries Act | 2006 | Governs fisheries in State of Arunachal |
| Asia | India (Gujarat) | Act No. 8 of 2003 | Gujarat Fisheries Act | 2003 | Governs fisheries in State of Gujarat |
| Asia | India (Haryana) | Punjab Act No. II of 1914 | Punjab Fisheries Act | 1914 | Governs fisheries in State of Punjab |
| Asia | India (Himachal Pradesh) | Act No. 16 of 1976 | Himachal Pradesh Fisheries Act | 1976 | Governs fisheries in State of Himachal |

*(Continued)*

| Region | Country/State | Citation | Title | Year Passed | Summary or Subsections Applicable to Regulatory Issue |
|---|---|---|---|---|---|
| Asia | India (Kerala) | Act No. 10 of 1981 | Kerala Marine Fishing Regulation Act | 1981 | Governs fisheries in State of Kerala |
| Asia | India (Manipur) | Manipur Act No. 1 of 1992 | Manipur Fisheries Act | 1992 | Governs fisheries in State of Manipur |
| Asia | India (Mizoram) | Act No. 12 of 2005 | Mizoram Fisheries Act | 2005 | Governs fisheries in State of Mizoram |
| Asia | India (Tamil Nadu) | Act No. 106 cf 5 March 1983 | Tamil Nadu Marine Fishing Regulation Act | 1993 | Governs fisheries in State of Tamil Nadu |
| Asia | India (West Bengal) | Act No. IX of 1993; *The Calcutta Gazette*, Extraordinary, 14 June 1993, pp. 1–11 | West Bengal Marine Fishing Regulation Act | 1993 | Governs fisheries in State of West Bengal |
| Asia | Israel | P.O. Suppl. I, 157 | Fisheries Ordinance | 1937 | Governs fisheries management and conservation; contains provisions on:<br>• restricting taking and landing of fish; and<br>• prohibiting use of poison or explosives Implemented by 1937 Fisheries Rules, which:<br>• prohibit trawling in shallow Mediterranean waters;<br>• place certain gear restrictions and net mesh size requirements; establish protected fishing areas; and<br>• call for notification requirements in case of fish disease outbreaks |

(*Continued*)

| Region | Country/State | Citation | Title | Year Passed | Summary or Subsections Applicable to Regulatory Issue |
|---|---|---|---|---|---|
| Asia | Israel | Law No. 5758-1998 | National Parks, Nature Reserves, National Sites and Memorial Sites Law | 1963 | Provides basic legal structure for protecting Israel's natural habitats and assets, including marine habitats; protects certain species of fish and aquatic plant and animal life |
| Asia | Japan | Act No. 267 of 1949 | Fisheries Act | 1949 | Establishes Japan's basic fisheries production system |
| Asia | Japan | Act No. 313 of 1951 | Act on the Protection of Fishery Resources | 1951 | Prescribes measures for aquatic animal and plant conservation, including:<br>• catch restrictions;<br>• fishing method restrictions;<br>• total allowable catch limits;<br>• fleet capacity limits;<br>• protected water surfaces; and<br>• import quarantines (to prevent spread of diseases) |
| Asia | Japan | Act No. 76 of 1996 | Act on the Exercise of the Sovereign Right for Fishery, etc. in the Exclusive Economic Zone | 1996 | Establishes Japan's EEZ and implements UNCLOS; calls for setting total allowable catches (TACs) |

*(Continued)*

| Region | Country/State | Citation | Title | Year Passed | Summary or Subsections Applicable to Regulatory Issue |
|---|---|---|---|---|---|
| Asia | Japan | Act No. 89 of 2001 | Fisheries Basic Act | 2001 | Governs conservation and management of living aquatic resources; aims to:<br>• control water quality;<br>• protect and provide nursery grounds for aquatic animals and plants; and<br>• preserve forests in order to conserve and improve the environment for aquatic animals and plants<br>Establishes a Fishery Policy Council; delegates most fisheries management to local fishery cooperative associations (FCAs) who set their own rules on harvest times, net mesh sizes, and sustainable yields |
| Asia | Russia | Law No. 166-FZ of 2004 (Dec 20, 2004) | Federal Law No. 166-FZ on Fisheries and Conservation of Aquatic Biological Resources | 2004 | Governs fishery management and conservation; provides for specific conservation measures, including a quota system with total allowable catch (TAC) limits set annually by the Federal Fisheries Institution Implemented by various decrees and regulations, including:<br>• Ministerial Decree No. 644 regarding the distribution of quotas for fish farming, stock enhancement, acclimatization, educational and training purposes;<br>• Ministerial Decree No. 694 validating the list of fishing gear prohibited for import to the Russian Federation; and<br>• Ministerial Decree No. 768 regarding the distribution of the total allowable catch in accordance with quotas |

*(Continued)*

| Region | Country/State | Citation | Title | Year Passed | Summary or Subsections Applicable to Regulatory Issue |
|---|---|---|---|---|---|
| Australia/ New Zealand | Australia | Act. No. 162 of 1991 | Fisheries Management Act | 1991 | Governs fisheries management; defines the Australian Fishing Zone; establishes the Australian Fishery Management Authority; calls for fishery management plans (which may allow for statutory fishing rights and fishery closures); places general ban on driftnet fishing<br>Implemented by:<br>• 1992 Fisheries Management Regulations, providing for catch limits and incidental seabird catch measures;<br>• 1995 Fisheries Management (Northern Prawn Fishery) Regulations;<br>• 1995 Fisheries Management (Southern Bluefin Tuna Fishery) Regulations, setting weight and season restrictions;<br>• 1998 Fisheries Management (South East Trawl Fishery) Regulations;<br>• 2002 Fisheries Management (Bass Strait Central Zone Scallop Fishery) Regulations;<br>• 2004 Fisheries Management (Southern and Eastern Scalefish and Shark Fishery) Regulations;<br>• 2006 Fisheries Management (Western Tuna and Billfish Fishery) Regulations;<br>• 2009 Fisheries Management (Eastern Tuna and Billfish Fishery) Regulations; and<br>• 2009 Fisheries Management (International Agreements) Regulations |
| Australia/ New Zealand | Australia | Act No. 91 of 1999 | Environment Protection and Biodiversity Conservation Act | 1999 | Implements CBD<br>Chapter II allows government to decide whether an action (including fisheries actions in marine areas) has, will have, or is likely to have a significant impact on certain aspects of the environment |

*(Continued)*

| Region | Country/State | Citation | Title | Year Passed | Summary or Subsections Applicable to Regulatory Issue |
|---|---|---|---|---|---|
| Australia/ New Zealand | New Zealand | Act No. 28 of 1977; *Regional Compendium of Fisheries Legislation (Western Pacific Region)*, FAO, June 1993, pp. 1018–1044; FAL No. 27, vol. 2, 1978, pp. 58–78 | Territorial Sea and Exclusive Economic Zone Act | 1977 | Defines New Zealand's EE Sections 11–13 call for setting total allowable catches (TACs) |
| Australia/ New Zealand | New Zealand | Act No. 88 of 1996 | Fisheries Act | 1996 | Seeks the "conservation, use, enhancement and development of fisheries resources while ensuring sustainability"; establishes the National Fisheries Advisory Council and a Catch History Review Committee<br>• Part 3 addresses sustainability measures<br>• Part 4 sets up a robust quota management system (including annual catch entitlements) Implemented by various regulations and orders, including:<br>• 2000 Fisheries (Southern Bluefin Tuna Quota) Regulations;<br>• 2005 Fisheries (Declaration of New Stocks Subject to Quota Management System) Notice; and<br>• 2005 Fisheries (Stocks Determined Not to be Subject to Quota Management System) Notice Amends the previous Fisheries Act of 1983 (No. 14 of 1983) |
| Australia/ New Zealand | New Zealand | Act No. 31 of 1997 | Fisheries (Quota Operations Validation) Act | 1997 | Confirm and validates overfishing and underfishing rights and entitlements |

(*Continued*)

| Region | Country/State | Citation | Title | Year Passed | Summary or Subsections Applicable to Regulatory Issue |
|---|---|---|---|---|---|
| Europe | European Union | O.J. L. 318, 27.11.1998, p. 63; Document 31998R0850R(01) | Corrigendum to Council Regulation (EC) No 850/98 of 30 March 1998 for the conservation of fishery resources through technical measures for the protection of juveniles of marine organisms | 1998 | Sets out technical measures Currently under review to accommodate new Common Fisheries Policy (CFP) reforms |
| Europe | European Union | O.J. L. 151, 11.6.2008, pp. 5–25 | Commission Regulation (EC) No 517/2008 of 10 June 2008 laying down detailed rules for the implementation of Council Regulation (EC) No 850/98 as regards the determination of the mesh size and assessing the thickness of twine of fishing nets | 2008 | Sets out technical measures Currently under review to accommodate new Common Fisheries Policy (CFP) reforms |

*(Continued)*

| Region | Country/State | Citation | Title | Year Passed | Summary or Subsections Applicable to Regulatory Issue |
|---|---|---|---|---|---|
| Europe | European Union | O.J. L. 354, 28.12.2013, pp. 22–61 | Regulation (EU) No 1380\2013 of the European Parliament and the Council of 11 December 2013 on the Common Fisheries Policy, amending Council Regulations (EC) No 1954/2003 and (EC) No 1224 and repealing Council Regulations (EC) no 2371 and (EC) No 639/2004 and Council Decision 2004/585/ EC. (EU Common Fisheries Policy or (CFP)) | 2013 | Sets out rules for managing European fishing fleets and conserving fish stocks as a common resource of the EU; demands an end to overfishing by the end of the decade and establishes a legally binding commitment to fish at sustainable levels; embraces the precautionary principle; calls for various overfishing and bycatch measures, including:<br>• total allowable catches (TACs);<br>• fleet capacity restrictions;<br>• minimum mesh sizes;<br>• gear restrictions; and<br>• landing obligations |

*(Continued)*

| Region | Country/State | Citation | Title | Year Passed | Summary or Subsections Applicable to Regulatory Issue |
|---|---|---|---|---|---|
| Europe | European Union | O.J. L. 198, 25.7.2019, pp. 105–201 | Regulation (EU) 2019/1241 of the European Parliament and of the Council of 20 June 2019 on the conservation of fisheries resources and the protection of marine ecosystems through technical measures, amending Council Regulations (EC) No 1967/2006, (EC) No 1224/2009 and Regulations (EU) No 1380/2013, (EU) 2016/1139, (EU) 2018/973, (EU) 2019/472 and (EU) 2019/1022 of the European Parliament and of the Council, and repealing Council Regulations (EC) No 894/97, (EC) No 850/98, (EC) No 2549/2000, (EC) No 254/2002, (EC) No 812/2004 and (EC) No 2187/2005 | 2019 | Sets out technical measures concerning:<br>• the taking and landing of marine biological resources;<br>• the operation of fishing gear; and<br>• the interaction of fishing activities with marine ecosystems. technical measures shall in particular contribute to achieving the following objectives:<br>• optimize exploitation patterns to provide protection for juveniles and spawning aggregations of marine biological resources;<br>• ensure that incidental catches of sensitive marine species are minimized and where possible eliminated;<br>• minimize the negative environmental impacts of fishing on marine habitats;<br>Sets out technical measures, including provisions on:<br>• prohibited fishing gear and methods;<br>• general restrictions on gear and conditions for their use;<br>• protection of sensitive species and their habitats;<br>• catches of marine mammals, seabirds and marine reptiles;<br>• protection of vulnerable marine ecosystems;<br>• minimum conservation reference sizes;<br>• measures to reduce discarding.<br>Further technical measures are laid down for specific regions. |

*(Continued)*

| Region | Country/State | Citation | Title | Year Passed | Summary or Subsections Applicable to Regulatory Issue |
|---|---|---|---|---|---|
| Europe | European Union—Germany | – | – | – | Regional fishing associations (*Landesfischereiverba€nde*) regulate and manage fishing in Germany, with each federal state having its own *Landesfischereiverband* and regulations [examples of such regional laws are included in this Appendix] |
| Europe | European Union—Germany (Bayern) | (August 15, 1908); FAOLEX No: LEX-FAOC072488 | Bavarian Fishery Law/ *Fischereigesetz fu€r Bayern* | 1908 | Governs fisheries in Bayern |
| Europe | European Union—Germany (Berlin | (June 19, 1995); FAOLEX No: LEX-FAOC073695 | Berlin Fishery Law/ *Berliner Landesfischereigesetz* | 1995 | Governs fisheries in Berlin |
| Europe | European Union—Germany (Brandenburg) | (May 13, 1993); FAOLEX No: LEX-FAOC073287 | Brandenburg Fishery Law/*Fischereigesetz fu€r das Land Brandenburg (BbgFischG)* | 1993 | Governs fisheries in Brandenburg |
| Europe | European Union—Germany (Hamburg) | (May 22, 1986); FAOLEX No: LEX-FAOC074736 | Hamburg Fishery Law/ *Hamburgisches Fischereigesetz* | 1986 | Governs fisheries in Hamburg |
| Europe | European Union—Germany (Hessen) | (Dec 19, 1990); FAOLEX No: LEX-FAOC075420 | Hessen Fishery Law/ *Fischereigesetz fu€r das Land Hessen (Hessisches Fischereigesetz—HFischG)* | 1990 | Governs fisheries in Hessen |

(*Continued*)

| Region | Country/State | Citation | Title | Year Passed | Summary or Subsections Applicable to Regulatory Issue |
|---|---|---|---|---|---|
| Europe | European Union—Germany (Niedersachsen) | (Feb 1, 1978); FAOLEX No: LEX-FAOC084524 | Lower Saxony Fishery Law/ *Niedersa€chsisches Fischereigesetz (Nds. FischG)* | 1978 | Governs fisheries in Niedersachsen |
| Europe | European Union—Germany (Rheinland-Pfalz) | (Dec 9, 1974); FAOLEX No: LEX-FAOC075265 | Rheinland-Pfalz Fishery Law/ *Landesfischereigesetz (LFischG)* | 1974 | Governs fisheries in Rheinland-Pfalz |
| Europe | European Union—Germany (Saarland) | (Jan 23, 1985); FAOLEX No: LEX-FAOC077116 | Saarland Fishery Law/ *Saarla€ndisches Fischereigesetz (SFischG)* | 1985 | Governs fisheries in Saarland |
| Europe | European Union—Germany (Sachsen) | (July 9, 2007); FAOLEX No: LEX-FAOC077045 | Saxony Fishery Law/ *Fischereigesetz fu€r den Freistaat Sachsen (Sa€chsisches Fischereigesetz—Sa€chsFischG)* | 2007 | Governs fisheries in Sachsen |
| Europe | European Union—Germany (Th€uringen) | (June 28, 2006); FAOLEX No: LEX-FAOC075395 | Th€uringen Fishery Law/*Thu€ringer Fischereigesetz (Thu€rFischG)* | 2006 | Governs fisheries in Th€uringen |

*(Continued)*

| Region | Country/State | Citation | Title | Year Passed | Summary or Subsections Applicable to Regulatory Issue |
|---|---|---|---|---|---|
| International | Parties electing to adopt the Code | In: *FAO Fisheries and Aquaculture Department* [online]. Rome. Updated 28 May 2014. | Code of Conduct for Responsible Fisheries (FAO Fisheries Code) | 1995 | Sets out voluntary, non-binding principles and standards for responsible fisheries practices;<br>Supplemented by various Technical Guidelines<br>Drafted in harmony with the UN Fish Stocks Agreement and the FAO Compliance Agreement |
| International | Parties to Agreement | 148 U.N.T.S. 1860 (Nov. 24, 1993) | Agreement to Promote Compliance with International Conservation and Management Measures by Fishing Vessels on the High Seas (FAO Compliance Agreement) | 1993 | Governs high seas fishing in international waters; aims to prevent bad faith vessel "re-flagging" in Regional Marine Fisheries Organization (RMFO) waters<br>Drafted in harmony with UN Fish Stocks Agreement and FAO Fisheries Code |
| International | Parties to Agreement | 2167 U.N.T.S. 3 (Dec. 4, 1995) | Agreement for the Implementation of the Provisions of the United Nations Convention on the Law of the Sea of 10 December 1982 relating to the Conservation and Management of Straddling Fish Stocks and Highly Migratory Fish Stocks (UN Fish Stocks Agreement) | 1995 | Lays out a legal conservation and management regime for straddling and highly migratory fish stocks; establishes RMFO networks<br>Drafted in harmony with the FAO Compliance Agreement and the FAO Fisheries Code |

(*Continued*)

| Region | Country/State | Citation | Title | Year Passed | Summary or Subsections Applicable to Regulatory Issue |
|---|---|---|---|---|---|
| International | Parties to Convention | 33 U.S.T. 3476 (May 20, 1980) | Convention on the Conservation of Antarctic Marine Living Resources (CCAMLR) | 1980 | Aims to preserve marine life and environmental integrity in and near Antarctica; establishes an Ecosystem Monitoring Program (CEMP)<br>Focuses on protecting the region's krill, an important foundation of the marine ecosystem food chain |
| International | Parties to Convention | S. Treaty Doc. No. 103-20 (1993), 31 I.L.M. 818 | Convention on Biological Diversity (CBD) | 1992 | Addresses all aspects of biodiversity; has yielded two workplans directly related to fisheries:<br>4. Jakarta Mandate on Marine and Coastal Ecosystems, which addresses the conservation and sustainable use of marine and coastal biological diversity; and<br>5. Workplan on Inland Water Ecosystems, which promotes ecosystem approach to freshwater fisheries |
| International | Parties to Convention | S. Treaty Doc. No. 103–20 (1993), 31 I.L.M. 818 | Convention on Biological Diversity (CBD) | 1992 | Addresses all aspects of biodiversity; has yielded two workplans directly related to fisheries:<br>6. Jakarta Mandate on Marine and Coastal Ecosystems, which addresses the conservation and sustainable use of marine and coastal biological diversity; and<br>7. Workplan on Inland Water Ecosystems, which promotes ecosystem approach to freshwater fisheries |
| International | Parties to UNCLOS | 1833 U.N.T.S. 397 (Montego Bay, Dec. 10, 1982) | United Nations Convention on the Law of the Sea (UNCLOS) | 1982 | Provides a global "constitution for the seas"; establishes a comprehensive governance regime for the world's oceans and fisheries; establishes the EEZ concept; prescribes specific fisheries management rules |
| International | Parties to Convention | S. Treaty Doc. No. 103-20 (1993), 31 I.L.M. 818 | Convention on Biological Diversity (CBD) | 1992 | Addresses all aspects of biodiversity; has yielded two workplans directly related to fisheries:<br>8. Jakarta Mandate on Marine and Coastal Ecosystems, which addresses the conservation and sustainable use of marine and coastal biological diversity; and<br>9. Workplan on Inland Water Ecosystems, which promotes ecosystem approach to freshwater fisheries |

(Continued)

| Region | Country/State | Citation | Title | Year Passed | Summary or Subsections Applicable to Regulatory Issue |
|---|---|---|---|---|---|
| International (Bilateral) | Norway and Iceland | (May 28, 1980) | Agreement between Norway and Iceland concerning Fishery and Continental Shelf Questions | 1980 | Calls for measures (including total allowable catch (TAC) limits) to promote the conservation, rational exploitation, and sound reproduction of capelin fish stocks; establishes a Fisheries Commission |
| International (Bilateral) | United States and Canada | T.I.A.S. 11091 (Jan 28, 1985) | Treaty between the Government of the United States of America and the Government of Canada Concerning Pacific Salmon (Pacific Salmon Treaty) | 1985 | Commits Parties to prevent the overfishing of Pacific salmon; ensures that both countries receive benefits equal to the salmon produced in their waters; allows for creating salmon enhancement programs |
| International (Regional) | Australia, Japan, New Zealand | 1819 U.N.T.S. 360 (May 10, 1993) | Convention for the Conservation of Southern Bluefin Tuna (CCSBT) | 1993 | Aims to conserve and optimally utilize Southern Bluefin tuna; calls for total allowable catch (TAC) and bycatch measures; establishes the Commission for the Conservation of Southern Bluefin Tuna |
| International (Regional) | Parties to Convention (Atlantic coastal nations in Europe, Americas, Africa, Asia) | 673 U.N.T.S. 63, 20 U.S.T. 2887, T.I.A.S. 6767 (May 14, 1966) | International Convention for the Conservation of Atlantic Tunas (ICCAT) | 1966 | Sets up a program for international cooperation in research and conservation; calls for setting maximum sustainable yield (MSY) levels for Atlantic tuna populations; establishes the International Commission for the Conservation of Atlantic Tunas |

*(Continued)*

| Region | Country/State | Citation | Title | Year Passed | Summary or Subsections Applicable to Regulatory Issue |
|---|---|---|---|---|---|
| International (Regional) | Parties to Convention (nations in Americas, Asia, and Europe) | 80 U.N.T.S. 3, 1 U.S.T. 230, T.I.A.S. No. 2044 | Convention for the Strengthening of the Inter-American Tropical Tuna Commission (Antigua Convention) | 2010 | Aims to conserve and manage tropical tuna Updates and strengthens the original 1949 Convention, which established the Inter-American Tropical Tuna Commission (IATTC) |
| International (Regional) | Parties to Convention (nations in Europe, North America, Northern Arctic) | 24 U.S.T. 1080, T.I.A.S. 7628 | Convention for the International Council for the Exploration of the Sea (ICES) | 1964 | Promotes scientific research and investigation programs on the sea and marine living resources; establishes International Council for the Exploration of the Sea |
| International (Regional) | Parties to Convention (North Atlantic coastal nations) | 1338 U.N.T.S. 33, T.I.A.S. No. 10789 (Mar 2, 1982) | Convention for the Conservation of Salmon in the North Atlantic (North Atlantic Salmon Treaty or NASCO) | 1982 | Aims to conserve North Atlantic salmon; prohibits salmon fishing beyond areas of fisheries jurisdiction of coastal states (and within certain areas of national jurisdiction); establishes the North Atlantic Salmon Conservation Organization |
| International (Regional) | Parties to Convention (Northwest Atlantic coastal nations in Americas, Europe, and Asia) | 157 U.N.T.S. 157, 1 U.S.T. 477, T.I.A.S. No. 2089 (Feb 8, 1949) | International Convention for the Northwest Atlantic Fisheries (Northwest Atlantic Fisheries Treaty) | 1949 | Aims to protect Northwest Atlantic fisheries; calls for: scientific research programs;<br>• total allowable catches (TACs); and<br>• bycatch measures<br>Establishes the Northwest Atlantic Fisheries Organization (NAFO) |

(*Continued*)

| Region | Country/State | Citation | Title | Year Passed | Summary or Subsections Applicable to Regulatory Issue |
|---|---|---|---|---|---|
| International (Regional) | Parties to Convention (Pacific coastal nations in Americas, Europe, Asia, and Australia/New Zealand) | 40 I.L.M. 227 (2001) | Convention on the Conservation and Management of Highly Migratory Fish Stocks in the Western and Central Pacific Ocean | 2000 | Aims to ensure the long-term conservation and sustainable use of highly migratory fish stocks in the western and central Pacific Ocean; establishes the Commission for the Conservation and Management of Highly Migratory Fish Stocks in the Western and Central Pacific Ocean |
| International (Regional) | Parties to Convention (Southeast Atlantic coastal nations in Africa, Europe, Asia, and Americas) | 2221 U.N.T.S. 189 (Apr. 20, 2001) | Convention on the Conservation and Management of Fishery Resources in the South-East Atlantic Ocean (SEAFO Convention) | 2001 | Aims to ensure the long-term conservation and sustainable use of fishery resources on the high seas of the South East Atlantic Ocean; embraces the precautionary principle; establishes the South East Atlantic Fisheries Organisation (SEAFO) |

# Index

Note: **Bold** page numbers refer to tables; *italic* page numbers refer to figures.

For Product Safety Concerns and Information please contact our EU
representative GPSR@taylorandfrancis.com
Taylor & Francis Verlag GmbH, Kaufingerstraße 24, 80331 München, Germany

www.ingramcontent.com/pod-product-compliance
Ingram Content Group UK Ltd.
Pitfield, Milton Keynes, MK11 3LW, UK
UKHW020023200926
475702UK00014B/2023
* 9 7 8 0 3 6 7 6 8 4 7 6 1 *